Offshore Renewables for a Transition to a Low Carbon Society

Offshore Renewables for a Transition to a Low Carbon Society

Editors

Liliana Rusu
Vicky Stratigaki

MDPI Basel • Beijing • Wuhan • Barcelona • Belgrade • Novi Sad • Cluj • Manchester

Editors
Liliana Rusu
Department of Mechanical Engineering
'Dunarea de Jos' University of Galati
Galati
Romania

Vicky Stratigaki
Department of Civil Engineering
Ghent University
Ghent
Belgium

Editorial Office
MDPI
St. Alban-Anlage 66
4052 Basel, Switzerland

This is a reprint of articles from the Special Issue published online in the open access journal *Journal of Marine Science and Engineering* (ISSN 2077-1312) (available at: www.mdpi.com/journal/jmse/special_issues/bz_low_carbon_society).

For citation purposes, cite each article independently as indicated on the article page online and as indicated below:

Lastname, A.A.; Lastname, B.B. Article Title. *Journal Name* **Year**, *Volume Number*, Page Range.

ISBN 978-3-0365-9231-2 (Hbk)
ISBN 978-3-0365-9230-5 (PDF)
doi.org/10.3390/books978-3-0365-9230-5

Contents

About the Editors

Liliana Rusu

Prof. Liliana Rusu has been a full professor at 'Dunarea de Jos' University of Galati, Romania, Department of Mechanical Engineering, since 2016. She received her first Ph.D. degree in Mechanical Engineering from 'Dunarea de Jos' University of Galati in 2006, and her second Ph.D. in Marine Engineering and Naval Architecture was awarded to her in 2009 by the University of Lisbon, Portugal (former Technical University of Lisbon), under the supervision of Prof. Carlos Guedes Soares. In 2015, she received her Habilitation degree. She has continued her collaboration with CENTEC in the framework of a post-doctoral fellowship (individual grant financed by the Portuguese Foundation for Science and Technology), where she is currently working as a Professor Collaborator. Since December 2017, she has been a member of the Marine Knowledge Expert group of the European Commission. Since 2016, she has also been a member of the National Ethics Council for Research Activities. Her research interests include wave modelling, assessment of the wave climate in various geographical areas, assessment of wave/wind energy, data assimilation, and climate change impacts.

Vicky Stratigaki

Dr. Vicky Stratigaki works within the Department of Civil Engineering, Ghent University, Technologiepark 60, 9052 Zwijnaarder (Ghent), Belgium. Her scientific interests are mainly related to marine renewable energy; wave energy converter arrays; experimental modelling; numerical coupling methodologies; WEC–WEC interactions; wave–structure interactions.

Journal of
Marine Science
and Engineering

MDPI

Editorial

Offshore Renewables for a Transition to a Low Carbon Society

Liliana Rusu [1,*] and Vicky Stratigaki [2]

1 Faculty of Engineering, "Dunarea de Jos" University of Galati, 47 Domneasca Street, 800008 Galati, Romania
2 Department of Civil Engineering, Ghent University, Technologiepark 60, 9052 Zwijnaarde, Belgium; vicky.stratigaki@ugent.be
* Correspondence: liliana.rusu@ugal.ro

The need to reduce CO_2 emissions is of utmost importance considering the climate changes that have become more evident and affect us through the significant impact they have. The development of energy production using clean resources represents one of the most effective measures that can lead to a reduction in greenhouse gas emissions, thus preventing the continuation of the warming process. Besides the fact that they are an inexhaustible energy source that nature offers us, renewable energy sources can help us to mitigate the greenhouse gas emissions [1]. Taking into consideration that the oceans and seas of the planet occupy a much larger area than the land and offer various sources of renewable energy insufficiently exploited until now, although they present a high potential, it is obvious that the exploitation of renewable resources from the marine environment must be developed faster and on wider geographical spaces. In this general context, the new strategy established in the framework of the "European Green Deal" highlighted the importance of offshore renewables in transforming the European Union into a low-carbon energy space [2].

From this perspective, *Journal of Marine Science and Engineering* pays special attention to promoting studies regarding the offshore renewable sector, including evaluation of the resources that can be exploited in these zones and their dynamics in the context of climate change. Some technological challenges related to optimization aspects of wave energy converters (WECs) are also discussed. In this context, some very recent works targeting these important issues, and which were published in the Special Issue of *JMSE* entitled "Offshore Renewables for a Transition to a Low-Carbon Society" are highlighted in this Editorial.

Among the renewable resources available in offshore areas, the exploitation of wind energy is probably the most advanced, with the faster growth certainly being influenced by the existence of mature technologies already developed for the implementation of wind farms on land. Offshore wind has the advantage that it is stronger and less turbulent than the one on land, and therefore, it is more energetic and stable and thus more efficient. This makes offshore wind exploitation cost effective and therefore competitive with low-cost energy sources, although maintenance in the harsh marine environment is often more difficult.

Since the energy extracted from wind depends on the wind speed and this is variable, the identification of the best locations is based on various sources of information, such as in situ and satellite measurements or the results obtained through simulations with numerical models. Of course, a global perspective of these resources, with a high spatial and temporal resolution, can only be provided by atmospheric models whose results can be improved by combining them with satellite data or even in situ measurements.

The assessment of wind energy potential in the Romanian coastal area shows that this area can be a viable candidate for the development of offshore wind projects that have good wind energy potential and a large continental shelf. Furthermore, Romania is a part of the European Union, which aims to significantly expand the offshore wind sector [3]. Using the data provided by one of the most credible databases regarding wind speed at 10 m (U_{10})

Citation: Rusu, L.; Stratigaki, V. Offshore Renewables for a Transition to a Low Carbon Society. *J. Mar. Sci. Eng.* **2023**, *11*, 1185. https://doi.org/10.3390/jmse11061185

Received: 2 June 2023
Accepted: 5 June 2023
Published: 7 June 2023

above sea level, namely ERA5 [4], a long-term analysis of the average and extreme wind speed values was carried out for the Romanian exclusive economic zone (EEZ) in the Black Sea. A comparison with satellite measurements indicates a good agreement between the data in the locations targeted. However, an accurate evaluation of the wind speed at the wind turbine hub height is of higher significance for the wind energy potential estimation. From this perspective, U_{100} data spanning twenty years are also retrieved from ERA5.

Another important aspect, which must be taken into account for the realization of new renewable energy projects, is represented by technical and economic evaluations. Such a study was carried out for the Philippines and is published in this Special Issue [5]. Two of the most used turbines were considered for the technical analyses, namely Siemens SWT-3.6-120 and Senvion 6.2M126. The results of this combined assessment are reflected in the calculated levelized cost of electricity (LCOE) of the offshore wind farms in the Philippines.

Considering that the oceans and seas of the planet occupy a much larger area than the land and that they offer various insufficiently exploited sources of energy, although they present a high potential, it is obvious that the exploitation of resources from the marine environment must be developed faster. Besides the exploitation of the offshore wind energy, in the marine environment there is also the possibility of exploiting wave energy, which has great potential in many coastal areas [6] and is not yet exploited to its true potential. At the same time, wave energy can be predicted with good accuracy, higher than wind or solar.

Another important aspect is to find the most suitable WEC for a coastal area, taking into consideration that there are currently a wide variety of WECs tested. Given the characteristics of the sea state conditions on the western coast of the Iberian Peninsula, this region is identified by various studies as having great potential for wave energy exploitation. Thus, the Galician coast (NW Spain) is one of the areas identified with a high potential for wave energy exploitation. In this context, for the Galician coast (NW Spain), an evaluation of the wave power resource was carried out together with an analysis of the expected electrical energy output that could have been produced by four WECs over the period from 2014 to 2021 [7]. For each device, the performance was investigated considering two elements: efficiency and power load factor. The results indicate that the Atargis device, a cycloidal WEC, appears to be the most efficient.

A very important issue is also the optimization of the WECs, and in the case of a floating flapping-panel device, numerical simulation studies and a laboratory physical model hydrodynamic study were performed to evaluate the feasibility of the device [8]. In this way, an optimal damping coefficient to give maximum captured energy was identified.

It is clear that commercial wave energy exploitation can be efficient by placing multiple WECs in an array configuration. The point absorber is an efficient device that captures energy from waves coming from various directions, and for this reason, it is very suitable to be placed in arrays. However, an optimization of the WEC array layout is necessary, together with the applied control strategy. An extensive literature review on the state of the art in physical modelling of the point-absorber WEC arrays is presented in an article published in the present Special Issue [9]. Some scientific gaps were also identified, and they were reformulated as design requirements for the experimental setup. The experimental setup of an array of two to five generic heaving point-absorber devices is presented. Wave basin testing in waves simulating real sea state conditions was performed in response to the urgent need for reliable data on the WEC array tests [10]. These data are used to perform the optimization of the WEC array and to validate new numerical models (non-linear). A model predictive control (MPC) method was also proposed for improving the efficiency in extracting the energy of a WEC array [11]. As a conclusion, the experimental tests using MPC with the interconnected system model indicate a better wave energy capture efficiency.

Offshore solar energy could be a viable option for making many coastal communities, islands, and isolated locations more sustainable. As with the other renewable energy sources presented here, the amount of energy that could be extracted in an area is influenced by the geographic location where the systems are deployed.

A new methodology to identify the most appropriate marine areas for the deployment of some offshore solar farms in the Aegean Sea is presented in [12]. Two phases are included in the methodology proposed. In the first one, exclusion criteria are applied for the marine zones based on the information collected from various sources and maps which were created using a GIS environment. Then, in the second phase, objective and subjective models were applied, and to establish the correlation between them, the Kendall rank correlation coefficient was used. The subjective model based on the value of the solar radiation encountered in an area indicates the offshore area near Crete Island as the most suitable site for an offshore solar farm deployment. On the other hand, the objective model considers the offshore area around Thasos Island as a better location.

Additionally, the installation of floating solar panels on the water surfaces of lakes and lagoons can represent a viable solution, and this is shown in the study focused on the western Black Sea and the Razim-Sinoe lagoon system [13]. In such locations, the environmental conditions in which the floating structures would work are milder than in open coastal areas. Based on the data retrieved from the ERA5 database regarding the solar radiation at four points (one point located on the Romanian coast, the other three in lakes), it was found that the energy produced at all points has similar values.

It is also well known that large quantities of greenhouse gases are produced in maritime transport, and possible solutions should aim to reduce these emissions. For this reason, a solution would be to provide reliable information based on a long-term analysis of the mean and extreme values of the weather conditions, especially wind and sea state conditions that affect navigation safety. Such a study was carried out for the main maritime routes in the European seas and where information can be found regarding the areas that should be avoided as they are more exposed to extreme events [14]. The seasonal analysis also indicated the periods of the year that present lower risks for maritime transport.

Finally, the editors are confident that the works included in this Special Issue offer useful information for both researchers and stakeholders.

Conflicts of Interest: The authors declare no conflict of interest.

References

1. Soukissian, T.; O'Hagan, A.M.; Azzellino, A.; Boero, F.; Brito e Melo, A.; Comiskey, P.; Gao, Z.; Howell, D.; Le Boulluec, M.; Maisondieu, C.; et al. *European Offshore Renewable Energy: Towards a Sustainable Future*; Future Science Brief No. 9; European Marine Board: Ostend, Belgium, 2023.
2. A European Green Deal. Available online: https://commission.europa.eu/strategy-and-policy/priorities-2019-2024/european-green-deal_en (accessed on 19 March 2023).
3. Onea, F.; Rusu, E.; Rusu, L. Assessment of the Offshore Wind Energy Potential in the Romanian Exclusive Economic Zone. *J. Mar. Sci. Eng.* **2021**, *9*, 531. [CrossRef]
4. Hersbach, H.; Bell, B.; Berrisford, P.; Hirahara, S.; Horányi, A.; Muñoz-Sabater, J.; Nicolas, J.; Peubey, C.; Radu, R.; Schepers, D.; et al. The ERA5 Global Reanalysis. *Q. J. R. Meteorol. Soc.* **2020**, *146*, 1999–2049. [CrossRef]
5. Maandal, G.L.D.; Tamayao-Kieke, M.-A.M.; Danao, L.A.M. Techno-Economic Assessment of Offshore Wind Energy in the Philippines. *J. Mar. Sci. Eng.* **2021**, *9*, 758. [CrossRef]
6. Rusu, L.; Rusu, E. Evaluation of the Worldwide Wave Energy Distribution Based on ERA5 Data and Altimeter Measurements. *Energies* **2021**, *14*, 394. [CrossRef]
7. Arguilé-Pérez, B.; Ribeiro, A.S.; Costoya, X.; deCastro, M.; Carracedo, P.; Dias, J.M.; Rusu, L.; Gómez-Gesteira, M. Harnessing of Different WECs to Harvest Wave Energy along the Galician Coast (NW Spain). *J. Mar. Sci. Eng.* **2022**, *10*, 719. [CrossRef]
8. Song, T.; Li, Z.; Zheng, H.; Liang, C.; Wan, Z. Optimization on Hydrodynamic Performance for First Level Energy-Capturing Enhancement of a Floating Wave Energy Converter System with Flapping-Panel-Slope. *J. Mar. Sci. Eng.* **2023**, *11*, 345. [CrossRef]
9. Vervaet, T.; Stratigaki, V.; De Backer, B.; Stockman, K.; Vantorre, M.; Troch, P. Experimental Modelling of Point-Absorber Wave Energy Converter Arrays: A Comprehensive Review, Identification of Research Gaps and Design of the WECfarm Setup. *J. Mar. Sci. Eng.* **2022**, *10*, 1062. [CrossRef]
10. Vervaet, T.; Stratigaki, V.; Ferri, F.; De Beule, L.; Claerbout, H.; De Witte, B.; Vantorre, M.; Troch, P. Experimental Modelling of an Isolated WECfarm Real-Time Controllable Heaving Point Absorber Wave Energy Converter. *J. Mar. Sci. Eng.* **2022**, *10*, 1480. [CrossRef]

11. Zhang, B.; Zhang, H.; Yang, S.; Chen, S.; Bai, X.; Khan, A. Predictive Control for a Wave-Energy Converter Array Based on an Interconnected Model. *J. Mar. Sci. Eng.* **2022**, *10*, 1033. [CrossRef]
12. Vagiona, D.G.; Tzekakis, G.; Loukogeorgaki, E.; Karanikolas, N. Site Selection of Offshore Solar Farm Deployment in the Aegean Sea, Greece. *J. Mar. Sci. Eng.* **2022**, *10*, 224. [CrossRef]
13. Manolache, A.I.; Andrei, G.; Rusu, L. An Evaluation of the Efficiency of the Floating Solar Panels in the Western Black Sea and the Razim-Sinoe Lagunar System. *J. Mar. Sci. Eng.* **2023**, *11*, 203. [CrossRef]
14. Chirosca, A.-M.; Rusu, L. Characteristics of the Wind and Wave Climate along the European Seas Focusing on the Main Maritime Routes. *J. Mar. Sci. Eng.* **2022**, *10*, 75. [CrossRef]

Journal of
Marine Science and Engineering

MDPI

Article

An Evaluation of the Efficiency of the Floating Solar Panels in the Western Black Sea and the Razim-Sinoe Lagunar System

Alexandra Ionelia Manolache, Gabriel Andrei and Liliana Rusu *

Department of Mechanical Engineering, Faculty of Engineering, "Dunarea de Jos" University of Galati, 47 Domneasca Street, 800008 Galati, Romania
* Correspondence: liliana.rusu@ugal.ro

Abstract: The development of novel solar power technologies is regarded as one of the essential solutions to meeting the world's rising energy demand. Floating photovoltaic panels (FPV) have several advantages over land-based installations, including faster deployment, lower maintenance costs, and increased efficiency. Romania is considered a country with enormous solar energy potential, which is one of the most exploited sectors of the renewable energy sector. With this in mind, the purpose of this work is to assess the energetic potential provided by the sun, taking into account three lakes in Romania's east and extending to the west of the Black Sea. In this context, we examine the hourly distribution of solar radiation for the year 2021. The solar radiation data were extracted using the ERA5 database, as well as data collected in situ near them. Following this research, we discovered that all of the chosen locations have a high energetic potential and could be used as locations for the exploitation of solar energy, thereby avoiding the use of land that could be used for agricultural purposes in these areas. We also noticed that there are minor differences between the solar radiation values obtained from the ERA5 database and the measured ones.

Keywords: solar radiation; marine renewable energy; floating solar panels (FPV); sustainability; Romanian nearshore

Citation: Manolache, A.I.; Andrei, G.; Rusu, L. An Evaluation of the Efficiency of the Floating Solar Panels in the Western Black Sea and the Razim-Sinoe Lagunar System. *J. Mar. Sci. Eng.* **2023**, *11*, 203. https://doi.org/10.3390/jmse11010203

Academic Editors: Bang-Fuh Chen and M. Dolores Esteban

Received: 15 November 2022
Revised: 10 December 2022
Accepted: 10 January 2023
Published: 12 January 2023

1. Introduction

Since the Framework Convention on Climate Change was adopted in 1992, the globe has altered dramatically. In 1990, industrialized countries accounted for two-thirds of global emissions; today, they account for roughly half, and by 2020, developing countries will account for two-thirds of global emissions. The Kyoto Protocol, which has governed the limitation of greenhouse gas emissions until now is no longer sufficient. To mitigate the worst effects of climate change, the Paris Agreement was adopted. By reducing global warming to well below 2 °C and pursuing efforts to restrict it to 1.5 °C [1], the Paris Agreement [2] lays out a worldwide framework to avoid severe climate change. It also aims to support countries in their efforts to improve their capacity to deal with the negative effects on the environment. Aside from tackling climate change, there is also a significant interest in reducing global greenhouse gas emissions. By 2030, it is desired to have cut greenhouse gas emissions by at least 55% [3]. To accomplish this, we must stop relying on fossil fuels [4] and start investing in reliable [5] clean, accessible, and affordable alternative energy sources. The sun, wind, water, waste, and heat from the Earth are all abundant sources of renewable energy that are renewed by nature and release little to no greenhouse gases or air pollution.

The energy sector is taking the lead in the decarbonization effort [6] because significant investments are made in mature and affordable renewable energy technologies, such as wind and photovoltaic (PV). Solar energy is regarded as the most promising source of renewable energy [7,8]. It is a free, clean, and ever-lasting source of energy. In ancient times, it met the necessity for cooking and warmth. Nowadays, it is employed in a variety of ways, including converting solar energy to electrical energy by means of solar panels (PVs). The sun produces more than enough energy to cover the entire world's energy requirements and, unlike fossil fuels, it will not run out anytime soon. The only constraint of solar power as a renewable energy source is our ability to convert it into electricity efficiently and cost-effectively. The harvesting and usage of light and/or heat energy generated by the sun, as well as the technologies (passive and active) involved in achieving such goals, are regarded as being essential to the solar energy concept [9,10]. Solar energy is harnessed using three primary technologies: photovoltaics (PV), which directly converts light to electricity; concentrating solar power (CSP) [11,12], which uses heat from the sun (thermal energy) to drive utility-scale electric turbines; and solar heating and cooling (SHC) systems [13], which collect thermal energy to provide hot water and air heating or conditioning. The most common technology is PVs, which are devices made of semiconductor materials that directly convert sunlight falling on them to electrical energy. PVs should be installed in such a way that they form a tilt angle with the horizontal plane to extract the most power from them and allow sunlight to fall at a steep angle [14].

According to global data, more solar photovoltaic capacity is being installed than any other generation technology, making solar power the world's favorite new type of electricity generation. Solar photovoltaics had an electrical capacity of 8,485,405 GW in 2021, out of a total of 30,683 GW of renewable energy [15]. The photovoltaic sector accounts for approximately 28% of total renewable energy. However, when we look at Romania's renewable energy sector, we can see that its capacity is not very large, with a capacity quantity of approximately 11,138 TW, of which the photovoltaic sector accounts for only 12.6% (approximately 1.4 TW). When we look at the map of solar radiation capacity around the world, we can see that countries such as the Netherlands, Poland, the United Kingdom, and Germany are in areas with solar energy resources similar to Romania, but their production is much higher. Until now, Romania has had a little well-established supporting program for the development of the renewable energy sector; however, in 2022, a program to increase wind and solar energy production will be developed for small and large enterprises that can help to improve the renewable sector.

Floating photovoltaics is a concept that has gained popularity in recent years, with no commercial deployments and only a few demonstrator projects deployed globally [16,17]. Many places around the world, primary islands such as Japan, Singapore, Korea, and the Philippines, do not have enough land for PV installations. There is already demand in this field in countries such as Japan, Australia, the United States, Brazil, Korea, India, and others, and it is expected to expand globally. Floating solar systems can be installed in bodies of water such as oceans, lakes, lagoons, reservoirs, fish farms, dams, canals, and so on. Far Niente wineries in California, USA, received the most media attention and was widely regarded as the first to develop a floating PV project (despite the fact that a research floating PV project had been installed the previous year in Aichi, Japan) [18]. A large number of studies have been conducted to examine the advantages that floating photovoltaics have over conventional ones on the ground [19,20]. Choi Y.K [21] also conducted research in this area, comparing empirical data from floating photovoltaic systems developed in Korea with those on the ground. Following studies that concluded that floating photovoltaic systems can be more efficient by over 11%, the development of larger projects could lead to a larger-scale approach to the floating solar energy sector. Another study that highlights the superior efficiency of FPV was conducted by Sasmanto A.A et al. [22].

The power output of solar cells varies in response to temperature changes. Because the efficiency of the PV module is temperature dependent, installing solar PV systems on the water's surface benefits from a significantly lower ambient temperature due to the cooling effect of water [23–25]. If aluminum frames are used to support the floating solar PV module, the cooler temperature from the water is also carried out, lowering the overall temperature of the modules, as determined by the work of Liu H. et al. [26]; this aspect is also studied by El Hammoumi A. et al. [27].

Another aspect that is being thoroughly researched is the environmental benefits. Floating photovoltaics reduce negative effects such as deforestation, bird death, erosion, microclimate change, and others [28]. In addition, photovoltaics can reduce the evaporation of lake water and prevent algae growth [29]. Studies in these areas are being conducted by Elshafei M. [30], where one of the main conclusions is related to the possibility of reducing water loss through floating photovoltaics.

Potential assessment is one of the most popular FPV-related topics since it has been demonstrated that there is technological potential for anthropogenic reservoirs all over the world. A study that investigates the potential of solar energy is the one in reference [31], which investigates the possibility of placing floating photovoltaic panels in existing hydropower reservoirs in Laos. Another study that focuses on the energy potential is the one from reference [32], which analyzes the possibility of placing FPV on 10% of the surfaces covered by water, which could generate 31% of the energy needs in Spain. The research in references [33,34] examines the same topic, but this time for India. However, they also integrate studies on the prevention of water evaporation, which is a significant issue in this region. A study that covers the whole of Europe is found in reference [35], which studies all the hydropower basins on the continent in the desire to increase attention on this industry with the idea in mind of achieving carbon neutrality by 2030. One of the nations mentioned is Romania, which has a surface area of 394 km^2 and ranks fourth among the countries with reservoir areas. This region might be used for the installation of FPV with a capacity of 44.1 TWh if the entire surface is used. For Romania, this topic is not addressed; there are some studies, such as the one in reference [35] which address the entirety of Europe and is primarily focused on hydropower reservoirs, but there are no studies that also analyze the lakes that are used for irrigation. A study that addresses this topic is carried out by Popa B. et al. [36], which analyzes the possibility of locating a 1 MW FPV farm on Morii Lake.

The need to find innovative strategies that may result in acquiring huge amounts of energy is required given the current state of Europe, which is dealing with a widespread energy crisis that has caused energy costs to rise exponentially. At the national level, Romania is now dealing with this issue and is working to find quick solutions, making recent investments in renewable energy stand out. The goal of this study is to provide insight into the Romanian market's unrealized potential for floating solar energy. It intends to demonstrate the advantages of FPV over those on land, enabling the growth of the solar energy industry. In this way, the importing of power from other countries will be reduced to the maximum.

This work aims to evaluate the solar radiation on three lakes in Romania: Lake Razim, Lake Sinoe, and Lake Golovita, but it also extends to locations with deeper waters. To perform this, data on solar radiation and temperature were extracted from the ERA5 database, and in situ data will be used to validate the satellite data. A 540 kWh PV will be used to outline an overview of the solar radiation from the chosen locations. The novelty of this work is that it focuses on water-covered areas in Romania, where there is a desire to use PVs on water. The first FVP farm in Romania was launched in 2022 in the port of Constanta and was evaluated at an annual production of 15,000 kWh provided by its 22 PVs. So, the values obtained from this study can be considered as benchmarks in future investigations for the identification of other locations for the exploration of this energy sector that is less addressed in Romania, and it could have a huge impact in combating polluting sources. However, there are many challenges that this sector must overcome, the major one is related to the sea waves, as the system developed for Romania is able to deal only with waves of a maximum height of one meter.

This work is divided into five sections, the first of which presents the literature review that was used to write the work to observe what has been studied over time and to structure the work. The second section discusses the research area and gives the data used to assess resources, technical specifications used to compute energy production, and methods. The third section presents the results acquired after processing the ERA5 database data. Section four contains data from in situ measurements, as well as a brief discussion of the results and comparisons to previous research. Finally, the study's conclusions and potential future research topics are given.

2. Materials and Methods

2.1. The Location of Interest

While early floating PV projects were typically located on landlocked bodies of water, such as lakes and reservoirs, organizations have recently begun to consider installing plants on offshore waters. Although this entails far more technical challenges, such as surviving heavy swells and overcoming saltwater corrosion, it also opens up vast new areas for floating PV, potentially in conjunction with aquaculture. To carry out the study, four points of interest were chosen (Table 1), three of which are in the Razim-Sinoie Lagoon Complex and one in the western part of the Black Sea. The location at P4 was chosen to determine whether there are better resources in open water than on the lake; their exact position is also illustrated in Figure 1.

Table 1. The location of the four chosen sites.

	Sites			
	P1	P2	P3	P4
Location	Lake Razim	Lake Golovita	Lake Sinoe	Black Sea
Latitude (°)	44°57′07.03″ N	44°43′17.40″ N	44°28′00.95″ N	44°20′35.93″ N
Longitude (°)	28°51′37.11″ E	28°47′32.28″ E	28°45′23.41″ E	28°41′39.11″ E

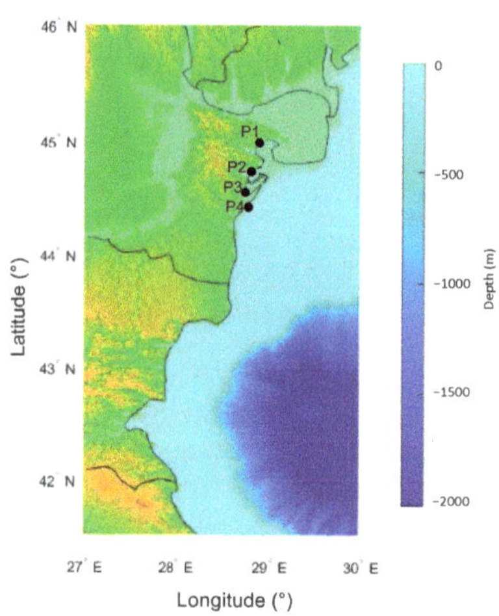

Figure 1. Map of the western side of the Black Sea and the location of the four sites considered.

The Razim-Sinoie Complex is part of the Danube Delta Biosphere Reserve and is located in the south of the Danube Delta. It consists primarily of lakes, sea beds, and some higher relief formations. Even though its total surface area is approximately 800 square kilometers, its depth does not exceed 3.5 m. These are low-salinity lakes formed by the mixing of fresh and saltwater due to their proximity to both the Danube and the Black Sea.

Point P4 is located in the Black Sea, an intercontinental sea located between South-Eastern Europe and Anatolia. Romania has a 245 km coastline that connects to the Black Sea in the southeast.

The average annual temperature in the lake area is around 10 °C, and the coastal area, which is a strip of 10–15 km west of the seashore, benefits from temperatures above 11 °C.

2.2. ERA5 Data Set

The ECMWF (European Centre for Medium-Range Weather Forecasts) is a non-governmental organization supported by 35 countries. It generates global numerical weather forecasts and other data for its member and cooperating countries, as well as the general public. The ECMWF's most recent reanalysis product is ERA5 [37,38]. Following several years of modeling and data assimilation advancements, a new model cycle for the Integrated Forecasting System (IFS Cycle 47r3) [39] was introduced into the reanalysis operations to ensure a substantial improvement in forecast accuracy and computational efficiency. The reanalysis combines the model with observations from around the world to create a globally complete and consistent data set that is constrained by physical laws. For each hour of the day, the ERA5 data set provides estimates of numerous atmospheric, land–surface, and sea-state parameters on $0.25^{\circ} \times 0.25^{\circ}$ latitude–longitude grids [40,41]. Surface solar radiation downwards (SSRD) [42] and total sky direct solar radiation at the surface (FDIR) [43] are the parameters used in this study to represent the amount of shortwave radiation (surface direct and diffuse solar radiation) and direct radiation reaching the Earth's surface, respectively. The year 2021 was chosen as the study's time frame.

2.3. The Mathematical Model Used

The entire quantity of shortwave radiation received from above by a surface horizontal to the ground is referred to as global horizontal irradiance (GHI). This value is especially important for solar installations since it combines both direct normal irradiance (DNI) and diffuse horizontal irradiance (DHI), and the relation between all three parameters can be expressed using the formula below [44,45]. DNI is solar radiation that travels in a straight line from the sun's current position in the sky. DHI is solar energy that has been scattered by molecules and particles in the atmosphere and comes from all directions equally. GHI represents all the light that arrives on a horizontal plane, from the sun, sky, and clouds. By subtracting DNI from GHI, DHI can be obtained. We should mention that direct solar radiation also includes the radiation that has been scattered by cloud particles by a fraction of a degree. All parameters are in W/m^2.

$$\mathrm{GHI} = \mathrm{DHI} + \mathrm{DNI}\cos\theta_z \tag{1}$$

$$\mathrm{DHI} = \mathrm{GHI} - \mathrm{DNI}\cos\theta_z \tag{2}$$

where θ_z represents the zenith angle in degrees. The angle formed by the sun's beams and the vertical direction is known as the solar zenith angle [46]. This means that the zenith angle decreases as the Sun rises higher in the sky. The formula is:

$$\cos\theta = \sin\phi \sin\delta + \cos\phi \cos\delta \cos h \tag{3}$$

where ϕ represents the local latitude, δ is the current declination of the sun, and h is the hour angle, in the local solar time. All mentioned parameters are in degrees.

The solar declination angle (Equation (4)) is the angle formed by the sun's beams and the Earth's equator [37]. The solar declination angle varies with time; it is not a fixed quantity. Every single day will be unique. However, the angle is limited to 23.44° and 23.44°.

$$\delta = 23.4 \sin\left(360^{\circ}\frac{n + 284}{365}\right) \tag{4}$$

where n denotes the day of the year (for example 1 for January 1, 32 for February 1, 60 for March 1, etc.).

The hour angle (h) is defined as the angular displacement of the sun east or west of the local meridian caused by the Earth's rotation and is stated in degrees as:

$$h = 15(\mathrm{AST} - 12) \tag{5}$$

where AST means the apparent solar time.

The difference between the two types of solar time is described by the equation of time (Equation (6)) [47,48]. The two periods that are different are the mean solar time, which follows a hypothetical mean sun with uniform motion down the celestial equator, and apparent solar time (AST), which directly tracks the diurnal motion of the sun [49]. By measuring the sun's present location (hour angle), as indicated (with varying degrees of precision) by a sundial, one can determine the apparent solar time. The time displayed by a reliable clock set up so that its variations from apparent solar time has a mean of zero over the course of a year, which would be the mean solar time for the same location. The conversion from local standard time (LST) to solar time is accomplished in two steps. First, the equation of time is applied to the local standard time, and then a longitude (LON) correction is applied. This longitude correction is four minutes of time per degree of difference between the local (site) longitude and the longitude of the time zone's local standard meridian (LSM) [50]; so, AST is connected to LST as follows:

$$\mathrm{AST} = \mathrm{LST} + \frac{\mathrm{ET}}{60} + \frac{\mathrm{LON} - \mathrm{LSM}}{15} \tag{6}$$

$$\mathrm{ET} = 2.2918(0.0075 + 0.1868\ \cos(\mathrm{B}) - 3.2077\ \sin(\mathrm{B}) - 1.4615\ \cos(2\mathrm{B}) - 4.089\ \sin(2\mathrm{B}) \\ \text{and} \\ \mathrm{B} = \frac{360(\mathrm{n}-1)}{365} \tag{7}$$

where TZ is the time zone, given in coordinated hours ahead or behind universal time (UTC).

The average total solar radiation for an inclined surface (W/m^2) can be calculated using the method developed by Liu and Jardon [51], as follows:

$$\mathrm{GHI_T} = \mathrm{R}\cdot\mathrm{GHI} \tag{8}$$

R is the monthly ratio of daily average radiation on a tilted surface to that on a level surface. R may be calculated by evaluating the direct, diffuse, and reflected components of radiation incident on the tilted surface separately [52,53]. Assuming that diffuse and reflected radiation is isotropic, R can be calculated using the formula [54]:

$$\mathrm{R} = \left(1 - \frac{\mathrm{DHI}}{\mathrm{GHI}}\right)\mathrm{R_b} + \mathrm{DHI}\left(\frac{1+\cos\beta}{2\,\mathrm{GHI}}\right) + \rho\left(\frac{1-\cos\beta}{2}\right) \tag{9}$$

where ρ is the ground reflectance and has a value of 0.2 [55] for hot and humid tropical locations and R_b is the beam conversion factor and for the northern hemisphere and can be expressed as:

$$\mathrm{R_b} = \frac{\cos(\phi-\beta)\cos\delta\sin\omega_h + \left(\frac{\pi}{180}\right)\omega_h\sin(\phi-\beta)\sin\delta}{\cos\phi\cos\delta\sin\omega_s + \left(\frac{\pi}{180}\right)\omega_s\sin\phi\sin\delta} \tag{10}$$

where ω_s' (°) is the sunrise or sunset hour angle for the inclined surface [53] and has the following equation:

$$\omega_s' = \min\left\{\omega_s, \cos^{-1}(\tan(\phi-\beta)\tan\delta)\right\} \tag{11}$$

The sunrise and sunset hour angles both have the same numerical value; the sunrise angle is negative and the sunset angle is positive. The following equation may be used to compute both:

$$\omega_s = \cos^{-1}(-\tan\phi\tan\delta) \tag{12}$$

The annual solar energy output of a photovoltaic system

$$\mathrm{P} = \mathrm{A_p}\cdot\mathrm{r}\cdot\mathrm{GHI_T}\cdot\mathrm{PR} \tag{13}$$

where A_p is the area of the PV in m^2, r is the panel yield in % (determined using Equation (14)), GHI_T is the average solar radiation on panels W/m^2, and PR is the performance ratio which has a value of 0.75 usually but can range between 0.5 and 0.9.

$$r = \frac{PE}{10A_p} \quad (14)$$

where PE is the electrical power kWp.

3. Results

Because the production of sun-based energy is dependent on the availability of the sun, it is necessary to determine the hours when solar radiation is present. The average hourly global, diffuse, and direct radiation values calculated as a mean for all days in 2021 are illustrated in Figure 2. The peak value for all four locations is recorded around 11 o'clock. The only difference is given by site P1, which has the maximum at 10 o'clock, but the difference between the global radiation from 11 o'clock and that from 10 o'clock is only 1 W/m^2. For these locations, no value for solar radiation was recorded in an 8 h interval.

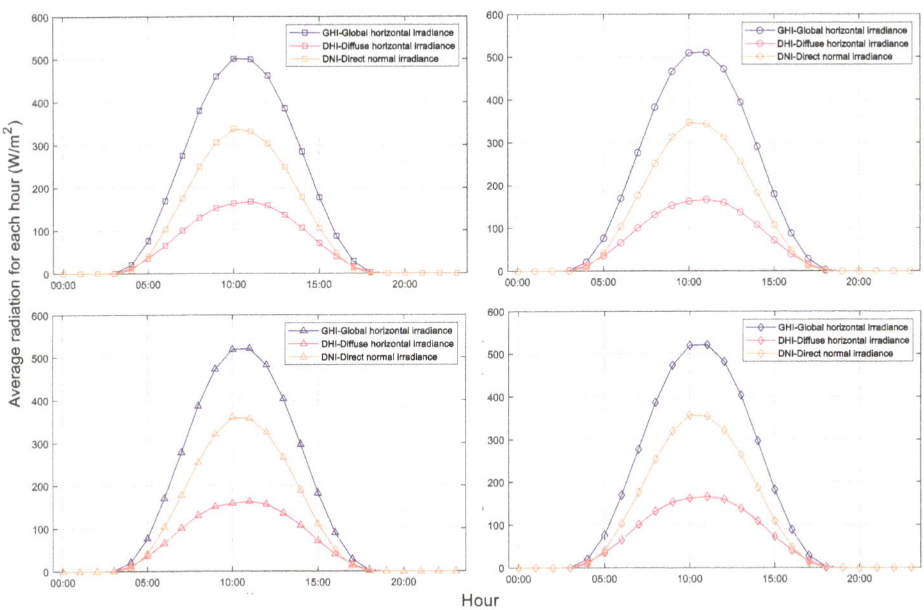

Figure 2. The curve of GHI, DHI, and DNI. The square symbol is for Lake Razim; the circle symbol is for Lake Golovita; the triangle symbol is for Lake SInoe; and the diamond symbol is for the Black Sea.

In Figures 3 and 4, it can be seen that the summer months in this region receive a large quantity of solar radiation, with the month of July having the highest value of global solar radiation and the P3 site also having the highest value of approximately 287 W/m^2, and the P1 site having the lowest value of 278 W/m^2. December receives the least amount of solar radiation with just 41 W/m^2 registered for site P1. The difference between the most productive site and the least productive is 3.2% for the maximum radiation and 9% for the minimum.

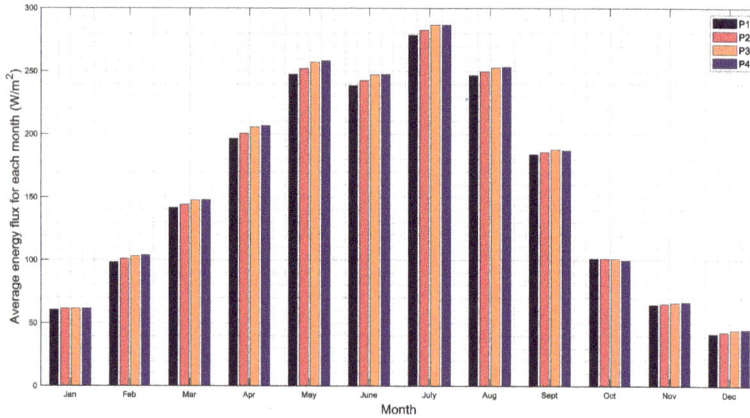

Figure 3. Monthly average GHI of solar irradiance (W/m^2).

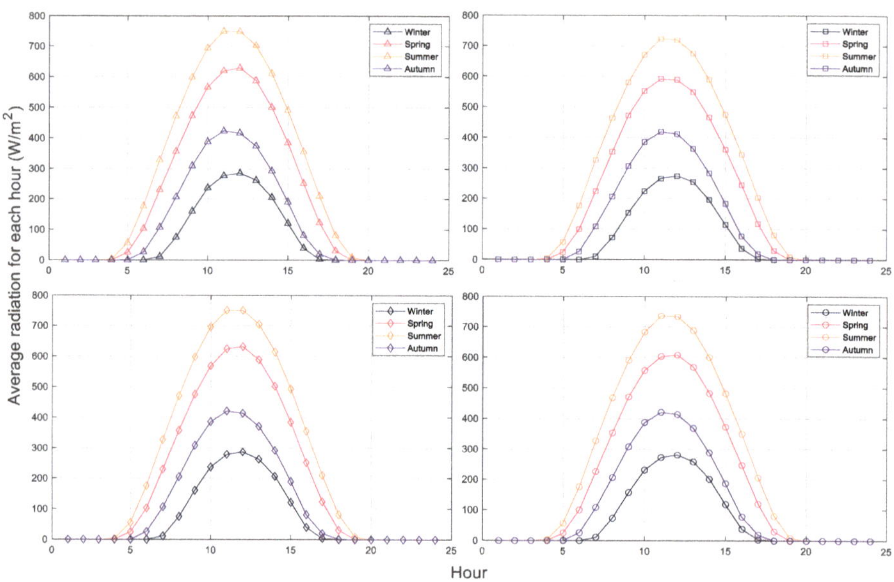

Figure 4. Solar radiation (in W/m^2) for each season and daily hour. The square symbol is for Lake Razim; the circle symbol is for Lake Golovita; the triangle symbol is for Lake SInoe; and the diamond symbol is for the Black Sea.

Figure 5 describes the tilt angle for each day of the year, and Table 2 represents the monthly one. As can be seen, the differences from one location to another is insignificant. According to the seasonal optimal tilt angle calculation based on solar angles, the tilt angle of the summer season is the smallest and the tilt angle of the winter season is the largest. The maximum tilt angle was observed for 20–21 December as 68.4°. For winter, the optimal tilt angle is considered to be 64° for all four locations, and for summer it is 24°. Additionally, for the cold half of the year, the optimal tilt angle can be obtained from the sum of the latitude and 15°, and for the warm half of the year by the difference between the latitude and 15°. This result can also be obtained by using Table 2, the average of the cold months being approximately 59° and the average of the warm months being approximately 29°.

The annual optimum tilt angle was calculated by averaging the value of optimum tilt angles for all months of a year and was found to be the exact latitude value for each location.

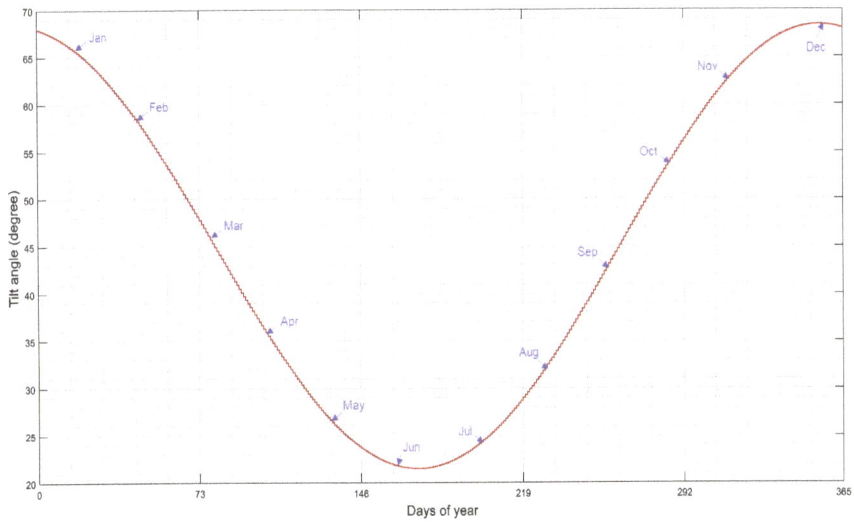

Figure 5. The annual variation in tilt angle.

Table 2. Monthly optimum fixed tilt angles.

Sites	January	February	March	April	May	June
P1	65.8	58.3	47.3	35.5	26.1	21.9
P2	65.6	58.0	47.1	35.2	25.9	21.6
P3	65.3	57.8	46.9	35.0	25.7	21.4
P4	65.2	57.7	46.7	34.9	25.5	21.3
	July	**August**	**September**	**October**	**November**	**December**
P1	23.9	31.7	43.0	54.8	64.0	68.0
P2	23.6	31.4	42.7	54.6	63.8	67.8
P3	23.4	31.2	42.5	54.3	63.5	67.6
P4	23.2	31.0	42.3	54.2	63.4	67.4

The graph below (Figure 6) depicts the change in the solar declination angle over time. The solar declination angle is positive from the vernal equinox to the autumnal equinox, as illustrated in the graph below, or from 20 March 20 to 22 September (or 23), and it is expected to be negative for the rest of the year. The declination angle must be estimated to calculate the solar elevation. The solar elevation increases with the declination angle and its peak is during the summer months. As winter approaches, the declination angle reduces and the solar elevation decreases. As a result, in winter, the sun descends toward the horizon. Thus, the tilt angle of the PV rises, and the PVs are virtually vertically aligned to optimize solar output, as can be seen in Figure 5.

To determine the annual energy production, the average solar radiation for the year 2021 was used, which has a value of 700 kWh. The PV used for this study was JRH 540 W, with a panel area of 2.584 m2 and a maximum power of 540 W (Table 3) with an adjustable tilt angle. Figure 7 shows that sites P3 and P4 have almost identical values for the annual energy production of 728 W/m^2, the difference between them being almost imperceptible. The lowest value is recorded in location P1 with approximately 700 W/m^2. The capacity factor for the year 2021 is around 18.2%; the lowest value is also recorded for the P1 site at only 17.7%.

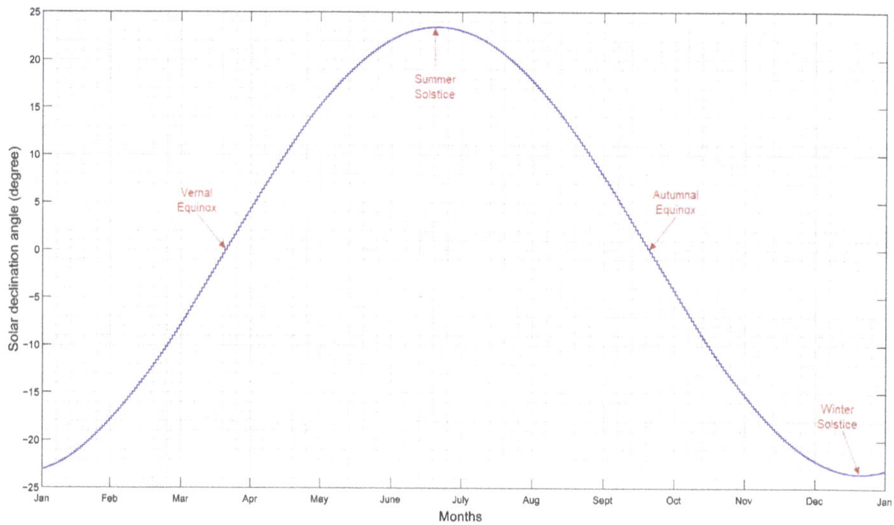

Figure 6. The annual variation in declination angle.

Table 3. PV catalog specifications.

Characteristics	
Power	540 W
Type	Monocrystalline
Area	2.584 m^2
Number of cells	144
Open Cct voltage	49.55 V
Short Cct current	13.89 A
Voltage, max power	41.62 V
Current, max power	12.98 A

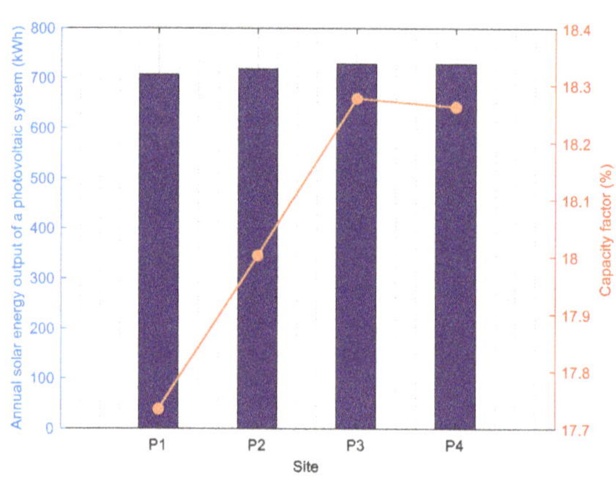

Figure 7. The annual energy production of the PV of 540 W and the capacity factor for the four locations.

The temperature coefficient, or more specifically, the open-circuit voltage temperature coefficient, given in either a percentage of VOC per degree C (%/°C) or volts per degree C (V/°C), is one of the factors that can affect the actual performance of a photovoltaic panel, causing it to vary away from its theoretical value. For the chosen panel, the temperature coefficient has a value of 0.275%/°C. This coefficient represents the amount by which its output voltage, current, or power varies as a result of a physical change in the ambient temperature conditions surrounding it before the array begins to warm up. As a result, the performance of the panel will decrease when the temperature increases compared to the reference one, which is 25 °C, and when the temperature decreases, it will improve. The performance of the 540 W PV can be seen in Figure 8.

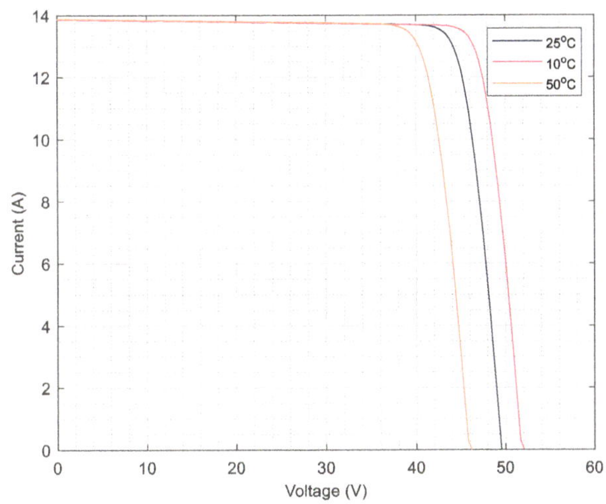

Figure 8. Temperature-dependent PV module current output.

Figures 9 and 10 show the average minimum and maximum temperatures for our locations. According to the graph, the average maximum temperature in summer is 25.7 °C, while the average minimum temperature in winter is around 7.5 °C for lake locations and 4 °C for sea locations. Figure 10 shows that the water temperature is more constant than the air temperature, so in winter the water temperature is higher than the air temperature, and in summer the opposite, because water heats up and cools down slower.

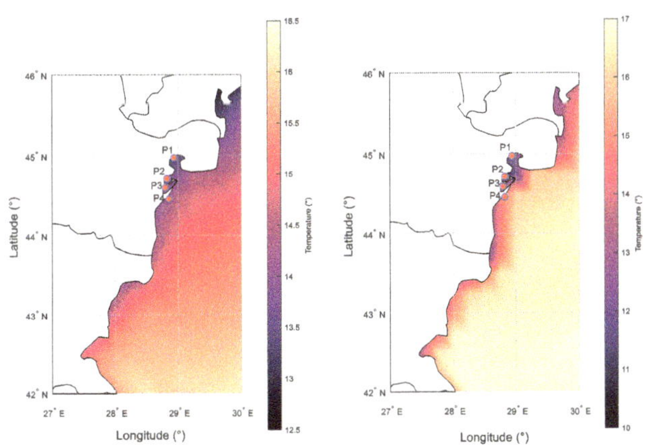

Figure 9. Map with the lakes of eastern Romania and the western Black Sea. The left side shows the air temperature and the right side shows the water temperature.

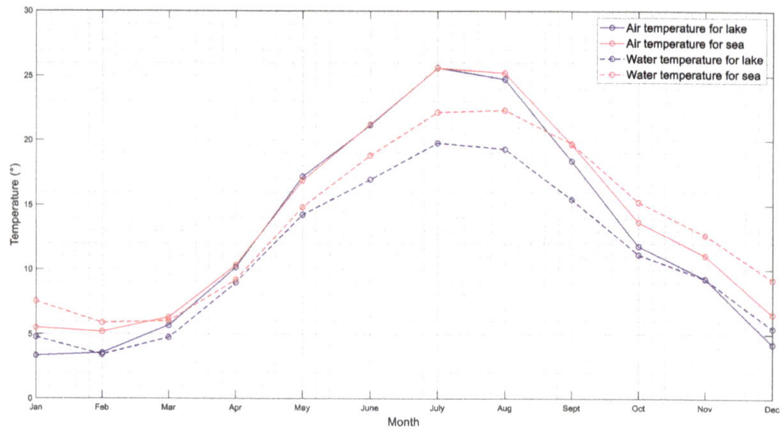

Figure 10. Average monthly temperature variations.

4. Discussion

Renewable energy from the sun is one of the most used for the production of green energy globally, ranking 3rd after hydropower and wind. In this case, solar radiation was examined for four locations in Romania that are located on the water, to evaluate the possibility of exploiting solar energy not only on land but also on the water. An important aspect to emphasize is the one related to the validation of the data obtained from the ERA5 database. Numerous studies evaluate the veracity of the data obtained from different databases, such as the studies performed by Jiang et al. [40], which analyze 98 locations in China using in situ data and data from the ERA5 database, after which they could observe that the values for GHI are close for the two, but large differences appear for DHI and DNI because the model cannot accurately evaluate aerosols, clouds, and their interaction. Another study that analyzed the correlation between the data obtained from the ERA5 database and measured data, for three locations in Germany, is the one in Ref. [56]. This paper showed that the best results were recorded in high-pressure situations, and in the rest of the cases, the ERA5 data overestimated the results. Similar studies were also performed in references [57,58], and following them, we could observe that the areas that are dominated by rains and clouds have the weakest results when it comes to the correlation of the ERA5 database with the in situ data. Other studies compare data from different reanalysis databases, such as the work by Ref. [59], which conducts a comparative study between the solar radiation obtained from the ERA-5, MERRA-2, ERA-Interim, JRA-55, NCEP-NCAR, NCEP-DOE, and CFSR databases, and measured data. The closest results are those of the ERA5 and ERA-Interim databases. The better results were observed for the ERA5-land database, which provides better results than ERA5 [60]. This can be attributed to the ERA5-land database's much better resolution of 9 km compared to ERA5's resolution of 31 km.

Figure 11 shows the GHI for four locations in Romania, all in areas where solar radiation is considered to have the highest values. The Constanta site is relevant in this case because it is located near the four locations used for this study. Since no measuring devices are installed in these locations, we used the in situ measurements in Constanta to further compare the results obtained to determine if they are relevant or not. As can be seen in Figures 3 and 12, the differences between the measured data and those from the database are relatively small. In this case, the root mean square error (RMSE) has a value of 32.29 W/m^2 for P1, 30.06 W/m^2 for P2, 27.73 W/m^2 for P3, and 27.58 W/m^2 for P4. In this case, the data obtained from the ERA5 database underestimates the real values of solar radiation. Additionally, in Figure 11, the other three locations were chosen to make a small comparison between the solar resources; all these locations exploit the soil resources, especially the areas in the vicinity of Bucharest, but we can see that the best resources are still in Constanta where the study in question is carried out.

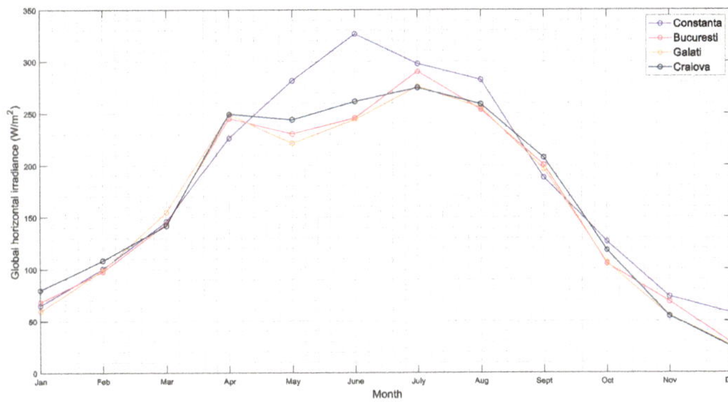

Figure 11. GHI for four locations in Romania from high-potential areas.

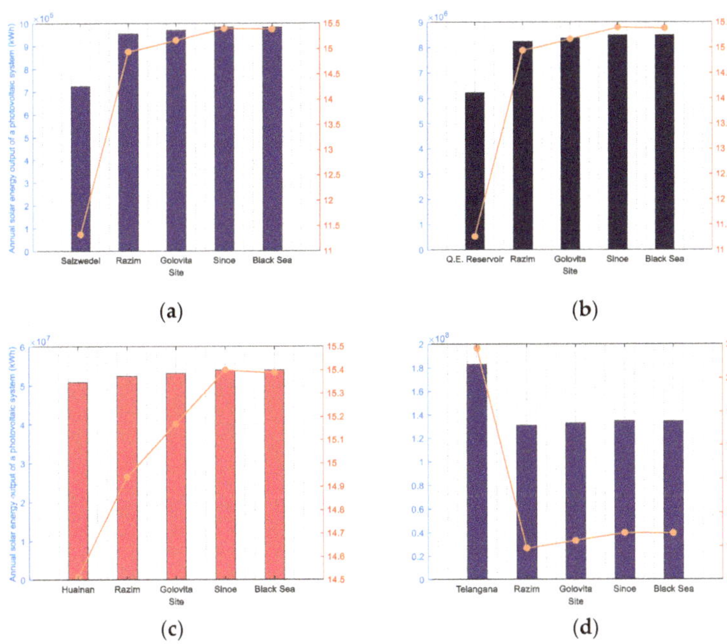

Figure 12. Four FPV farms annual energy output and their usage within the study's chosen locations: (**a**) Germany; (**b**) the United Kingdom; (**c**) China; and (**d**) India.

Another study focused on solar radiation in Romania is the one developed by AKTAG and YILMAZ [61] that analyzed the port cities on the shore of the Black Sea, among which is Constanta. In the reference mentioned above, the average solar radiation can be deduced as 14.02 $MJm^{-2}day^{-1}$, which would represent approximately 162.26 W/m^2 as an average for the year studied. The annual average for the four locations is 158.39 W/m^2 for P1, 160.9 W/m^2 for P2, 163.51 W/m^2 for P3, and 163.73 W/m^2 for P4. We can see that the values are close so that the profile of solar radiation can be said to be kept constant over time.

The more that FPV technology advances, the more farms of this kind appear all over the world. There is currently more than 3 MW of installed capacity. In this context, we have chosen to simulate four projects that are in the course of commissioning or are already in operation. These projects were selected based on size and installed capacity. If we refer to the size, they range from 6800 m^2 to 2,023,430 m^2. With these visible differences, the installed capacities will also be varied, from 729 kW to 0.10 GW projects. In terms of the regions, we picked both high- and low-radiation areas but also regions that were similar to the one in this study. Table 4 shows the locations of the chosen farms, as well as their properties.

Table 4. The location of the four chosen sites.

	Sites			
	Germany	**United Kingdom**	**China**	**India**
Location	Salzwedel	Queen Elizabeth II Reservoir	Huainan	Telangana
Capacity	729 kW	6.3 MW	40 MW	100 MW
Area	6800 m^2	57,000 m^2	800,000 m^2	2,023,430 m^2

The only farm that generates more energy at its original site is the one in India, as shown in Figure 12, because the solar radiation there is substantially greater than in the areas we selected for the research. Projects, such as those in Figure 12, may provide for around 280, 2425, 15,650, or 38,510 homes, given that the average household usage in Romania is roughly 283 kWh/month, or 3396 kWh/year. Even if the project with the lowest capacity (729 kW) may be regarded as weak, we can state that in this situation it may be useful. A project like this could be sufficient for the first three locations that are on the lake since they could supply the communities nearby that are not developed and where the population is declining.

Romania does not have high values of solar radiation on a global scale, as it is located in an area with medium to low solar radiation resources. However, countries such as Germany, which is located in more deficient areas, is among the top four countries with the most PVs installed (in fourth place), as is Japan, which is the third.

Another aspect worth analyzing is the one related to water and air temperatures. As already mentioned in this paper, water is an important factor/element that solar panels can use as a cooling system. As noted above, when the temperature increases, the performance of the panel decreases, so that in the summer months, the average monthly temperature for the selected areas is about 25.7 °C, with a maximum monthly temperature of 35 °C. When the temperature exceeds 25 °C, the PV will no longer perform as expected by the manufacturer. In this case, the temperature of the water can play an important role in cooling the panels, with temperatures 3 or 5 degrees lower than that of the air. The idea of reducing the temperature of the panels has been intensively studied over time and is described extensively in reference papers [62–64]. All these studies have shown that energy production improves by 3–6% compared to conventional PVs, and water has a positive effect even when its temperature is higher than that of the air.

5. Conclusions

The development of new projects that will result in the generation of power is crucial given the energy crisis Romania is facing. By implementing these projects, we can reduce the purchase of electricity from external grids. Since there are currently no studies that provide information about possible locations on the water where solar energy can be exploited, this study presents an overview of the benefits that can come from the implementation of FPV farms.

The present research highlights the importance of floating PV systems installed on bodies of still water such as ponds, lakes, dams, and reservoirs but also the sea. In this study, the GHI, DNI, and DHI components of surface solar irradiance observed on the west side of Romania have been analyzed, based on 1 h data obtained from the ERA5 database, and the analyzed period is between January 2021 and 31 December 2021. It also compares energy production by floating PV plants in different locations. The study yielded the following conclusions:

- The data obtained from the ERA5 database are similar to the measured ones; those from ERA5 underestimate solar radiation.
- The energy production for the chosen locations is high, reaching 700 kWh for a 540 W PV. These results are similar to the first PV farm located on the water in Romania, which is estimated at 15,000 kWh for 22 panels.
- Because the PVs will be floating on water, they will be cooler and thus will produce more power than those installed on land. Overheating can lead to component damage, and by placing them on water, maintenance can be significantly reduced. Furthermore, rain and wind help to clean the surface of the PV, minimizing the amount of maintenance necessary.
- The tilt angle has a significant impact on solar energy production. In our case, the value of this angle should be set at 64° and for summer at 24°. The use of PVs without an angle reduces energy production by about 10–15%. This tilt angle is a challenge for FPVs because the technology has not yet been sufficiently developed.
- FPV technology could be an innovative solution to the problem of insufficient land. The majority of land in Romania is used for agricultural purposes, and the country is positioned among the first countries in agriculture in Europe.
- The lakes used for this study have two main economic purposes, one being tourism, having beautiful fauna and flora, and the main one being for irrigation. Taking into account the remark about how the panels help to reduce water evaporation, as was also concluded in reference [65], we can say that the economic impact of irrigation is increased by the location of PVs on the water.
- By integrating several FPV farms into the four sites, we were able to see that, in three out of the four cases, our locations were able to produce more energy than their existing positions. Dobrogea nowadays is characterized by undeveloped, largely uninhabited settlements, and renewable energy sources are advantageous to these communities. Many of these villages use wind turbines to provide electricity. As a result, it is possible to assure the nearby communities' access to electricity by building even the smallest FVP project.
- The Black Sea location provided the best results, but it is also the most difficult in terms of environmental conditions, as it will be affected by waves, and the structure must be designed to last implying higher costs.

Previous research has revealed good qualities of wind energy for the Black Sea [66,67]. We can conclude that the Black Sea has enormous potential for renewable energy, whether it is solar or wind energy.

As future research directions, we intend to direct this study to a practical case either carried out in the laboratory or carried out in situ at one of the locations in the study.

Author Contributions: A.I.M. drafted this manuscript, processed the numerical data, and assembled the manuscript, G.A. checked and revised the manuscript and supervised it, L.R. came up with the idea of floating solar panels and was in charge of funding acquisition and project administration. All authors have read and agreed to the published version of the manuscript.

Funding: This work was carried out in the framework of the research project DREAM (Dynamics of the Resources and Technological Advance in Harvesting Marine Renewable Energy), supported by the Romanian Executive Agency for Higher Education, Research, Development, and Innovation Funding-UEFISCDI (grant number PN-III-P4-ID-PCE-2020-0008).

Institutional Review Board Statement: Not applicable.

Informed Consent Statement: Not applicable.

Data Availability Statement: Not applicable.

Conflicts of Interest: The authors declare no conflict of interest.

References

1. IRENA. *Global Hydrogen Trade to Meet the 1.5 °C Climate Goal: Part II—Technology Review of Hydrogen Carriers*; IRENA: Abu Dhabi, United Arab Emirates, 2022; ISBN 978-92-9260-431-8.
2. United Nations. *Report of the Conference of the Parties to the United Nations Framework Convention on Climate Change (21st Session)*; United Nations: Paris, France, 2015; Volume 4, p. 2017.
3. Wang, C.; Wang, Y.; Tong, X.; Ulgiati, S.; Liang, S.; Xu, M.; Wei, W.; Li, X.; Jin, M.; Mao, J. Mapping Potentials and Bridging Regional Gaps of Renewable Resources in China. *Renew. Sustain. Energy Rev.* **2020**, *134*, 110337. [CrossRef]
4. Singh, D.P.; Dhami, S.S.; Banwait, S.S.; Goyal, D. Performance Comparison of Fixed and Tracking Type Solar Plants. *Int. J. Innov. Technol. Explor. Eng.* **2020**, *9*, 612–620. [CrossRef]
5. Ravichandran, N.; Ravichandran, N.; Panneerselvam, B. Comparative Assessment of Offshore Floating Photovoltaic Systems Using Thin Film Modules for Maldives Islands. *Sustain. Energy Technol. Assess.* **2022**, *53*, 102490. [CrossRef]
6. Olabi, A.G.; Abdelkareem, M.A. Renewable Energy and Climate Change. *Renew. Sustain. Energy Rev.* **2022**, *158*, 112111. [CrossRef]
7. Kannan, N.; Vakeesan, D. Solar Energy for Future World—A Review. *Renew. Sustain. Energy Rev.* **2016**, *62*, 1092–1105. [CrossRef]
8. Essak, L.; Ghosh, A. Floating Photovoltaics: A Review. *Clean Technol.* **2022**, *4*, 752–769. [CrossRef]
9. Devabhaktuni, V.; Alam, M.; Depuru, S.S.S.R.; Green, R.C., II; Nims, D.; Near, C. Solar Energy: Trends and Enabling Technologies. *Renew. Sustain. Energy Rev.* **2013**, *19*, 555–564. [CrossRef]
10. Claus, R.; López, M. Key Issues in the Design of Floating Photovoltaic Structures for the Marine Environment. *Renew. Sustain. Energy Rev.* **2022**, *164*, 112502. [CrossRef]
11. Weinstein, L.A.; Loomis, J.; Bhatia, B.; Bierman, D.M.; Wang, E.N.; Chen, G. Concentrating Solar Power. *Chem. Rev.* **2015**, *115*, 12797–12838. [CrossRef]
12. Santos, J.J.C.S.; Palacio, J.C.E.; Reyes, A.M.M.; Carvalho, M.; Freire, A.J.R.; Barone, M.A. Concentrating Solar Power. In *Advances in Renewable Energies and Power Technologies*; Elsevier: Amsterdam, The Netherlands, 2018; pp. 373–402.
13. Ge, T.S.; Wang, R.Z.; Xu, Z.Y.; Pan, Q.W.; Du, S.; Chen, X.M.; Ma, T.; Wu, X.N.; Sun, X.L.; Chen, J.F. Solar Heating and Cooling: Present and Future Development. *Renew Energy* **2018**, *126*, 1126–1140. [CrossRef]
14. Hairong, X.; Hao, G.; Yusuke, Y.; Takayoshi, S.; Renzhi, M. Photo-Enhanced Rechargeable High-Energy-Density Metal Batteries for Solar Energy Conversion and Storage. *Nano Res. Energy* **2022**, *1*, e9120007. [CrossRef]
15. International Renewable Energy Agency. *Renewable Capacity Statistics 2022*; IRENA: Abu Dhabi, United Arab Emirates, 2022; ISBN 978-92-9260-428-8.
16. Sahu, A.; Yadav, N.; Sudhakar, K. Floating Photovoltaic Power Plant: A Review. *Renew. Sustain. Energy Rev.* **2016**, *66*, 815–824. [CrossRef]
17. Vo, T.T.E.; Ko, H.; Huh, J.; Park, N. Overview of Possibilities of Solar Floating Photovoltaic Systems in the OffShore Industry. *Energies* **2021**, *14*, 6988. [CrossRef]
18. Kumar, M.; Niyaz, H.M.; Gupta, R. Challenges and Opportunities towards the Development of Floating Photovoltaic Systems. *Sol. Energy Mater. Sol. Cells* **2021**, *233*, 111408. [CrossRef]
19. Trapani, K.; Millar, D.L. Floating Photovoltaic Arrays to Power the Mining Industry: A Case Study for the McFaulds Lake (Ring of Fire). *Env. Prog. Sustain. Energy* **2016**, *35*, 898–905. [CrossRef]
20. Trapani, K.; Millar, D.L.; Smith, H.C.M. Novel Offshore Application of Photovoltaics in Comparison to Conventional Marine Renewable Energy Technologies. *Renew Energy* **2013**, *50*, 879–888. [CrossRef]
21. Choi, Y.K. A Study on Power Generation Analysis of Floating PV System Considering Environmental Impact. *Int. J. Softw. Eng. Its Appl.* **2014**, *8*, 75–84. [CrossRef]
22. Sasmanto, A.A.; Dewi, T. Eligibility Study on Floating Solar Panel Installation over Brackish Water in Sungsang, South Sumatra. *EMITTER Int. J. Eng. Technol.* **2020**, *8*, 240–255. [CrossRef]
23. Majid, Z.A.A.; Ruslan, M.H.; Sopian, K.; Othman, M.Y.; Azmi, M.S.M. Study on Performance of 80 Watt Floating Photovoltaic Panel. *J. Mech. Eng. Sci.* **2014**, *7*, 1150–1156. [CrossRef]
24. Cazzaniga, R.; Cicu, M.; Rosa-Clot, M.; Rosa-Clot, P.; Tina, G.M.; Ventura, C. Floating Photovoltaic Plants: Performance Analysis and Design Solutions. *Renew. Sustain. Energy Rev.* **2018**, *81*, 1730–1741. [CrossRef]
25. Micheli, L. The Temperature of Floating Photovoltaics: Case Studies, Models and Recent Findings. *Sol. Energy* **2022**, *242*, 234–245. [CrossRef]
26. Liu, H.; Krishna, V.; Lun Leung, J.; Reindl, T.; Zhao, L. Field Experience and Performance Analysis of Floating PV Technologies in the Tropics. *Prog. Photovolt. Res. Appl.* **2018**, *26*, 957–967. [CrossRef]
27. El Hammoumi, A.; Chalh, A.; Allouhi, A.; Motahhir, S.; El Ghzizal, A.; Derouich, A. Design and Construction of a Test Bench to Investigate the Potential of Floating PV Systems. *J. Clean. Prod.* **2021**, *278*, 123917. [CrossRef]
28. Pimentel Da Silva, G.D.; Branco, D.A.C. Is Floating Photovoltaic Better than Conventional Photovoltaic? Assessing Environmental Impacts. *Impact Assess. Proj. Apprais.* **2018**, *36*, 390–400. [CrossRef]

29. Haas, J.; Khalighi, J.; De La Fuente, A.; Gerbersdorf, S.U.; Nowak, W.; Chen, P.-J. Floating Photovoltaic Plants: Ecological Impacts versus Hydropower Operation Flexibility. *Energy Convers. Manag.* **2020**, *206*, 112414. [CrossRef]
30. Elshafei, M.; Ibrahim, A.; Helmy, A.; Abdallah, M.; Eldeib, A.; Badawy, M.; AbdelRazek, S. Study of Massive Floating Solar Panels over Lake Nasser. *J. Energy* **2021**, *2021*, 6674091. [CrossRef]
31. Nhiavue, Y.; Lee, H.S.; Chisale, S.W.; Cabrera, J.S. Prioritization of Renewable Energy for Sustainable Electricity Generation and an Assessment of Floating Photovoltaic Potential in Lao PDR. *Energies* **2022**, *15*, 8243. [CrossRef]
32. López, M.; Soto, F.; Hernández, Z.A. Assessment of the Potential of Floating Solar Photovoltaic Panels in Bodies of Water in Mainland Spain. *J. Clean. Prod.* **2022**, *340*, 130752. [CrossRef]
33. Mamatha, G.; Kulkarni, P.S. Assessment of Floating Solar Photovoltaic Potential in India's Existing Hydropower Reservoirs. *Energy Sustain. Dev.* **2022**, *69*, 64–76. [CrossRef]
34. Agrawal, K.K.; Jha, S.K.; Mittal, R.K.; Vashishtha, S. Assessment of Floating Solar PV (FSPV) Potential and Water Conservation: Case Study on Rajghat Dam in Uttar Pradesh, India. *Energy Sustain. Dev.* **2022**, *66*, 287–295. [CrossRef]
35. Kakoulaki, G.; Sanchez, R.G.; Amillo, A.G.; Szabo, S.; de Felice, M.; Farinosi, F.; de Felice, L.; Bisselink, B.; Seliger, R.; Kougias, I. Benefits of Pairing Floating Solar Photovoltaics with Hydropower Reservoirs in Europe. *Renew. Sustain. Energy Rev.* **2023**, *171*, 112989. [CrossRef]
36. Popa, B.; Vuta, L.I.; Dumitran, G.E.; Picioroaga, I.; Calin-Arhip, M.; Porumb, R.-F. FPV for Sustainable Electricity Generation in a Large European City. *Sustainability* **2021**, *14*, 349. [CrossRef]
37. Hersbach, H. *The ERA5 Atmospheric Reanalysis*; American Geophysical Union: Washington, DC, USA, 2016; p. NG33D-01.
38. Hersbach, H.; Bell, B.; Berrisford, P.; Hirahara, S.; Horányi, A.; Muñoz-Sabater, J.; Nicolas, J.; Peubey, C.; Radu, R.; Schepers, D. The ERA5 Global Reanalysis. *Q. J. R. Meteorol. Soc.* **2020**, *146*, 1999–2049. [CrossRef]
39. Savazzi, A.C.M.; Nuijens, L.; Sandu, I.; George, G.; Bechtold, P. The Representation of Winds in the Lower Troposphere in ECMWF Forecasts and Reanalyses during the EUREC4A Field Campaign. *Atmos. Chem. Phys. Discuss.* **2022**, *22*, 1–29.
40. Jiang, H.; Yang, Y.; Bai, Y.; Wang, H. Evaluation of the Total, Direct, and Diffuse Solar Radiations from the ERA5 Reanalysis Data in China. *IEEE Geosci. Remote Sens. Lett.* **2019**, *17*, 47–51. [CrossRef]
41. Gleixner, S.; Demissie, T.; Diro, G.T. Did ERA5 Improve Temperature and Precipitation Reanalysis over East Africa? *Atmosphere* **2020**, *11*, 996. [CrossRef]
42. Hogan, R. *Radiation Quantities in the ECMWF Model and MARS*; ECMWF: Reading, UK, 2015.
43. Jiang, H.; Yang, Y.; Wang, H.; Bai, Y.; Bai, Y. Surface Diffuse Solar Radiation Determined by Reanalysis and Satellite over East Asia: Evaluation and Comparison. *Remote Sens.* **2020**, *12*, 1387. [CrossRef]
44. Stein, J.S.; Hansen, C.W.; Reno, M.J. *Global Horizontal Irradiance Clear Sky Models: Implementation and Analysis*; Sandia National Laboratories (SNL): Albuquerque, NM, USA; Livermore, CA, USA, 2012.
45. Yang, D.; Sharma, V.; Ye, Z.; Lim, L.I.; Zhao, L.; Aryaputera, A.W. Forecasting of Global Horizontal Irradiance by Exponential Smoothing, Using Decompositions. *Energy* **2015**, *81*, 111–119. [CrossRef]
46. Fatemi, S.A.; Kuh, A. Solar Radiation Forecasting Using Zenith Angle. In Proceedings of the 2013 IEEE Global Conference on Signal and Information Processing, Austin, TX, USA, 3–5 December 2013; IEEE: Piscataway, NJ, USA, 2013; pp. 523–526.
47. Deceased, J.A.D.; Beckman, W.A. *Solar Engineering of Thermal Processes*; Wiley: Hoboken, NJ, USA, 1982; Volume 3, ISBN 9780470873663.
48. Duffie, J.A.; Beckman, W.A.; Blair, N. *Solar Engineering of Thermal Processes, Photovoltaics and Wind*; John Wiley & Sons: Hoboken, NJ, USA, 2020; ISBN 11195-40283.
49. Sproul, A.B. Derivation of the Solar Geometric Relationships Using Vector Analysis. *Renew. Energy* **2007**, *32*, 1187–1205. [CrossRef]
50. *ASHRAE Handbook of Fundamentals*; ASHRAE Inc.: Atlanta, GA, USA, 2009; ISBN 6785392187.
51. Liu, B.Y.H.; Jordan, R.C. Daily Insolation on Surfaces Tilted towards the Equator. *Trans. ASHRAE* **1962**, *67*, 526–541.
52. Islam, M.A.; Alam, M.S.; Sharker, K.K.; Nandi, S.K. Estimation of Solar Radiation on Horizontal and Tilted Surface over Bangladesh. *Comput. Water Energy Environ. Eng.* **2016**, *5*, 54–69. [CrossRef]
53. Bakirci, K. General Models for Optimum Tilt Angles of Solar Panels: Turkey Case Study. *Renew. Sustain. Energy Rev.* **2012**, *16*, 6149–6159. [CrossRef]
54. Berisha, X.; Zeqiri, A.; Meha, D. Solar Radiation–The Estimation of the Optimum Tilt Angles for South-Facing Surfaces in Pristina. *Preprints* **2017**, 2017080010. [CrossRef]
55. Muneer, T. *Solar Radiation and Daylight Models*; Routledge: Boca Raton, FL, USA, 2007; ISBN 00804-74411.
56. Mierzwiak, M.; Kroszczyński, K.; Araszkiewicz, A. On Solar Radiation Prediction for the East–Central European Region. *Energies* **2022**, *15*, 3153. [CrossRef]
57. Jiao, B.; Li, Q.; Sun, W.; Martin, W. Uncertainties in the Global and Continental Surface Solar Radiation Variations: Inter-Comparison of in-Situ Observations, Reanalyses, and Model Simulations. *Clim. Dyn.* **2022**, *59*, 2499–2516. [CrossRef]
58. Trolliet, M.; Walawender, J.P.; Bourlès, B.; Boilley, A.; Trentmann, J.; Blanc, P.; Lefèvre, M.; Wald, L. Downwelling Surface Solar Irradiance in the Tropical Atlantic Ocean: A Comparison of Re-Analyses and Satellite-Derived Data Sets to PIRATA Measurements. *Ocean. Sci.* **2018**, *14*, 1021–1056. [CrossRef]
59. Tahir, Z.R.; Azhar, M.; Mumtaz, M.; Asim, M.; Moeenuddin, G.; Sharif, H.; Hassan, S. Evaluation of the Reanalysis Surface Solar Radiation from NCEP, ECMWF, NASA, and JMA Using Surface Observations for Balochistan, Pakistan. *J. Renew. Sustain. Energy* **2020**, *12*, 23703. [CrossRef]

60. Galanaki, E.; Emmanouil, G.; Lagouvardos, K.; Kotroni, V. Long-Term Patterns and Trends of Shortwave Global Irradiance over the Euro-Mediterranean Region. *Atmosphere* **2021**, *12*, 1431. [CrossRef]
61. Aktağ, A.; Yilmaz, E. A Suitable Model to Estimate Global Solar Radiation in Black Sea Shoreline Countries. *Energy Sources Part A Recovery Util. Environ. Eff.* **2012**, *34*, 1628–1636. [CrossRef]
62. Elminshawy, N.A.S.; El-Damhogi, D.G.; Ibrahim, I.A.; Elminshawy, A.; Osama, A. Assessment of Floating Photovoltaic Productivity with Fins-Assisted Passive Cooling. *Appl. Energy* **2022**, *325*, 119810. [CrossRef]
63. Dörenkämper, M.; Wahed, A.; Kumar, A.; de Jong, M.; Kroon, J.; Reindl, T. The Cooling Effect of Floating PV in Two Different Climate Zones: A Comparison of Field Test Data from the Netherlands and Singapore. *Sol. Energy* **2021**, *219*, 15–23. [CrossRef]
64. Kjeldstad, T.; Lindholm, D.; Marstein, E.; Selj, J. Cooling of Floating Photovoltaics and the Importance of Water Temperature. *Sol. Energy* **2021**, *218*, 544–551. [CrossRef]
65. Abdelal, Q. Floating PV; an Assessment of Water Quality and Evaporation Reduction in Semi-Arid Regions. *Int. J. Low-Carbon Technol.* **2021**, *16*, 732–739. [CrossRef]
66. Diaconita, A.I.; Rusu, L.; Andrei, G. A Local Perspective on Wind Energy Potential in Six Reference Sites on the Western Coast of the Black Sea Considering Five Different Types of Wind Turbines. *Inventions* **2021**, *6*, 44. [CrossRef]
67. Diaconita, A.; Andrei, G.; Rusu, L. New Insights into the Wind Energy Potential of the West Black Sea Area Based on the North Sea Wind Farms Model. *Energy Rep.* **2021**, *7*, 112–118. [CrossRef]

Journal of
Marine Science and Engineering

MDPI

Article

Experimental Modelling of an Isolated WECfarm Real-Time Controllable Heaving Point Absorber Wave Energy Converter

Timothy Vervaet [1,*], Vasiliki Stratigaki [1], Francesco Ferri [2], Louis De Beule [1,†], Hendrik Claerbout [1,†], Bono De Witte [1,†], Marc Vantorre [1] and Peter Troch [1]

1 Department of Civil Engineering, Ghent University, Technologiepark 60, B-9052 Zwijnaarde, Belgium
2 Department of the Built Environment, Aalborg University, Thomas Manns Vej 23, 9220 Aalborg, Denmark
* Correspondence: timothy.vervaet@ugent.be; Tel.: +32-9-264-54-89
† These authors contributed equally to this work.

Abstract: To offer point absorber wave energy converters (WECs) as a bankable product on the marine renewable energy market, multiple WECs will be installed together in an array configuration. The wave energy community (research and industrial) has identified the urgent need for available realistic and reliable data on WEC array tests in order to perform a better WEC array optimization approach and in order to validate recently developed (non-linear) numerical models. The 'WECfarm' project is initiated to cover this scientific gap on necessary experimental data. The 'WECfarm' experimental setup consists of an array of five generic heaving point-absorber WECs. The WECs are equipped with a permanent magnet synchronous motor (PMSM), addressing the need for WEC array tests with an accurate and actively controllable power take-off (PTO). The WEC array control and data acquisition are realized with a Speedgoat Performance real-time target machine, offering the possibility to implement advanced WEC array control strategies in the MATLAB-Simulink environment. The presented article describes the experimental setup, the performed tests and the results of the test campaign using a single, isolated 'WECfarm' WEC in April 2021 at the wave basin of Aalborg University (AAU), Denmark. A Coulomb and viscous friction model is determined to partly compensate for the drivetrain (motor, gearbox, rack and pinion) friction. A system identification (SID) approach is adopted considering the WEC system to be composed of two single input single output (SISO) models, the radiation and the excitation model. Radiation tests yield the intrinsic impedance. Excitation tests yield the excitation frequency response function. Adopting an impedance matching approach, the control parameters for the resistive and reactive controller are determined from the complex conjugate of the intrinsic impedance. Both controllers are tested for a selection of regular wave conditions. The performed experimental test campaign using an isolated 'WECfarm' WEC allows a full evaluation of the WEC design prior to extending the setup to five WECs. Within the 'WECfarm' project, an experimental campaign with a five-WEC array in the Coastal and Ocean Basin (COB) in Ostend, Belgium, is under preparation.

Keywords: wave energy converter (WEC); heaving point absorber WEC; WECfarm; physical modeling; system identification (SID); real-time control; MATLAB-Simulink

Citation: Vervaet, T.; Stratigaki, V.; Ferri, F.; De Beule, L.; Claerbout, H.; De Witte, B.; Vantorre, M.; Troch, P. Experimental Modelling of an Isolated WECfarm Real-Time Controllable Heaving Point Absorber Wave Energy Converter. *J. Mar. Sci. Eng.* **2022**, *10*, 1480. https://doi.org/10.3390/jmse10101480

Academic Editor: Constantine Michailides

Received: 9 September 2022
Accepted: 7 October 2022
Published: 11 October 2022

1. Introduction

A point absorber wave energy converter (WEC) consists of a floating or submerged body to capture energy from different wave directions. The point absorber diameter should preferably be in the range of 5–10% of the prevailing wavelength [1]. Due to its ability to absorb energy from different directions, this WEC type is particularly suitable to put in arrays. In a WEC array, hydrodynamic interactions between the WECs occur through radiation and diffraction of waves. Both constructive and destructive interactions will occur between individual WECs within a WEC array called near-field interactions.

Göteman et al. define directions for future research necessary for a better WEC array optimization approach [2]. There is a need for available real life data for the validation of WEC array modeling and optimization. This is a research gap for the full wave energy sector, not only in the optimization of WEC arrays. In computational fluid dynamics-based numerical wave tanks (CNWT), the Power Take-Off (PTO) system is mostly modeled as a linear spring–damper system, not representing realistic PTO dynamics and inefficiencies and undermining the overall model fidelity [3]. To validate CNWT considering WECs, it is desirable to incorporate a realistic, nonlinear PTO model. As the computational power capabilities increase yearly, so do the numerical model's capabilities, stressing the need for experimental data to validate the model. However, publicly available databases from WEC array experiments are scarce. Vervaet et al. identified 17 experimental campaigns on point absorber WEC arrays, carried out during the last decades [4]. This limited number of experimental campaigns is due to the high cost of constructing and testing in wave basin facilities, as well as due to the complexity of the experiments and related instrumentation [5]. Therefore, the 'WECfarm' project aims to deliver a dataset to cover the research gap on the need for publicly available real life and reliable data to validate these new advanced numerical models. Vervaet et al. discuss the state of the art in physical modeling of point absorber WEC arrays and the identification of research gaps, resulting in design specifications of the 'WECfarm' experimental setup [4].

The 'WECfarm' experimental setup consists of an array of five heaving point-absorber WECs, designed as a unique test bench for future innovative WEC array research, able to address the current requirements and research gaps on physical WEC array testing. Given the limited number of five WECs, the WECfarm WEC array is not classified as a large WEC array, as the Manchester Bobber 25-WEC array [6], the PerAWaT project 24-WEC array [7] and the WECwakes project 25-WEC array [8]. Vervaet et al. discuss the features of the experimental setup, for which the most important ones are summarized below [4]. The WEC buoy is designed to be generic, being a truncated cylinder with a draft of 0.16 m and a radius of 0.30 m. The high diameter-to-draft ratio of 3.75 yields a flat Response Amplitude Operator (RAO) response and high resonance bandwidth, enhancing WEC–WEC interactions. The WECs are equipped with a Permanent Magnet Synchronous Motor (PMSM), addressing the need for WEC array tests with an accurate and actively controllable PTO. The air bushings linear guiding system excludes guiding friction in the power absorption measurements. The WEC array control and data acquisition are realized with a Speedgoat (Speedgoat, Köniz, Switzerland) Performance real-time target machine, offering the possibility to implement advanced WEC array control strategies in the MATLAB-Simulink environment. Wave basin testing with 'WECfarm' WEC arrays targets to include long- and short-crested waves and extreme wave conditions, representing real sea conditions. Within the 'WECfarm' project, two experimental campaigns have been performed at the Aalborg University (AAU) wave basin: (a) testing of the first WEC in April 2021, addressed in the presented article; (b) testing of a two WEC array in February 2022. An experimental campaign with a five WEC array, in the new wave basin; the Coastal and Ocean Basin (COB) in Ostend (Belgium) [9], is scheduled in 2023.

Friction characterization tests are performed to quantify the drivetrain (motor, gearbox, rack and pinion) introduced friction, whereafter a Coulomb and viscous based friction model for partial compensation of the drivetrain friction is implemented in the MATLAB-Simulink control model. Beatty et al. used a Proportional Integral (PI) force control to minimize the error between the target and measured forces for the physical model of a Wavestar WEC [10]. However, Bacelli et al. stress that closing the loop around a force sensor may induce negative consequences for the design of higher level control loops [11]. The presented friction compensation methodology provides an alternative for closing the PTO force feedback loop around the force sensor.

Coe et al. present a WEC control design based on the principle of impedance matching [12]. The control parameters yield from the complex conjugate of the intrinsic impedance, determined by radiation system identification (SID) tests, where the WEC is excited by a torque noise signal in calm water. Bacelli et al. performed SID tests for an isolated heaving point absorber WEC [13]. This article discusses the application of SID tests on an isolated 'WECfarm' WEC. The SID methodology will be extended to arrays with two to five WECs. Application of the impedance matching methodology for WEC arrays will yield valuable data and insights on WEC-WEC interactions and WEC array control optimization. Within the presented test campaign, the resistive and reactive controller are tested for a selection of regular wave conditions. This testing campaign allows a full evaluation of the WEC prior to extending the setup to five WECs.

Section 2 provides a detailed overview of the experimental setup, with a focus on the 'WECfarm' WEC, instrumentation and wave basin setup. The experimental results of the drivetrain friction model characterization tests, the SID tests and the power absorption tests are discussed in Section 3. A summary of the findings and conclusions are presented in Section 4.

2. Experimental Setup

2.1. WECfarm WEC and Instrumentation

For a detailed discussion on the design of the 'WECfarm' five-WEC array, the reader is referred to [4]. In this article, we consider a single, isolated 'WECfarm' WEC. Figure 1 shows a 3D rendering of the final design of the device. The used right-handed coordinate system has its origin at the intersection of the still water level (SWL) with the vertical axis through the center of the WEC buoy. This allows us to express displacements of the WEC buoy relative to the SWL. The x-axis corresponds with the positive wave propagation direction. The y-axis follows from the motor sign convention: a positive torque results in a downward motion of the WEC buoy. Therefore, the z-axis is pointed downwards to define positive forces, displacements, velocities and accelerations.

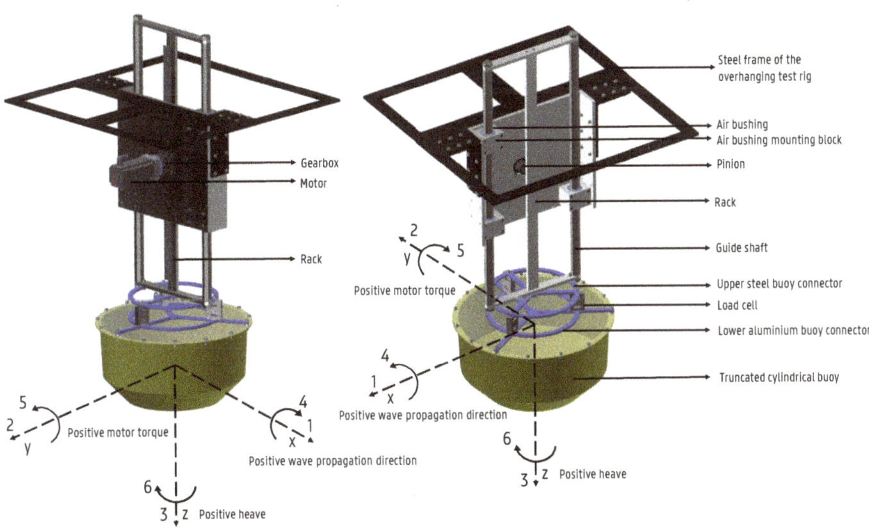

Figure 1. Rendering of the 'WECfarm' WEC, made with Autodesk Inventor (Autodesk, San Rafael, CA, USA).

To exclude friction in the linear guiding, 40 mm OAV (OAV Air Bearings, Princeton, NJ, USA) air bushings are used. The air bushings are characterized by a load versus pressure curve, where one air bushing can cope with a maximum radial load of 720 N, for a nominal pressure of 5.5 bar [14]. A configuration of three OAV 40 mm air bushings guarantees a permanent layer of air between the guide shafts and the bushings for the most extreme wave conditions, resulting in zero-friction linear guiding on the condition of proper alignment. The PTO system of the WEC is designed as a PMSM connected to a gearbox powering a rack and pinion system. The pinion pitch circle radius R_{pinion} is equal to 0.0212205 m. A Wittenstein (Wittenstein, Igersheim, Germany) single-stage gearbox 'NPR 025S-MF1-4 -2E1-1S' with ratio i = 4 is connected to a Beckhoff (Beckhoff Automation, Verl, Germany) PMSM 'AM8542-2E11-0000' with an inertia of 6.17 kg cm^2, a rated torque of 3.97 Nm and a rated speed of 1200 RPM, for 230 V AC power supply [15,16]. The velocity on the pinion will be four times less than on the motor shaft, while the torque on the pinion will be four times more than on the motor shaft. The Beckhoff PMSM is powered and controlled by a Beckhoff motor drive type 'AX5103-0000-0212'. The hydrodynamic part of the WEC consists of an Acrylonitril-Butadieen-Styreen (ABS) thermofolded truncated cylindrical buoy, covered with a Polymethylmethacrylate (PMMA) plate. Figure 2a shows a 2D rendering of the WEC buoy with its dimensions. The WEC buoy is 0.32 m high and designed with a draft of 0.16 m. This draft corresponds with a submerged volume of 0.03683 m^3. Therefore, the mass of the WEC buoy and hydrodynamically activated parts on top of it is 36.83 kg.

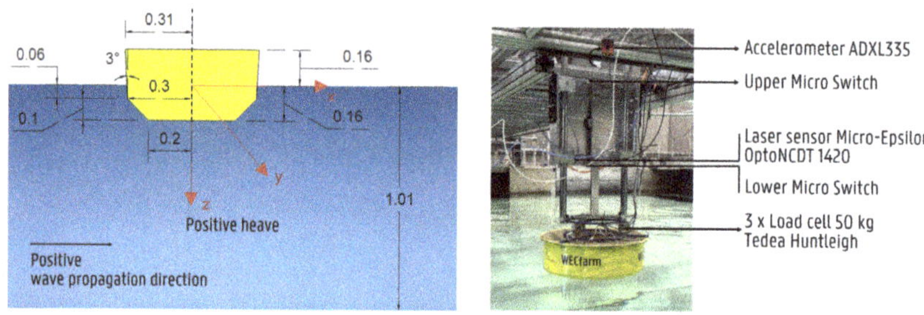

(a) (b)

Figure 2. 'WECfarm' WEC: (**a**) 2D rendering of the WEC buoy, dimensions in m; (**b**) sensors and their respective location on the WEC.

Figure 3 shows a scheme of the data acquisition and control flow for the 'WECfarm' setup with the isolated WEC, with a legend indicating the signal type. The MATLAB-Simulink real-time control model is built on the host PC and loaded on the Speedgoat Performance real-time target machine by Ethernet communication. This target machine runs the Simulink model and processes the input/output (I/O) at a sample frequency of 1000 Hz. In this context, the high sample frequency corresponds with the defined 'real-time' terminology. For each test, the various time series of each logged Simulink signal are saved within a single MATLAB structure.

A scheme on the bottom of Figure 3 shows the sensor input for the Speedgoat IO133 terminal board. The accelerometer ADXL335 (Analog Devices, Norwood, MA, USA) is used to measure the acceleration of the WEC buoy in the heave direction and is attached on top of the rack, the furthest position on the WEC from the water. The accelerometer has a linearity of $\pm$0.3 % of the Full Scale Output (FSO) [17]. Three Tedea Huntleigh (Vishay Precision Group, Malvern, PA, USA) 50 kg load cells are placed between the hydrodynamic part (=the buoy) and the electromechanical part (=the motor) to measure the actual applied forces. The load cells with accuracy class C3 have a total error (per OIML R60) of 0.020% of the rated output [18]. A configuration of at least three load cells is required to avoid torsion and bending influencing the measurements. The mass of the three load cells together is equal to 0.682 kg, the mass above the load cells m_{top} is equal to 27.610 kg and the mass below the load cells is equal to 8.534 kg, which results in a total hydrodynamically activated mass m of 36.83 kg. A TLE analog weight transmitter (Laumas Elettronica, Montechiarugolo, Italy) is used to amplify these three analog signals and to sum them to one analog signal. In case the WEC is locked, the wave heave excitation force F_e can be measured. In case the motor is active, the load cells measure the PTO force F_{PTO}. The upper micro switch and the lower micro switch are used as safety limit switches. It is necessary to limit the amplitude of the WEC buoy displacement to prevent the guiding system damaging the structure. The laser sensor Micro-Epsilon optoNCDT 1420-500 (Micro-Epsilon, Ortenburg, Germany) is installed as a backup for the motor encoder to measure the displacement of the WEC buoy relative to the SWL. The laser sensor has a linearity of ±500 μm, equivalent to ±0.1% of the FSO [19]. Moreover, the laser sensor can be used for displacement measurements for tests without the motor. Three pneumatic indicators, one for each air bushing, are used as a visual safety indicator in the pneumatic circuit. As long as the air bushing is provided with a certain air pressure, the red balloon in the indicator stays inflated, confirming the air bushings are pressurized. Figure 2b shows a picture of the WEC as installed at the AAU wave basin with indication of the location of the sensors.

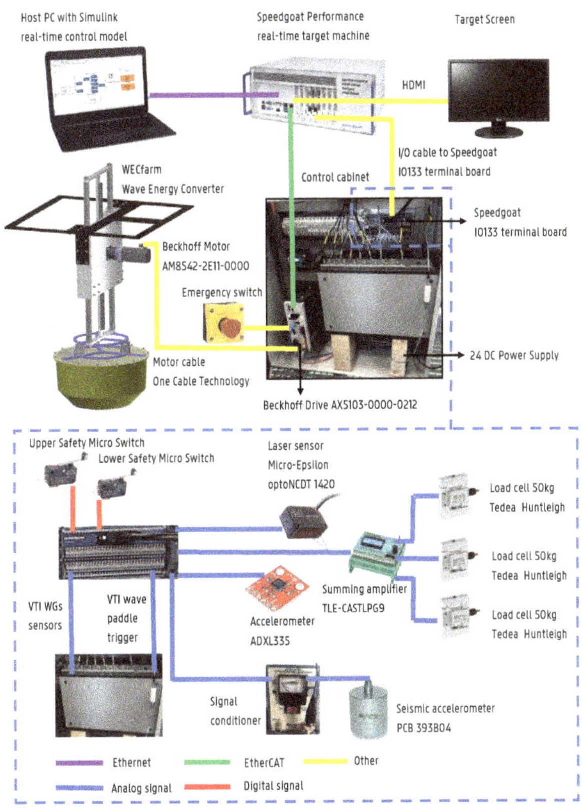

Figure 3. General data acquisition and control flow for the isolated 'WECfarm' WEC.

The motor drive, the Speedgoat IO133 terminal board, DC power supply and loss current switches are centralized in the control cabinet. Figure 3 shows a picture of the inside of the control cabinet. During the experimental campaign, the VTI (VTI Instruments Corporation, Irvine, CA, USA) wave gauge sensor system was put inside the control cabinet to establish the connection of the VTI wave gauge sensor analog output with the Speedgoat IO133 terminal board analog input. The VTI wave paddle trigger is used for synchronization by providing a constant voltage signal from the moment the wave paddles are activited. The seismic accelerometer (PCB 393B04) is placed on top of the steel frame to quantify possible vibrations of this frame. Vibrations of the frame and resonance in particular are to be avoided, since these affect the measurement quality of the other sensors.

The torque request in the Simulink model is sent by the EtherCAT (Ethernet for Control Automation Technology) communication protocol to the Beckhoff motor drive as a Master Data Telegramm (MDT) process parameter. On the other hand, the Speedgoat target machine can receive by EtherCAT communication Amplifier Telegramm (AT) process parameters from the Beckhoff motor drive. The Beckhoff PMSM input and output signals are sent from and to the Beckhoff motor drive by the One Cable Technology (OCT), which allows to power the motor and process feedback. The motor drive receives the absolute position within one revolution at an 18 bit resolution from the single-turn absolute encoder. This encoder allows real-time determination of the state (position and velocity) of the WEC buoy. The drive provides the motor with a certain current, corresponding to a torque by multiplication with the torque constant of 1.91 Nm/A [16].

The uncertainty in the measured values with the given instrumentation is minimized by the selection of sensors with high resolution, high accuracy and low linearity error. It is important that the sensors are correctly calibrated, with a zero offset for the equilibrium position with draft 0.16 m, represented in Figure 2a. Besides the uncertainty related to sensor measurements, it is important to quantify the uncertainty of the physical testing results. Lamont-Kane et al. identified five distinct sources of uncertainty for physical testing of WEC-arrays [20]: Spatial variation of the wave-field within the wave basin; temporal variation of the wave-field from one repeat to another; the repeatability of model response for any single individual WEC; the reproducibility of model response between various nominally identical WECs (not applicable for the test campaign with a single, isolated WEC in the presented article) and the variation in the time-series of an incident irregular wave train. The quantification of these sources of uncertainty for the performed experimental campaign is not addressed in the presented article.

2.2. Wave Basin Setup

The experimental campaign took place at the AAU wave basin of the Ocean and Coastal Engineering Laboratory of the Department of the Built Environment [21]. The wave basin measures 14.60 m × 19.30 m × 1.50 m (length × width × depth) with an active test area of 8.00 m × 13.00 m (length × width). The wave generation system is 13 × 1.5 m (width × height) with 30 individually controlled wave paddles (snake type configuration). The system allows accurate generation of 3D waves due to narrow vertically hinged paddles (0.43 m segment width) with maximum wave height up to 0.45 m (at 3.0 s period) and typical maximum significant wave height H_s in the range of 0.25–0.30 m. The wave basin is equipped with passive wave absorber elements. The AwaSys wave generation software is able to generate regular, irregular, solitary waves, execute 2-D and 3-D active wave absorption (reflection compensation) and generate 2nd order irregular unidirectional and multidirectional waves [22].

Seven resistive wave gauges (WGs) are installed in the wave basin to measure incident, diffracted, radiated and reflected waves during the tests. Figure 4a shows the planview layout of the wave basin with the central location of the WEC buoy, WGs, wave generation system and passive absorption. The interdistance between the different WGs and the WEC buoy are indicated in m and waves are generated from the bottom of Figure 4a. These seven WGs are also displayed in Figure 5a. The numbering is based on the used analog input ports of the Speedgoat IO133 terminal board.

Figure 4b shows a 'bird's-eye' perspective of the experimental setup at the AAU wave basin. The WEC is attached to the bridge over the wave basin, complying with the spirit level requirements. Waves are generated from the top right corner in Figure 4b. The host PC for the Simulink control and the PC with the AwaSys software are located in line with the bridge, on the bottom right corner in Figure 4b. The water level is set equal to 1.010 m, which results in an equal positive and negative heave stroke with a magnitude of 0.25 m. This water level should be kept constant during the experimental campaign, since changing water level yields an offset for the position measurements from the laser sensor and motor encoder. A water density ρ of 1000 kg/m^3 is taken into account.

Figure 5a shows a picture of the overview of the setup as schematized in Figure 3. Figure 5b shows the coordinate system and sign convention on a picture of the experimental setup, as adopted in Figures 1 and 2a. During this test campaign, waves are generated in the x-direction, corresponding to the direction of the greatest stiffness of the WEC.

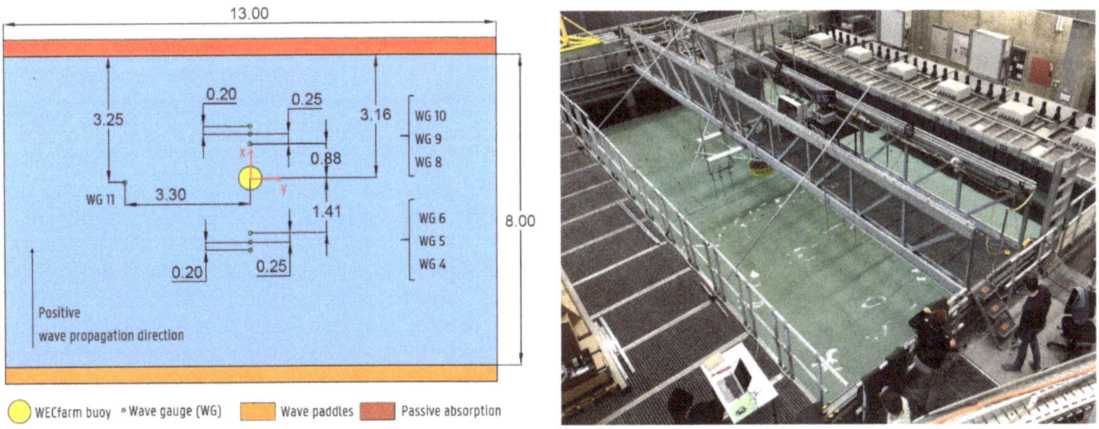

(**a**) (**b**)

Figure 4. Experimental setup of the isolated 'WECfarm' WEC at the AAU wave basin: (**a**) wave basin planview layout with dimensions in meters; (**b**) bird's-eye perspective picture towards the wave paddles.

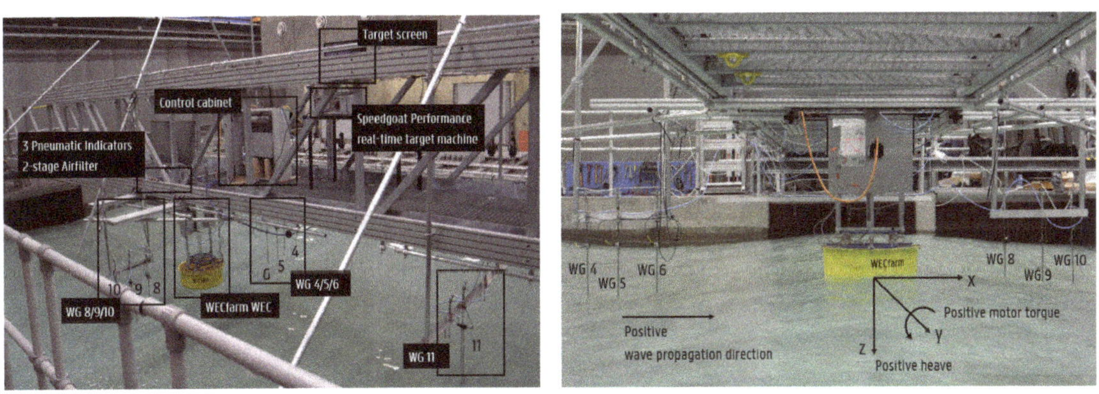

(**a**) (**b**)

Figure 5. Experimental setup of the isolated 'WECfarm' WEC at the AAU wave basin: (**a**) picture with indication of the subsystems; (**b**) picture with indication of the coordinate system and sign convention.

3. Results

3.1. Test Matrix

Table 1 gives an overview of the different types of tests that are performed. Based on their underlying purpose, they are subdivided in three categories. The next sections discuss the different test categories.

Table 1. Overview of the types of performed tests.

Category	Test Type
Friction model	Friction characterization tests
System Identification	Radiation tests Excitation tests Free decay tests
Power absorption	Resistive control tests Reactive control tests

3.2. Friction Model

The air bushings coefficient of friction is a function of air shear from motion, not from surface contact. Therefore, a friction coefficient of 0.00008 in log-scale due to the contribution from air molecules and gravitation can be taken into account [23]. To obtain the air gap between the shaft and the air bushing, a compressor with a two stage air filter supplies these bushings with clean and dry air under a nominal pressure of 5.5 bar. The gearbox break-away torque T_{01} is expected to be around 0.20–0.40 Nm, with convergence towards lower values for a longer operational lifetime [24]. The motor static friction M_R is reported to be equal to 0.02 Nm [15]. As a result, the total drivetrain friction experienced by the WEC will be mainly determined by the Coulomb friction attributed to the gearbox. Moreover, additional Coulomb and viscous friction attributed to the rack and pinion will occur. Based on the empirical characterization of the actual friction, a static friction model is constructed, discussed in Section 3.2.1 [25].

3.2.1. Friction Characterization Tests

The friction of the WEC with the motor and gearbox installed is characterized by a zero torque command on the motor. The WEC buoy is manually moved up and down from below the load cells with following elements in the test sequence: slowly at a targeted constant velocity close to 0.0 m/s and an accelerated motion with amplitude up to 0.40 m/s. Figure 6a shows a picture of a friction characterization test in the empty AAU wave basin. The tests are repeated in the filled AAU wave basin to benefit from buoyancy forces.

(a) (b)

Figure 6. Friction characterization tests (**a**) in the empty AAU wave basin; (**b**) with application of lateral loading in the-x-direction.

During the execution of the lifting procedure, the force acting on the loadcells $F_{loadcells}$, the position z, the velocity $\dot{z}$ and the acceleration $\ddot{z}$ are measured. The first element of the test sequence allows us to determine the Coulomb damping coefficient C_{Cou}, while the second element allows us to determine the viscous damping coefficient C_{Vis}.

The friction force $F_{Friction}$ to compensate for equals $F_{loadcells}$ reduced by the acceleration force F_{acc} caused by m_{top} and by the rotational inertia of the motor and the gearbox. To take this rotational inertia of the motor and the gearbox into account in F_{acc}, it is expressed as a mass in the heave direction, i.e., the mass attributed to the motor and gearbox inertia m_{MGI}:

$$m_{MGI} = \frac{J_{motor} \cdot i^2 + J_{gearbox}}{R^2_{pinion}} = \frac{6.17\,\text{kg cm}^2 \cdot 4^2 + 0.71\,\text{kg cm}^2}{(2.12205\,\text{cm})^2} = 22.08\,\text{kg} \quad (1)$$

with J_{motor} the motor inertia equal to 6.17 kg cm^2 [16] and $J_{gearbox}$ the gearbox inertia equal to 0.71 kg cm^2 [26]. It is anticipated that the actual inertia will be higher, as the inertia of the rack and pinion is not taken into account in Equation (1). The resulting $F_{Friction}$ is:

$$F_{Friction} = F_{loadcells} - F_{acc} = F_{loadcells} - (-\ddot{z} \cdot (m_{top} + m_{MGI})) \quad (2)$$

In case the motor and the gearbox are not installed and the air bushings work properly, F_{acc} should equal $F_{loadcells}$, resulting in $F_{Friction}$ = 0 N. Figure 7a shows the z and $\dot{z}$ time series for the friction characterization Test_054. Figure 7b shows the time series of $F_{loadcells}$, F_{acc}, $F_{Friction}$ and the fitting of C_{Vis} and C_{Cou}. $F_{loadcells}$ exceeds F_{acc} due to the addition of the drivetrain (motor, gearbox and rack and pinion) friction.

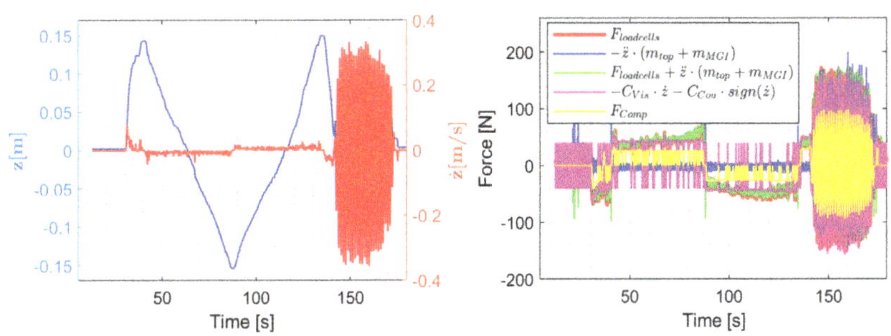

(a) (b)

Figure 7. Friction characterization WEC with gearbox and motor (**a**) z and $\dot{z}$ time series (Test_054); (**b**) Force time series (Test_054).

Fitting a model to $F_{Friction}$ results in the friction compensation model F_{Comp}, given by Equation (3).

$$F_{Comp} = \begin{cases} 0 & \text{for} \quad -\dot{z}_{Bou} \leq \dot{z} \leq \dot{z}_{Bou} \\ -(-C_{Vis} \cdot \dot{z} - C_{Cou} \cdot sign(\dot{z}))C_C & \text{for} \quad \dot{z} < -\dot{z}_{Bou} \,||\, \dot{z} > \dot{z}_{Bou} \end{cases} \quad (3)$$

Note that an additional minus sign is necessary to have F_{Comp} in the same direction as $\dot{z}$. The model is composed of a Coulomb part proportional to the sign of the velocity and a viscous part proportional to the velocity. A velocity boundary $\dot{z}_{Bou}$ is imposed to avoid rapid sign switches for the Coulomb friction term at low velocities. Consequently, no friction compensation is taken into account between $-\dot{z}_{Bou}$ and $\dot{z}_{Bou}$. These boundaries correspond approximately with the noise range on the velocity feedback from the motor encoder. Moreover, it is preferred to partly compensate for the friction to preserve a realistic PTO, expressed by the compensation factor C_C. Figure 8a shows the $F_{loadcells}$ to $\dot{z}$ mapping for the friction characterization Test_054, which allows us to fit the parameters of F_{Comp}. Table 2 provides the obtained values for C_{Vis}, C_{Cou}, C_C, and $\dot{z}_{Bou}$.

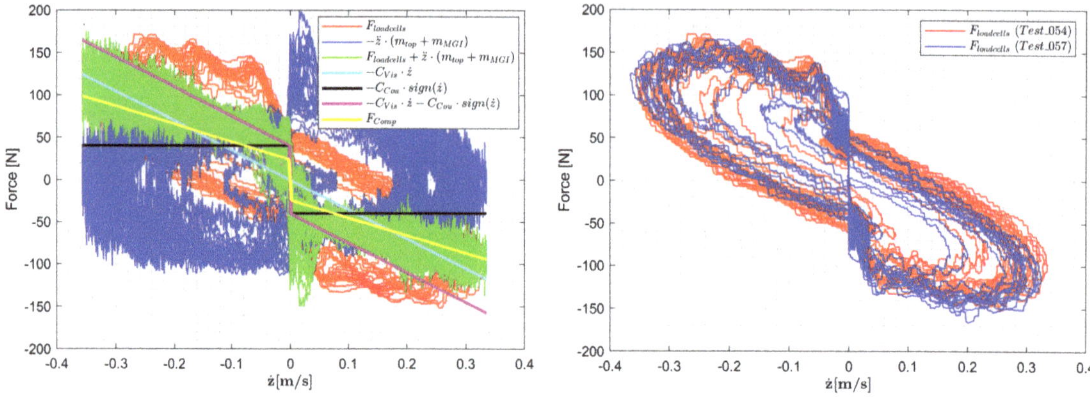

(**a**) (**b**)

Figure 8. Force to velocity mapping from the WEC friction characterization tests: (**a**) No lateral loading (Test_054); (**b**) No lateral loading (Test_054) versus −20 kg lateral loading in the x-direction (Test_057).

Table 2. C_{Vis}, C_{Cou}, C_C and $\dot{z}_{Bou}$ for the resulting F_{Comp}.

C_{Vis} [Ns/m]	C_{Cou} [N]	C_C [-]	$\dot{z}_{Bou}$ [m/s]
350	40	0.60	0.0012

Figure 7b shows the time series of F_{Comp}. The proper functioning of the velocity boundaries can be observed as less sign switches occuring for F_{Comp} (yellow curve) compared to the fitting of C_{Vis} and C_{Cou} (magenta curve). The difference between $F_{Friction}$ (green curve) and F_{Comp} (yellow curve) is the remaining PTO friction.

The friction characterization tests are repeated with different lateral loading conditions, as an approximation for the surge or sway wave excitation force. In the wave propagation direction (x-direction) −20 kg is applied with a tension spring and rope around the WEC buoy, shown in Figure 6b (Test_057). This loading of 196 N corresponds approximately with a surge wave excitation force for a wave with period T = 1.0 s and wave height H = 0.40 m, according to linear potential flow simulations with the open-source software package openWEC [27], with the integration of the Boundary Element Method (BEM) code Nemoh. The $F_{loadcells}$ to $\dot{z}$ mapping in Figure 8b shows no increased $F_{loadcells}$ for Test_057 compared to Test_054, demonstrating the proper functioning of the air bushings. In the Simulink model, the uncompensated input motor torque $\tau_{motor,uncomp}$ augmented with $F_{Comp}(R_{pinion}/i)$ yields the compensated input motor torque $\tau_{motor,comp}$:

$$\tau_{motor,comp} = \tau_{motor,uncomp} + F_{Comp}\frac{R_{pinion}}{i} \tag{4}$$

Note that the friction compensation in Equation (4) is implemented as a feedback control structure, since the velocity output of the WEC is used. Commonly, a friction compensation feedforward control structure is adopted when tracking a reference is the objective, which is not the case here [28]. Table 3 provides an overview of all the executed friction characterization tests, with indication of the filling of the AAU wave basin, lateral loading force, lateral loading direction and applied air pressure. The Test_ID is the identification number of each performed test.

Table 3. Overview of the friction characterization tests.

Test_ID	Filled Basin	Loading Force [kg]	Loading Direction	Air Pressure [bar]
Test_007	no	-	-	5.0
Test_008	no	-	-	5.0
Test_009	no	-	-	0
Test_024	yes	-	-	5.0
Test_025	yes	10	-x	5.0
Test_026	yes	15	-x	5.5
Test_054	yes	-	-	5.5
Test_055	yes	10	-x	5.5
Test_056	yes	15	-x	5.5
Test_057	yes	20	-x	5.5
Test_058	yes	15	y	5.5

3.3. System Identification

3.3.1. Linear Decomposed Wave–WEC Interaction Model

To perform a SID of the WEC, a linear decomposition model formulation for the wave–WEC interaction is adopted. The problem is separated into radiation and excitation components. Distinct tests are performed for both the radiation and the excitation components to determine the WEC system response, whereafter radiation and excitation frequency response functions (FRFs) can be constructed from the discrete frequency components. At a high level of abstraction, the WEC can be considered as a system with two inputs: the surface elevation η and F_{PTO}, as displayed in Figure 9. When the WEC system, consisting of the PTO and the WEC buoy, is assumed to be linear for small motions and small waves, superposition can be applied. The WEC system is considered to be composed of two single input single ouput (SISO) models, the radiation and the excitation model, as displayed in Figure 9 [12,13,29]:

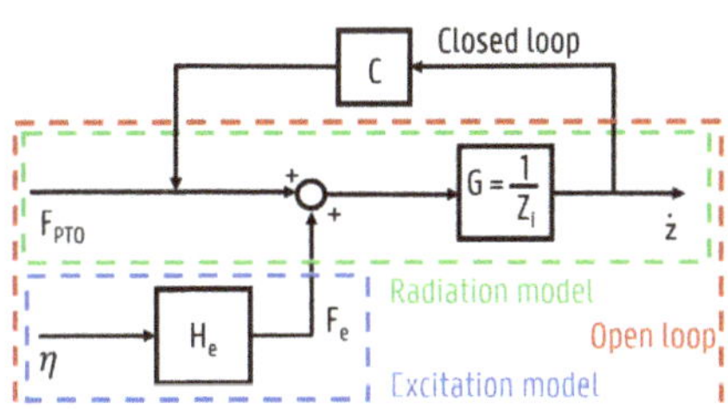

Figure 9. WEC block diagram based on the dual single input single ouput (SISO) radiation/diffraction model.

The radiation model is obtained by computing the ratio of the FRF of the output $\dot{z}$ to the FRF of the input F_{PTO}, resulting in the admittance $G(\omega)$:

$$G(\omega) = \frac{\hat{\dot{X}}(\omega)}{\hat{F}_{PTO}(\omega)} \tag{5}$$

The quantities are expressed as a function of the angular frequency ω. The upper case indicates that these variables are all in the frequency-domain, while the hat symbol ^ denotes that these variables are complex quantities. The control parameters for the impedance matching controller are determined from the intrinsic impedance $Z_i(\omega)$, defined as the inverse of $G(\omega)$:

$$Z_i(\omega) = (G(\omega))^{-1} = \frac{\hat{F}_{PTO}(\omega)}{\hat{\dot{X}}(\omega)} \tag{6}$$

Mechanical impedance is a measure of the opposition to motion from a source when a potential is applied, defined as the ratio of force (potential) to velocity (flow) in Equation (6). Note that $\widehat{X}$ is used instead of $\widehat{Z}$ for the FRF of $\dot{z}$, to avoid confusion with the defined Z_i. The excitation model is obtained by computing the ratio of the FRF of the output excitation force F_e, measured as $F_{loadcells}$, to the FRF of the input η at the location of the WEC, resulting in the excitation force coefficients $H_e(\omega)$:

$$H_e(\omega) = \frac{\widehat{F}_e(\omega)}{\widehat{\eta}(\omega)} \tag{7}$$

In the first instance, the above-discussed SID tests are carried out in "open loop", which means no output feedback is considered, represented in Figure 9 by no feedback arrow. A "closed loop" is obtained at any time the WEC is controlled and F_{PTO} is calculated based on an output measurement. In Figure 9, F_{PTO} depends on $\dot{z}$, as $F_{PTO} = C\dot{z}$, with C representing the control system dynamics. In case of resistive damping, C is a negative constant in this formulation. As a result, the frequency-domain WEC equation of motion, as displayed in Figure 9 as the dual SISO model, is given by:

$$\widehat{F}_{PTO} + H_e(\omega)\widehat{\eta} = \widehat{F}_{PTO} + \widehat{F}_e = Z_i(\omega)\widehat{X} \tag{8}$$

By applying the superposition principle, the decomposition in the radiation and excitation model is given by:

$$\widehat{X} = \frac{1}{Z_i(\omega)}(\widehat{F}_{PTO} + H_e(\omega)\widehat{\eta}) = \frac{1}{Z_i}\widehat{F}_{PTO} + \frac{H_e(\omega)}{Z_i(\omega)}\widehat{\eta} \tag{9}$$

3.3.2. Impedance Formulation and Radiation Tests

When the Fourier transform $\mathscr{F}$ differentiation property is applied to write $\ddot{z}$ in terms of $\dot{z}$:

$$\mathscr{F}[\ddot{z}(t)] = i\omega\widehat{X}(\omega) \tag{10}$$

the point absorber WEC equation of motion can be written in terms of $\widehat{X}(\omega)$ [12,30]:

$$m\, i\omega\, \widehat{X}(\omega) = -(B(\omega) + i\omega A(\omega))\widehat{X}(\omega) + \frac{K}{i\omega}\widehat{X}(\omega) + \widehat{F}_{PTO} + \widehat{F}_e \tag{11}$$

where m = 36.83 kg, $A(\omega)$ is the added mass coefficient and $B(\omega)$ is the hydrodynamic damping coefficient. The hydrostatic stiffness coefficient K is given by:

$$K = \rho g S \tag{12}$$

where S is the cross-sectional area of the WEC buoy at the SWL equal to 0.283 m^2, ρ = 1000 kg/m^3 and g the gravitational acceleration equal to 9.81 N/kg. Rearrangement of Equation (11) based on Equation (9), results in $Z_i(\omega)$:

$$Z_i(\omega) = \frac{\widehat{F}_{PTO} + \widehat{F}_e}{\widehat{X}} = B(\omega) + i\left(\omega(m + A(\omega)) - \frac{K}{\omega}\right) \tag{13}$$

$Z_i(\omega)$ is experimentally determined by executing a forced oscillation test in the AAU wave basin, without waves generated by the wave paddles. In Equation (13), F_e equals zero and F_{PTO} is a chirp signal with a frequency spectrum covering the bandwidth of interest [13]. Since it is experimentally more convenient to use F_{PTO} as an input and measure the output $\dot{z}$, Z_i is obtained as the inverse of G, defined by Equations (5) and (6).

Table 4 shows an overview of the performed radiation tests. The chirp-up PTO input torque is defined with an initial frequency of 0.0 Hz, a target time of 220 s and a frequency at target time of 4.0 Hz. The chirp-down PTO input torque is defined with an initial frequency of 4.0 Hz, a target time of 220 s and a frequency at target time of 0.0 Hz. The noise signal is defined with an amplitude of 1.0 Nm, multiplied by the defined PTO gain to control the absolute maximum heave amplitude of the WEC buoy z_{max}. Most of the test were executed with $C_C = 0.0$, without the friction compensation model discussed in Section 3.2 included. Test_112 and Test_113 consider $C_C = 0.6$, which is the WEC system to build the resistive and reactive controller on.

Table 4. Overview of the performed radiation tests.

Test_ID	PTO Input	Target Time [s]	Frequency Range [Hz]	PTO Gain	C_C [-]	z_{max} [m]
Test_018	Chirp-up	220	0.0–4.0	1.0	0.0	0.115
Test_021	Chirp-down	220	4.0–0.0	1.0	0.0	0.122
Test_053	Chirp-up	220	0.0–4.0	1.0	0.0	0.123
Test_059	Chirp-up	220	0.0–4.0	1.0	0.0	0.116
Test_066	Chirp-up	220	0.0–4.0	0.5	0.0	0.041
Test_109	Chirp-up	220	0.0–4.0	1.0	0.0	0.127
Test_112	Chirp-up	220	0.0–4.0	0.3	0.6	0.152
Test_113	Chirp-down	220	4.0–0.0	0.3	0.6	0.162
Test_163	Chirp-up	220	0.0–4.0	1.0	0.0	0.128

Figure 10a displays a radiation test for the 'WECfarm' WEC, where the circular radiated waves can be observed. Figure 10b shows the F_{PTO} and $\dot{z}$ time series, for which $\dot{z}$ displays a resonance. According to Equation (4), a $\tau_{motor,uncomp}$ of 1.0 Nm corresponds with a F_{PTO} of 188.5 N.

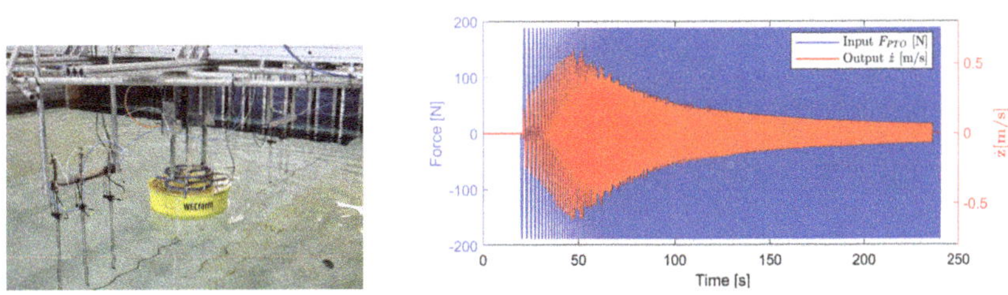

(**a**) (**b**)

Figure 10. Radiation test: (**a**) Picture of the setup in the AAU wave basin; (**b**) F_{PTO} and $\dot{z}$ time series (Test_059).

The calculated Z_i can be displayed as a bode plot with gain and phase, given in Figure 11 for Test_021, Test_059, Test_066, Test_112 and Test_113 [12]. Comparing Test_021 with Test_059 confirms that the chirp-up and chirp-down signal yields the same $Z_i(\omega)$. Test_066 with a gain of 0.5 and z_{max} = 0.041 m results in a higher identified Z_i compared to Test_059 with a gain of 1.0 and z_{max} = 0.116 m, stressing the importance of covering motion amplitudes representative for the WEC during operation. Test_112 and Test_113 demonstrate how the implementation of F_{Comp} alters the WEC system dynamics. As expected, $\dot{z}$ is more amplified when the friction is compensated. The WEC resonance frequency f_n is equal at 0.84 Hz, corresponding to a natural period T_n = 1.19 s.

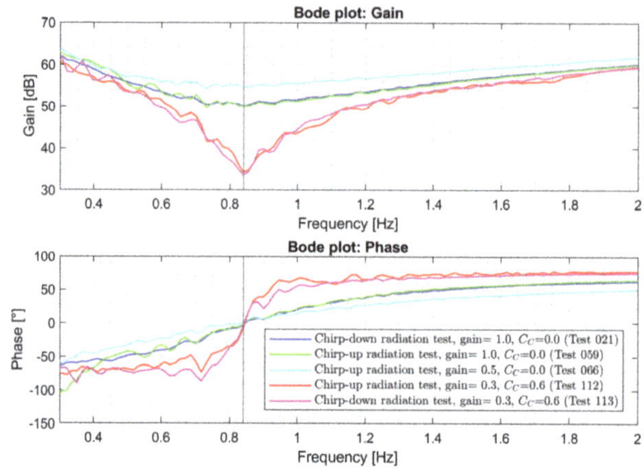

Figure 11. Bode plot of the experimentally identified Z_i for Test_021 , Test_059, Test_066, Test_112 and Test_113.

3.3.3. Excitation Tests

$H_e(\omega)$ is experimentally determined by locking the WEC in equilibrium position, as defined in Figure 2a, imposing a frequency rich η signal and measuring $F_{loadcells}$. The WEC is fixed by deactivating the motor drive, resulting in an active holding brake. Application of Equation (7) results in $H_e(\omega)$.

Table 5 shows an overview of the performed excitation tests. The wave input generated by the wave generation system is a JONSWAP wave spectrum, defined by a peak enhancement factor γ of 3.3 and the mentioned significant wave height H_s and peak period T_p. In the case of Test_133 with regular waves, H and T are mentioned. Using a start signal from the wave paddles and recording the wave paddle motion, allow for a deterministic comparison of the same sea state over different control inputs and types.

Table 5. Overview of the performed excitation tests.

Test_ID	Wave Input	H_s [m]	T_p [s]	Test_ID	Wave Input	H_s [m]	T_p [s]
Test_048	JONSWAP	0.05	1.0	Test_076	JONSWAP	0.07	2.0
Test_049	JONSWAP	0.05	1.0	Test_077	JONSWAP	0.09	2.0
Test_050	JONSWAP	0.07	1.0	Test_078	JONSWAP	0.11	2.0
Test_071	JONSWAP	0.09	1.0	Test_133	Regular	0.09	2.0
Test_072	JONSWAP	0.05	1.5	Test_144	JONSWAP	0.07	2.0
Test_073	JONSWAP	0.07	1.5	Test_145	JONSWAP	0.13	2.0
Test_074	JONSWAP	0.09	1.5	Test_167	JONSWAP	0.20	1.5
Test_075	JONSWAP	0.05	2.0	Test_168	JONSWAP	0.30	2.0

Figure 12a displays an excitation test for the WECfarm WEC, where the incoming and diffracted waves can be observed. Figure 12b shows the $F_{loadcells}$ and η_{WG11} time series for Test_074. When a wave crest (negative η) passes the WEC buoy, the load cells are compressed, resulting in a positive $F_{loadcells}$. When a wave trough (positive η) passes the WEC buoy, the load cells are under tension, resulting in a negative $F_{loadcells}$.

Figure 13a shows the resulting $\widehat{F}_{loadcells}$, equivalent to $\widehat{F}_e$, and $\widehat{\eta}$ for Test_074 and Test_077. The calculated H_e can be displayed with a gain and phase, given in Figure 13b for Test_074 and Test_077. Frequency smoothing has been performed on $\widehat{F}_{loadcells}$, $\widehat{\eta}$ and the gain and phase of H_e. The Gaussian-weighted moving average over a window of 30 frequency intervals of 0.0042 Hz has been taken. Figure 13b confirms the higher H_e for lower wave frequencies, as noticed in Figure 13a.

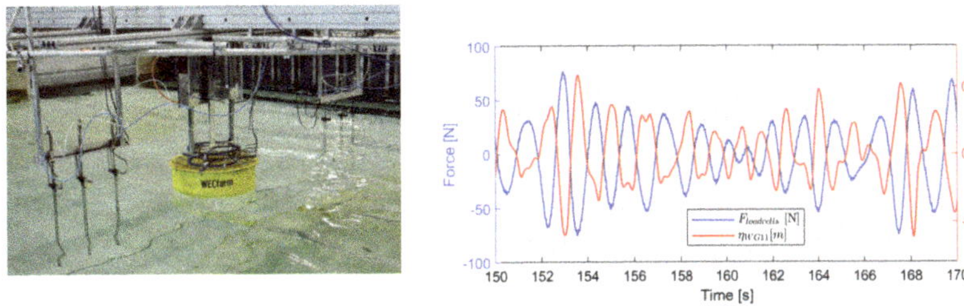

(**a**) (**b**)

Figure 12. Excitation test: (**a**) Picture of the setup in the AAU wave basin; (**b**) $F_{loadcells}$ and η_{WG11} time series (Test_074).

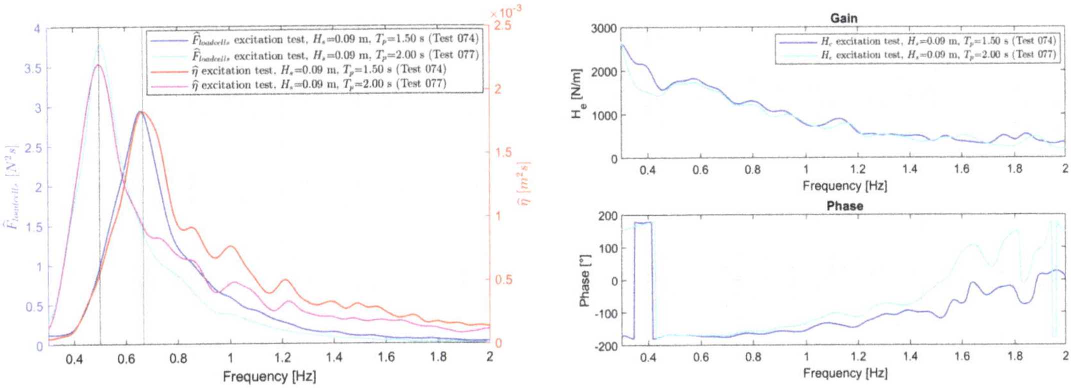

(**a**) (**b**)

Figure 13. Wave excitation Test_074 and Test_077: (**a**) $\widehat{F}_{loadcells}$ and $\widehat{\eta}$, frequency smoothened over a window of 30 frequencies; (**b**) H_e, frequency smoothened over a window of 30 frequencies.

To calculate $H_e(\omega)$ based on the most straightforward conceptual definition, a WG is placed on the location of the WEC buoy when the WEC buoy is not present and then record the η time series. The WEC buoy is then put in place and the same wave time series are run again, this time measuring the force on the WEC buoy caused by the waves. For control purposes, this procedure is of little use, since it is clearly not possible to measure η at the point where the WEC is located, once the WEC is in place. Therefore, η obtained with WG 11 is used as an approximation for η on the location of the WEC buoy. It is shown by Bacelli et al. that if the distance between the WEC buoy and the WG is increased, the term describing the diffracted waves at the WG becomes small enough and the original model given by Equation (9) can be adopted [13].

To verify the assumption that the data of WG 11 of these excitation tests could be used for this purpose, at the end of the test campaign some tests with wave spectra are executed with the WEC buoy removed and WG11 moved to the position of the WEC buoy, equilinear with the other WGs. For these tests, the interdistance WG 6 to WG 11 is 1.31 m and the interdistance WG 11 to WG 8 is 0.98 m. Figure 14 shows a picture of this layout of the WGs.

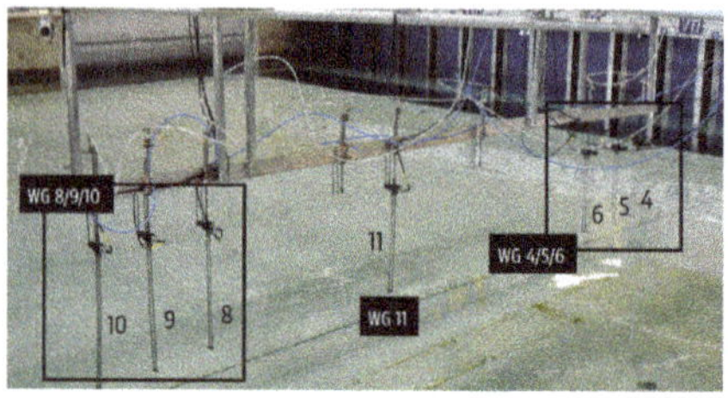

Figure 14. Picture of WGs layout for η measurements with WG 11 on the location of the WEC buoy.

Three JONSWAP spectra and one regular wave are tested, defined in Table 6.

Table 6. Overview of η measurement tests with WG11 on the location of the WEC buoy.

Test_ID	Wave Input	H_s [m]	T_p [s]
Test_172	JONSWAP	0.09	1.5
Test_173	JONSWAP	0.07	1.0
Test_174	JONSWAP	0.05	2.0
Test_175	Regular	0.09	2.0

The top Figure 15a shows the time series comparison between η measured by WG11 in excitation Test_074 and η measured by WG11 in Test_172. The bottom Figure 15a shows this comparison for Test_050 and Test_173.

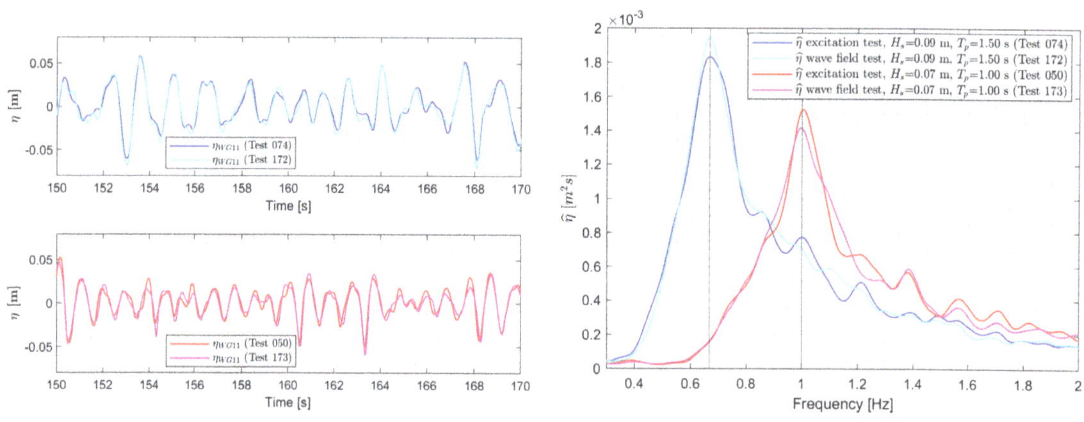

(**a**) (**b**)

Figure 15. Wave excitation tests (Test_074 and Test_050) and wave field accuracy tests (Test_172 and Test_173): (**a**) η_{WG11} time series; (**b**) $\widehat{\eta}$, frequency smoothened over a window of 30 frequencies.

For both wave conditions, a minor difference between the two time series can be observed. Figure 15b shows $\widehat{\eta}$ for Test_074 compared to Test_172 and $\widehat{\eta}$ for Test_050 compared to Test_173. Figure 15a,b confirm that η obtained with WG 11, according to the layout of the WGs in Figure 4a, can be used as an approximation for η on the location of the WEC buoy.

3.3.4. Free Decay Tests

Free decay tests are executed to determine the decay response. The heave displacement z allows us to calculate the logarithmic decrement Λ, the corresponding damping ratio

ξ_d and the natural period T_n. This T_n should correspond with the one obtained from the radiation tests discussed in Section 3.3.2. Conceiving the WEC as a mass-spring-damper system, ξ_d follows from Λ [31]:

$$\Lambda = \frac{1}{n-1} ln\left(\frac{x_1}{x_n}\right) = \frac{2\pi\xi_d}{\sqrt{1-\xi_d^2}} \tag{14}$$

in which x_1 and x_n are the values of the first and the n-th peak of the free decay motion z. For un underdamped system ($0 < \xi_d < 1$), the oscillations of the WEC buoy fade exponentially over time and tend toward zero, yielding the envelope:

$$z_e(t) = z_A \cdot exp(-\xi_d \omega_n t) \tag{15}$$

in which z_A is the amplitude of the free WEC response [31]. The angular natural frequency ω_n of the WEC buoy can be calculated from the damped natural frequency ω_d:

$$\omega_n = \frac{2\pi}{T_n} = \frac{\omega_d}{\sqrt{1-\xi_d^2}} \tag{16}$$

Table 7 gives an overview of the executed free decay tests, with indication of the start position of the WEC buoy z_{start}, C_C, n used Equation (14), ξ_d and T_n. A positive z_{start} corresponds to a submerged start position and a negative value corresponds to an elevated start position. The WEC buoy is submerged or elevated with a torque command. Once the torque is set equal to zero, the decay motion is initiated. Test_089 and Test_129 yield a T_n equal to 1.19 s, corresponding to the value obtained by the bode plots of the radiation tests displayed in Figure 11. When the motor and gearbox with pinion are removed in Test_170, a T_n equal to 1.00 s is obtained. This lower T_n can be attributed to the removed inertia within the PTO drivetrain since removing inertia results in a lower T_n. Test_089 and Test_129 yield a ξ_d of 0.13 and 0.12, respectively, corresponding to an underdamped system. Given this low ξ_d, ω_n will closely approximate ω_d.

Table 7. Overview of the performed free decay tests.

Test_ID	z_{start} [m]	C_C [-]	n	ξ_d [-]	T_n [s]
Test_089	−0.089	0.6	2	0.13	1.19
Test_129	0.078	0.6	3	0.12	1.19
Test_161	−0.202	0.6	n.a.	n.a.	n.a.
Test_162	0.222	0.0	n.a.	n.a.	n.a.
Test_170 (no motor, gearbox and pinion)	0.082; 0.140; 0.113; −0.204; −0.214; 0.039	n.a.	4	0.14	1.00

The limited draft in combination with the flat bottom will introduce important bottom slamming effects for tests where the WEC buoy re-enters the water after being lifted out [32]. These bottom slamming forces are assessed in Test_161 and Test_162, for which the start position of the WEC buoy is completely lifted out of the water. In addition, nonlinear viscous drag effects occur. The semi-empirical Morison equation describes the viscous force F_{vis} in function of a viscous drag coefficient C_D [33]:

$$F_{vis} = -0.5\,\rho\,S\,C_D\,\dot{X}|\dot{X}| \tag{17}$$

This C_D can be identified from experiments or from fully viscous modeling methods based on the Navier–Stokes equations [34–36]. The experimental quantification of bottom slamming effects and nonlinear viscous drag effects is not addressed in the presented article.

Test_170 considers multiple free decay tests for the WEC buoy with the motor and gearbox with pinion removed. While z is obtained by both the laser and encoder for the other tests, only laser measurements are available when the motor is removed. The WEC buoy is manually pushed down or lifted from the measurement bridge.

Figure 16a displays the WEC buoy displacing a water volume, equivalent to the volume of the WEC buoy, after being dropped from −0.202 m in Test_161. The concentric circular radiated waves can be observed. Bottom slamming forces are obtained from $F_{loadcells}$. Figure 16b shows the free decay z time series and corresponding exponential decay envelopes according to Equation (15) for Test_089, Test_129 and Test_170.

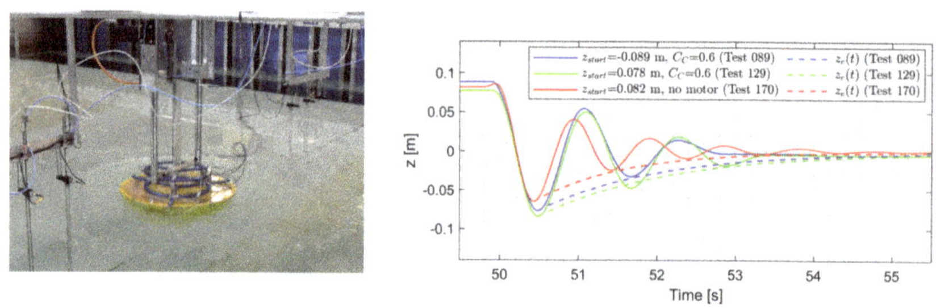

(a) (b)

Figure 16. Free decay tests: (**a**) Picture of Test_161; (**b**) z time series of the free decay tests and corresponding exponential decay envelopes (Test_089, Test_129 and Test_170).

Test_089 with an elevated start position of -0.089 m results in a quasi-equivalent decay response as Test_078 with a submerged start position of 0.078 m. In order to allow a comparison of the symmetry of the WEC system, Test_089 is plotted with inversed sign.

3.4. Power Absorption

Figure 17 displays the four quadrants where the WEC PTO system acts as a motor or generator, according to the adopted sign convention [37].

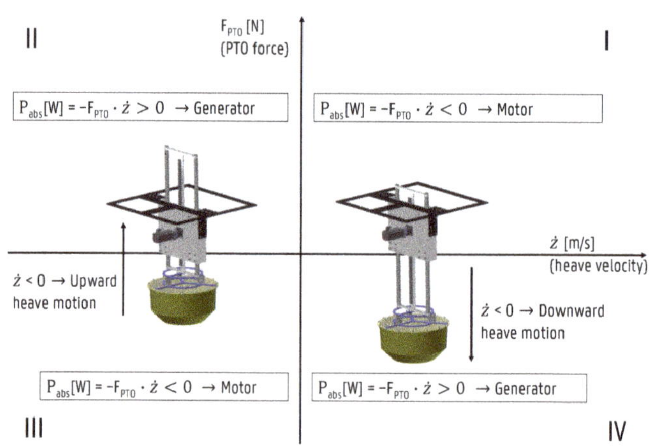

Figure 17. Power generation quadrant.

When F_{PTO} and the resultant $\dot{z}$ have an opposite sign, energy will be extracted from the waves and the PTO acts as a generator, occurring in quadrant II and IV. When F_{PTO} and the resultant $\dot{z}$ have an identical sign, energy will be consumed and the PTO acts as a motor, occurring in quadrant I and III. The defined convention in the Simulink model is that a net positive absorbed power value P_{abs} corresponds to mechanical power absorption and a net negative P_{abs} corresponds to adding power to the system. Therefore, a minus sign is added to the multiplication of F_{PTO} with $\dot{z}$ in order to obtain P_{abs}:

$$P_{abs} = -F_{PTO} \cdot \dot{z} \tag{18}$$

Note that this calculation results in instantaneous values. For regular waves, averaging over a number n of wave periods T results in the averaged absorbed power value $\overline{P}_{abs}$:

$$\overline{P}_{abs} = \frac{1}{nT} \int_0^{nT} P_{abs}(t)dt \tag{19}$$

3.4.1. Impedance Matching

Impedance matching can be applied to enable maximum power transfer between two oscillatory systems, the ocean waves and the WEC PTO system. Since impedance is defined as a complex value, the PTO generally has a resistance component (real part) and a reactance component (imaginary part). The maximum power transfer theorem says that the maximum possible power is delivered to the PTO when the PTO impedance Z_{PTO} (load impedance or input impedance) is equal to the complex conjugate (represented by *) of the impedance of the source Z_i (intrinsic impedance or output impedance). For two impedances to be complex conjugates their resistances must be equal and their reactances must be equal in magnitude and opposite in sign. With the WEC impedance model given by Equation (13), an optimal PTO is obtained by $Z_{PTO}(\omega)$ [12,38]:

$$Z_{PTO}(\omega) = Z_i^*(\omega) = B(\omega) - i\left(\omega(M + A(\omega)) - \frac{K}{\omega}\right) \tag{20}$$

The frequency response of the Proportional Integral Derivative (PID) controller is given by [12]:

$$FRF_{PID}(\omega) = \frac{(i\omega)^2 K_D + i\omega K_P + K_I}{i\omega} \tag{21}$$

K_P, K_I and K_D are the proportional, integral and derivative gains of the controller, respectively. The rearrangement of Equation (21) results in the PID controller impedance:

$$Z_{PID}(\omega) = i\omega K_D + K_P - i\frac{K_I}{\omega} = K_P + i(\omega K_D - \frac{K_I}{\omega}) \tag{22}$$

Given Equations (20) and (22), the proportional gain for the Proportional (P) controller, equivalent to the damping coefficient $C_{PTO,P}$ for the resistive control strategy, is given by:

$$C_{PTO,P}(\omega) = |Z_i^*(\omega)| = \sqrt{B^2(\omega) + \left(\omega(M + A(\omega)) - \frac{K}{\omega}\right)^2} \tag{23}$$

Given Equations (20) and (22), the proportional and integral gain for the Proportional Integral (PI) controller, equivalent to $C_{PTO,PI}$ and the spring coefficient $K_{PTO,PI}$ for the reactive control strategy, are given by:

$$C_{PTO,PI}(\omega) = Re\{Z_i^*(\omega)\} = B(\omega) \tag{24}$$

$$K_{PTO,PI}(\omega) = \omega \cdot Im\{Z_i^*(\omega)\} = \omega\left(-\left(\omega(M + A(\omega)) - \frac{K}{\omega}\right)\right) = -\omega^2(M + A(\omega)) + K \tag{25}$$

The presented study considers only P and PI control. Gu et al. present the implementation of a PID controller in a frequency domain model for a heaving point absorber WEC [39]. In this case the derivative controller acts on the acceleration term, corresponding to mass control.

3.4.2. Causal Impedance Matching P and PI Controller

For any causal system, the complex conjugate of the impedance will produce an acausal system. Since a WEC is a causal system, an acausal system is obtained for $Z_i^*(\omega)$. Therefore, to implement Equation (23) or Equations (24) and (25) perfectly across all wave frequencies, the future velocity of the WEC has to be known, resulting in an acausal controller implementation [12]. However, the controller does not need to operate perfectly at all wave frequencies simultaneously and can be designed to work well in a restricted range of frequencies. The energy in a typical sea state ranges over at most a single decade of frequencies, with the sea state changing appreciably only over the course of hours. This band-limited and slowly varying nature of ocean waves allows to utilize a causal realization of the impedance matching approach. Approximating $Z_i^*(\omega)$ in the peak frequency ω_p of the design sea state is straightforward to implement in the Simulink model. The impedance of the causal impedance matching P controller is equal to C_{PTO} defined in Equation (24) [12,38]:

$$Z_P(\omega) = C_{PTO}(\omega_p) \tag{26}$$

The impedance of the causal impedance matching PI controller is an interpolation of the impedance given by Equation (22) in ω_p [12,38]:

$$Z_{PI}(\omega) = C_{PTO}(\omega_p) - i \cdot \frac{K_{PTO}(\omega_p)}{\omega} \tag{27}$$

Based on the experimentally determined Z_i in Test_112, the impedance of a causal impedance matching P controller and PI controller can be determined, according to Equations (26) and (27), respectively. Figure 18a shows the gain and phase of these controllers, for an interpolation point of T = 1.50 s. Since in the power absorption tests various wave conditions are considered, Figure 18b plots $C_{PTO,P}(\omega)$ (Equation (23)), $C_{PTO,PI}(\omega)$ (Equation (24)) and $K_{PTO,PI}(\omega)$ (Equation (25)) in function of the wave frequency. The resulting control parameters for T = 1.50 s and T = 2.00 s are indicated. As expected, $K_{PTO,PI}(\omega) = 0\ \mathrm{kg/s^2}$ for T_n = 1.19 s.

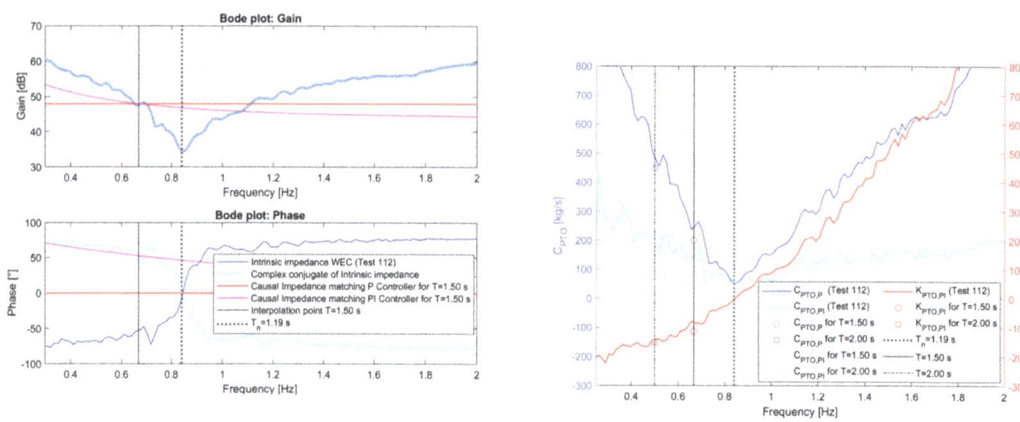

(a) (b)

Figure 18. Radiation Test_112: (**a**) Bode plot intrinsic impedance and causal impedance matching P and PI controller; (**b**) Coefficients causal impedance matching P and PI controller.

3.4.3. Resistive Control

The baseline control strategy is resistive control, equivalent to causal impedance matching P control as defined in Equation (26). $F_{PTO,P}$ is equal to $\dot{z}$ multiplied with a positive C_{PTO}:

$$F_{PTO,P} = -C_{PTO} \cdot \dot{z} \tag{28}$$

A minus sign is added, since $F_{PTO,P}$ should oppose $\dot{z}$. Tables 8 and 9 provide an overview of the performed resistive control tests, for regular and irregular waves, respectively. The applied C_C is given. For the regular waves, characterized by T and H, a range of C_{PTO} is tested during a single test. For the irregular waves, characterized by a JONSWAP wave spectrum with γ = 3.3, H_s and T_p, a fixed C_{PTO} is applied.

Table 8. Overview of the resistive control strategy tests for regular waves.

Test_ID	H [m]	T [s]	C_C [-]	C_{PTO} [kg/s]	Test_ID	H [m]	T [s]	C_C [-]	C_{PTO} [kg/s]
Test_080	0.09	2.0	0.0	0	Test_100	0.09	2.0	0.6	[200;800]
Test_081	0.07	2.0	0.0	0	Test_101	0.05	1.5	0.6	[50;550]
Test_085	0.09	2.0	0.0	[0;800]	Test_102	0.07	1.5	0.6	[50;450]
Test_087	0.09	1.0	0.6	0	Test_103	0.09	1.5	0.6	[50;550]
Test_090	0.09	2.0	0.6	[100;900]	Test_104	0.11	2.0	0.6	[200;800]
Test_091	0.05	1.0	0.6	0	Test_105	0.13	2.0	0.6	[200;800]
Test_092	0.09	1.0	0.6	0	Test_106	0.15	2.0	0.6	[200;800]
Test_093	0.09	1.0	0.6	[25;125]	Test_115	0.09	2.0	0.6	[200;800]
Test_094	0.09	1.0	0.6	[40;220]	Test_116	0.09	2.0	0.6	[200;800]
Test_095	0.07	1.0	0.6	[40;220]	Test_139	0.07	1.17	0.6	[50;500]
Test_096	0.05	1.0	0.6	[40;180]	Test_140	0.07	1.17	0.6	[40;100]
Test_097	0.05	2.0	0.6	[200;1400]	Test_141	0.20	2.0	0.6	[250;800]
Test_098	0.05	2.0	0.6	[200;800]	Test_164	0.20	2.0	0.6	[0;800]
Test_099	0.07	2.0	0.6	[200;800]					

Table 9. Overview of the resistive control strategy tests for irregular waves.

Test_ID	H_s [m]	T_p [s]	C_C [-]	C_{PTO} [kg/s]	Test_ID	H_s [m]	T_p [s]	C_C [-]	C_{PTO} [kg/s]
Test_079	0.05	1.0	0.0	0	Test_158	0.09	1.5	0.6	200
Test_130	0.09	1.5	0.6	300	Test_159	0.09	1.5	0.6	100
Test_131	0.09	1.5	0.6	200	Test_160	0.20	1.5	0.6	200
Test_132	0.09	1.5	0.6	400	Test_169	0.30	2.0	0.6	500

Figure 19a shows a snapshot of the z and $\dot{z}$ time series for Test_100. Figure 19b shows the corresponding C_{PTO} input, with a stepwise increase from 400 kg/s to 450 kg/s, and the P_{abs} output.

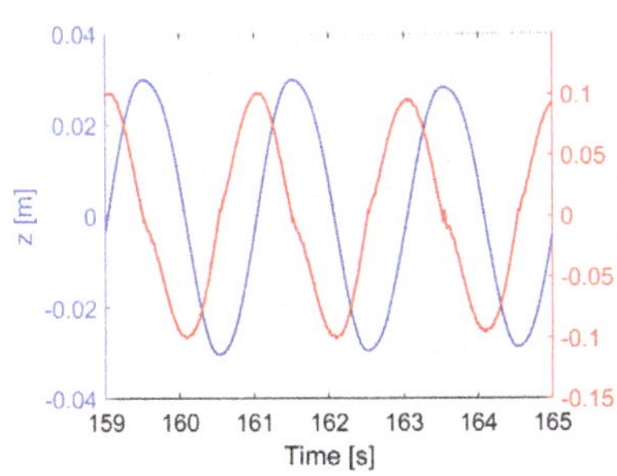

(a)

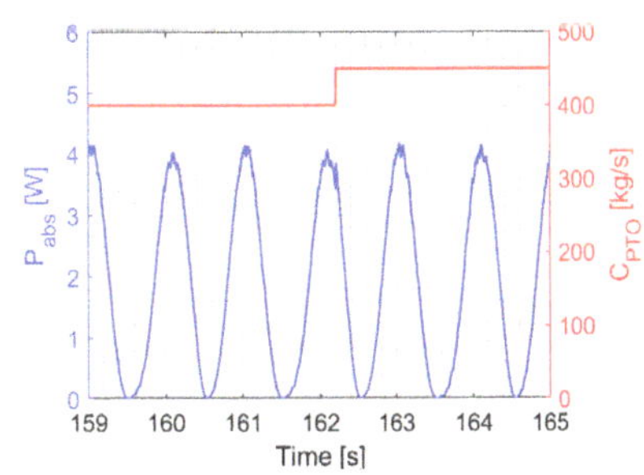

(b)

Figure 19. Resistive control test (Test_100): (**a**) z and $\dot{z}$ time series; (**b**) P_{abs} and C_{PTO} time series.

3.4.4. Reactive Control

Reactive control, equivalent to causal impedance matching PI control as defined in Equation (27), aims to bring the WEC into resonance by the addition of a spring. $F_{PTO,PI}$ is composed of $\dot{z}$ multiplied with a positive C_{PTO} and z multiplied with negative K_{PTO}.

$$F_{PTO,PI} = -C_{PTO} \cdot \dot{z} - K_{PTO} \cdot z \qquad (29)$$

Again, a minus sign is added, since $F_{PTO,PI}$ should oppose $\dot{z}$. Table 10 gives an overview of the performed reactive control tests for regular waves, characterized by H and T. C_C is fixed at 0.6 and a range of C_{PTO} and K_{PTO} is tested during each test.

Table 10. Overview of the reactive control strategy tests for regular waves.

Test_ID	H [m]	T [s]	C_C [-]	C_{PTO} [kg/s]	K_{PTO} [kg/s^2]
Test_137	0.09	2.0	0.6	[320;480]	[−1710;−1140]
Test_138	0.09	2.0	0.6	[48;72]	[−2280;−1520]
Test_146	0.09	1.5	0.6	[277;415]	[−978;−652]
Test_147	0.09	1.5	0.6	[70;104]	[−1680;−1120]

Figure 20a shows a snapshot of the z and $\dot{z}$ time series for Test_138. The Simulink control model did not impose constraints on z, $\dot{z}$, $\ddot{z}$, allowing these high values for z and $\dot{z}$. The stroke is mechanically limited to 0.25 m by the micro switches, displayed in Figure 3. Figure 20b shows the corresponding C_{PTO} input, with a stepwise increase from 60 kg/s to 72 kg/s, and the P_{abs} output. The net $\overline{P}_{abs}$ is calculated as the period averaged difference between the integrated generated power (positive P_{abs}) and the integrated added power (negative P_{abs}), according to Equation (19) and the sign convention as defined in Figure 17. K_{PTO} is equal to -1520 kg/s^2 for the displayed time window.

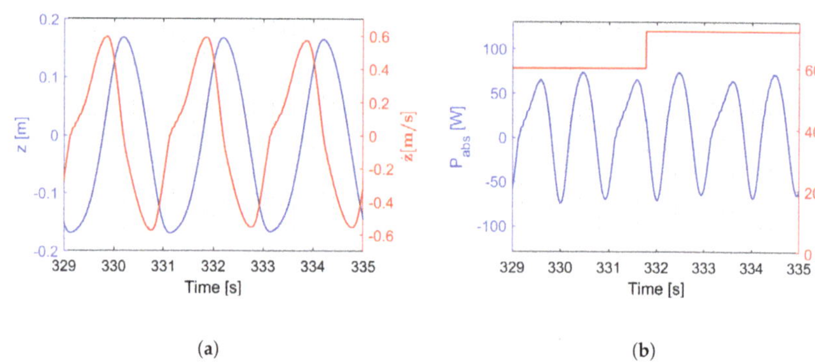

Figure 20. Reactive control test (Test_138): (**a**) z and $\dot{z}$ time series; (**b**) P_{abs} and C_{PTO} time series.

3.4.5. Power Absorption Comparison between the Resistive and Reactive Controller

Two regular wave conditions, characterized by H = 0.09 m, T = 2.00 s and by H = 0.09 m, T = 1.50 s, respectively, are considered to compare the presented resistive and reactive control strategy. $\overline{P}_{abs}$ is calculated with n = 5 in Equation (19). For resistive control, $C_{PTO,P}$ is obtained according to Equation (23) and for reactive control, $C_{PTO,PI}$ and $K_{PTO,PI}$ are obtained according to Equations (24) and (25), respectively. Figure 18b displays the obtained control parameters for T = 2.00 s and T = 1.50 s. For $C_{PTO,PI}$, lower values than experimentally indentified are adopted. For the regular wave with H = 0.09 m and T = 2.00 s, resistive control (Test_100) with C_{PTO} = 450 kg/s yields $\overline{P}_{abs}$ = 1.89 W and reactive control (Test_138) with C_{PTO} = 72 kg/s and K_{PTO} = -1520 kg/s yields $\overline{P}_{abs}$ = 10.26 W. For the regular wave with H = 0.09 m and T = 1.50 s, resistive control (Test_103) with C_{PTO}

= 200 kg/s yields $\overline{P}_{abs}$ = 3.09 W and reactive control (Test_147) with C_{PTO} = 70 kg/s and $K_{PTO} = -1120$ kg/s yields $\overline{P}_{abs}$ = 11.75 W. Figure 21 visualizes these results.

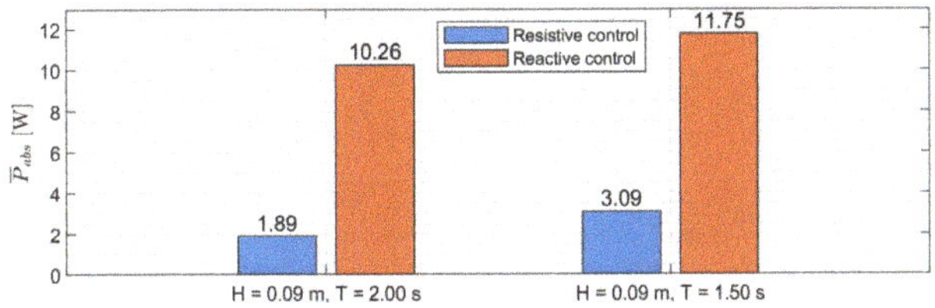

Figure 21. $\overline{P}_{abs}$ for Test_100, Test_103, Test_138 and Test_147.

For the reactive controller, a significant increase in power absorption compared to the resistive controller is observed, which is according to the literature [40]. Figure 20a shows that resonance occurs as z_{max} = 0.165 m is significantly higher than H/2 = 0.045 m. To properly assess the power performance of the reactive controller compared to the resistive controller, irregular waves should be considered, as discussed in Section 3.4.2.

The amplification on z requires the motor to add energy, which has a certain efficiency. Apart from larger peaks in F_{PTO} and P_{abs}, a major drawback of reactive control is the energy loss by dissipative processes inherent to the back-and-forth energy exchange between the PTO and the WEC buoy, especially when the magnitude of the exchanged energy is comparable to, or even significantly larger than, the net absorbed energy [41]. The presented research does not consider the PTO efficiency, nor the damping force, nor the reactive force. Strager et al. present a method to determine the optimal reactive control parameters for a given combination of non-ideal PTO efficiency and monochromatic wave frequency [42].

4. Discussion and Conclusions

Within the 'WECfarm' project, two test campaigns are performed at the AAU wave basin: (a) a testing of the first WEC in April 2021 and (b) a testing of a two-WEC array in February 2022, in preparation of five-WEC array tests. The main objective of the 'WECfarm' project is to cover the scientific gap on experimental data necessary for the validation of recently developed (non-linear) numerical models. The presented article discusses experimental testing of an isolated WEC, being the first test campaign within the 'WECfarm' project. The primary objective of evaluating the hydrodynamics, electromechanics, control platform, DAQ and structural performance of the WEC is to allow extending the setup to a five-WEC array.

The WEC buoy is a truncated cylinder with a high diameter to draft ratio to increase radiation and WEC–WEC interactions. The PTO system of the WEC is a PMSM connected to a gearbox powering a rack and pinion system. A configuration of three air bushings guarantees a permanent layer of air between the guide shafts and the bushings, resulting in zero-friction linear guiding. The WEC control and data acquisition are realized with a Speedgoat Performance real-time target machine, offering the possibility to implement advanced WEC array control strategies in the MATLAB-Simulink model. This unique PTO system in combination with the real-time target machine makes active and accurate PTO control possible.

The drivetrain (motor, gearbox, rack and pinion) friction is assessed by manually moving the WEC buoy up and down, from below the loadcells and under a zero torque command. Based on the empirical relationship between $F_{loadcells}$ and $\dot{z}$, a simplified model based on Coulomb and viscous friction is determined. The resulting friction compensation model F_{Comp} is implemented in Simulink as a torque augmentation on the torque command.

A SID approach is adopted, considering the WEC system to be composed of two SISO models, the radiation and the excitation model. The radiation tests in calm water, with F_{PTO} chirp-up and chirp-down noise signals as input and $\dot{z}$ as output, yield the intrinsic impedance Z_i. The excitation tests with the WEC buoy fixed, with as input various JONSWAP spectra and as output $F_{loadcells}$, yield the wave excitation force coefficients $H_e(\omega)$. The assumption that WG measurements at sufficient distance from the WEC buoy may be used to characterize η on the location of the WEC is confirmed. Free decay tests characterize the WEC buoy decay response, confirming that T_n = 1.19 s, obtained by the radiation tests.

Power absorption tests are executed with the resistive and reactive control strategy. The tests with regular waves are executed with 'real-time' tuning of the control parameters in the Simulink model. Adopting an impedance matching approach, the optimal $C_{PTO,P}$ for resistive control and the optimal $C_{PTO,PI}$ and $K_{PTO,PI}$ for reactive control are calculated from Z_i. For the two selected regular wave conditions, characterized by H = 0.09 m, T = 2.00 s and by H = 0.09 m, T = 1.50 s, the reactive controller results in a significant higher averaged absorbed power $\overline{P}_{abs}$ compared to the resistive controller. However, the presented research does not consider PTO efficiency, which is detrimental for reactive control.

The implementation of the friction compensation model F_{Comp} proved to be a good methodology to partly compensate Coulomb and viscous friction attributed to the drivetrain. The experimentally determined radiation and excitation model yields a simple, though accurate, model of the WEC system. The intrinsic impedance, resulting from the radiation test, is used to design a causal impedance matching P and PI controller, equivalent to a resistive and reactive controller. Although this approach was only tested limitedly, mainly for regular waves, the successful extension to a multiple-WEC array, considering irregular waves, is confirmed. The testing of the isolated 'WECfarm' WEC proved to be successful and the extension of the setup to a five-WEC array will allow it to comply with the future research objectives.

Author Contributions: Conceptualization, methodology and analysis, T.V.; execution of the experiments, T.V., F.F., L.D.B., H.C. and B.D.W.; data curation, T.V.; writing—original draft preparation, T.V.; writing—review and editing, V.S., F.F., M.V. and P.T.; funding acquisition for the experimental setup, V.S. and P.T. All authors have read and agreed to the published version of the manuscript.

Funding: This work is supported by the the FWO (Fonds Wetenschappelijk Onderzoek-Research Foundation Flanders), Belgium, through the following funding: (1) Timothy Vervaet is Ph.D. fellow (fellowship 11A6919N); (2) Vasiliki Stratigaki is a FWO postdoctoral researcher (fellowship 1267321N) and has been granted the 'FWO Research Grant' for constructing the WEC experimental set-up (FWO-KAN-DPA376). The travel expenses of Timothy Vervaet, Louis De Beule, Hendrik Claerbout and Bono De Witte for conducting the experimental campaign at the AAU wave basin were funded by four WECANet COST Action CA17105 Short Term Scientific Missions (STSMs). COST (European Cooperation in Science and Technology) is supported by the EU Framework Programme Horizon 2020. COST is a funding agency for research and innovation networks.

Institutional Review Board Statement: Not applicable.

Informed Consent Statement: Not applicable.

Data Availability Statement: The datasets resulting from the WECfarm project will be made available in due time on https://www.awww.ugent.be, accessed on 23 August 2022.

Acknowledgments: Michiel Herpelinck and Brecht De Backer were involved in the design of the WECfarm WEC in the framework of a master dissertation supervised by Prof. Kurt Stockman from the Department of Electromechanical, Systems and Metal Engineering of Ghent University. Brecht De Backer initiated the Autodesk Inventor drawings of the WEC. OAV (https://www.oavco.com/, accessed on 23 August 2022) is acknowledged for providing educational discount on the OAV 40 mm air bushings used in the experimental setup as well as for providing technical support [14]. Beckhoff, Wittenstein and Speedgoat are acknowledged for providing technical support. Aalborg University, Denmark is acknowledged for making the wave basin available from 12–18 April 2021 for the presented test campaign.

Conflicts of Interest: The authors declare no conflict of interest. The funders had no role in the design of the study; in the collection, analyses, or interpretation of data; in the writing of the manuscript, or in the decision to publish the results.

Abbreviations

The following abbreviations are used in this manuscript:

AAU	Aalborg University
ABS	Acrylonitril-Butadieen-Styreen
AT	Amplifier Telegramm
BEM	Boundary Element Method
CFD	Computational fluid dynamics
CNWT	CFD-based numerical wave tank
COB	Coastal and Ocean Basin
DAQ	Data Acquisition System
DOFs	Degrees of Freedom
EtherCAT	Ethernet for Control Automation Technology
FFT	Fast Fourier Transform
FRF	Frequency response function
FSO	Full Scale Output
I/O	Input and Output
JONSWAP	Joint North Sea Wave Project
MDT	Master Data Telegramm
MWL	Mean Water Level
MISO	Multiple input single output
OCT	One Cable Technology
P	Proportional
PI	Proportional Integral
PID	Proportional Integral Derivative
PMMA	Polymethylmethacrylate
PMSM	Permanent Magnet Synchronous Motor
PTO	Power Take-Off
RAO	Response Amplitude Operator
SID	System Identification
SISO	Single input single output
SWL	Still Water Level
WEC	Wave Energy Converter
WG	Wave Gauge

References

1. Falnes, J.; Lillebekken, P.M. Budal's latching-controlled-buoy type wave-power plant. In Proceedings of the European Wave Energy Conference, Cork, Ireland, 17–20 September 2003; pp. 233–244.
2. Göteman, M.; Giassi, M.; Engström, J.; Isberg, J. Advances and Challenges in Wave Energy Park Optimization—A Review. *Front. Energy Res.* **2020**, *8*, 26. [CrossRef]
3. Windt, C.; Davidson, J.; Ringwood, J. High-fidelity numerical modelling of ocean wave energy systems: A review of computational fluid dynamics-based numerical wave tanks. *Renew. Sustain. Energy Rev.* **2018**, *93*, 610–630. [CrossRef]
4. Vervaet, T.; Stratigaki, V.; De Backer, B.; Stockman, K.; Vantorre, M.; Troch, P. Experimental modelling of point absorber wave energy converter arrays: A comprehensive review, identification of research gaps and design of the WECfarm setup. *J. Mar. Sci. Eng.* **2022**, *10*, 1062. [CrossRef]
5. Nader, J.R.; Fleming, A.; Macfarlane, G.; Penesis, I.; Manasseh, R. Novel experimental modelling of the hydrodynamic interactions of arrays of wave energy converters. *Int. J. Mar. Energy* **2017**, *20*, 109–124. [CrossRef]
6. Stallard, T.; Stansby, P.K.; Williamson, A.J. An experimental study of closely spaced point absorber arrays. In Proceedings of the International Offshore and Polar Engineering Conference, Vancouver, BC, Canada, 6–11 July 2008.
7. Folley, M.; Whittaker, T. Preliminary Cross-Validation of Wave Energy Converter Array Interactions. In Proceedings of the International Conference on Offshore Mechanics and Arctic Engineering—OMAE, Nantes, France, 8–13 June 2013; Volume 10837.
8. Stratigaki, V.; Troch, P.; Stallard, T.; Forehand, D.; Kofoed, J.P.; Folley, M.; Benoit, M.; Babarit, A.; Kirkegaard, J. Wave basin experiments with large wave energy converter arrays to study interactions between the converters and effects on other users in the sea and the coastal area. *Energies* **2014**, *7*, 701–734. [CrossRef]
9. Troch, P.; Stratigaki, V.; Devriese, P.; Kortenhaus, A.; De Maeyer, J.; Monballiu, J.; Toorman, E.; Rawoens, P.; Vanneste, D.; Suzuki, T.; et al. Design features of the upcoming coastal and ocean basin in Ostend, Belgium. In Proceedings of the Coastal Structures Conference, Hannover, Germany, 30 September–2 Oktober, 2019; Goseberg, N., Schlurmann, T., Eds.; pp. 1191–1199.

10. Beatty, S.; Ferri, F.; Bocking, B.; Kofoed, J.P.; Buckham, B. Power Take-Off Simulation for Scale Model Testing of Wave Energy Converters. *Energies* **2017**, *10*, 973. [CrossRef]
11. Bacelli, G.; Ringwood, J. Constrained control of arrays of wave energy devices. *Int. J. Mar. Energy* **2013**, *3–4*, 53–69. [CrossRef]
12. Coe, R.G.; Bacelli, G.; Forbush, D. A practical approach to wave energy modeling and control. *Renew. Sustain. Energy Rev.* **2021**, *142*, 110791. [CrossRef]
13. Bacelli, G.; Coe, R.G.; Patterson, D.; Wilson, D. System Identification of a Heaving Point Absorber: Design of Experiment and Device Modeling. *Energies* **2017**, *10*, 472. [CrossRef]
14. OAV Air Bearings. OAV Air Bearing Product Book & Design Guide: Where Aerospace Techonology Meets Air Bearing Systems. 2021. Available online: https://www.oavco.com/brochures (accessed on 7 July 2021).
15. Beckhoff Automation. Beckhoff New Automation Technology, Operation Instructions, AM8000 and AM8500: Synchronous Servomotors, Version 4.9. 2022. Available online: https://download.beckhoff.com/download/document/motion/am8000_am8500_ba_en.pdf (accessed on 5 October 2022).
16. Beckhoff Automation. AM8542-wEyz | Servomotor with Increased Moment of Inertia 4.10 Nm (M0), F4 (87 mm). 2022. Available online: https://www.beckhoff.com/nl-be/products/motion/rotary-servomotors/am8500-servomotors-with-increased-rotor-moment-of-inertia/am8542-weyz.html (accessed on 2 June 2022).
17. Analog Devices. ADXL335: Small, Low Power, 3-Axis ±3 g Accelerometer. 2022. Available online: https://www.analog.com/media/en/technical-documentation/data-sheets/ADXL335.pdf (accessed on 5 October 2022).
18. Vishay Precision Group. Model 614, Tedea-Huntleigh: Tension Compression Load Cell. 2022. Available online: http://www.vishaypg.com/docs/12040/614.pdf (accessed on 5 October 2022).
19. Micro-Epsilon. Operating Instructions optoNCDT 1420. 2022. Available online: https://www.micro-epsilon.com/download/manuals/man--optoNCDT-1420--en.pdf (accessed on 5 October 2022).
20. Lamont-Kane, P.; Folley, M.; Whittaker, T. Investigating Uncertainties in Physical Testing of Wave Energy Converter Arrays. In Proceedings of the 10th European Wave and Tidal Energy Conference Series—EWTEC, Aalborg, Denmark, 2–5 September 2013.
21. Aalborg University. Ocean and Coastal Engineering Laboratory. 2022. Available online: https://www.en.build.aau.dk/laboratories/ocean-and-coastal-engineering/ (accessed on 25 June 2021).
22. Aalborg University. AwaSys. 2022. Available online: https://www.hydrosoft.civil.aau.dk/awasys/ (accessed on 2 August 2022).
23. OAV Air Bearings. *Friction Coefficient 40 mm OAV Air Bushings*; e-mail communication, from 24 June 2021; OAV Air Bearings: Princeton, NJ, USA, 2021.
24. Wittenstein Alpha. *Gearbox 'NPR 025S-MF1-4 -2E1-1S' Break-Away Torque*; e-mail communication, from 16 December 2021; Wittenstein Alpha: Igersheim, Germany, 2021.
25. Armstrong-Helouvry, B.; Dupont, P.; Canudas de Wit, C. A Survey of Models, Analysis Tools and Compensation Methods for the Control of Machines with Friction. *Automatica* **1994**, *30*, 1083–1138. [CrossRef]
26. Wittenstein Alpha. Alpha Basic Line & Alpha Value Line Product Catalog. 2022. Available online: https://alpha.wittenstein.de/en-en/alpha-value-line/ (accessed on 8 September 2022).
27. Verbrugghe, T. openWEC: Open Source Wave Energy Converter (WEC) Simulation Tool, Ghent University. 2018. Available online: https://users.ugent.be/~tverbrug/ (accessed on 2 June 2022).
28. Yang, P.; Zhang, Z.; Zhao, J.; Zhou, D. Improved PID Friction Feed-forward Compensation Control Based on Segment Friction Model. *Open Autom. Control. Syst. J.* **2014**, *6*, 1620–1628. [CrossRef]
29. Ringwood, J.; Davidson, J.; Giorgi, S. Identifying Models Using Recorded Data. In *Numerical Modelling of Wave Energy Converters*; Academic Press: Cambridge, MA, USA, 2016; pp. 123–147.
30. Falcão, A.F.d.O. Wave energy utilization: A review of the technologies. *Renew. Sustain. Energy Rev.* **2009**, *14*, 899–918. [CrossRef]
31. Stratigaki, V. Experimental Study and Numerical Modelling of Intra-Array Interactions and Extra-Array Effects of Wave Energy Converter Arrays. Ph.D. Thesis, Ghent University, Ghent, Belgium, 2014.
32. De Backer, G.; Vantorre, M.; Frigaard, P.; Beels, C.; De Rouck, J. Bottom slamming on heaving point absorber wave energy devices. *J. Mar. Sci. Technol.* **2010**, *15*, 119–130. [CrossRef]
33. Morison, J.R.; O'Brien, M.P.; Johnson, J.W.; Schaaf, S. The force exerted by surface waves on piles. *Pet. Trans. AIME* **1950**, *2*, 149–154. [CrossRef]
34. Penalba, M.; Giorgi, G.; Ringwood, J. A Review of Non-Linear Approaches for Wave Energy Converter Modelling. In Proceedings of the 11th European Wave and Tidal Energy Conference, Nantes, France, 6–11 September 2015.
35. Giorgi, G.; Ringwood, J.V. Nonlinear Froude-Krylov and viscous drag representations for wave energy converters in the computation/fidelity continuum. *Ocean Eng.* **2017**, *141*, 164–175. [CrossRef]
36. Quartier, N.; Ropero-Giralda, P.; MDomínguez, J.; Stratigaki, V.; Troch, P. Influence of the drag force on the average absorbed power of heaving wave energy converters using smoothed particle hydrodynamics. *Water* **2021**, *13*, 384. [CrossRef]
37. Antony, M.S.; Praveen Raj, R.S. Four Quadrant Operation of Vector Control of PMSM with Dynamic Braking. In Proceedings of the International Conference on Control, Communication & Computing India (ICCC), Trivandrum, India, 19–21 November 2015.
38. Faedo, N.; Carapellese, F.; Pasta, E.; Mattiazzo, G. On the principle of impedance-matching for underactuated wave energy harvesting systems. *Appl. Ocean. Res.* **2022**, *118*, 102958. [CrossRef]
39. Gu, Y.; Ding, B.; Sergiienko, N.Y.; Cazzolato, B.S. Power maximising control of a heaving point absorber wave energy converter. *IET Renew. Power Gener.* **2021**, *15*, 3296–3308. [CrossRef]

40. Coe, R.G.; Bacelli, G.; Wilson, D.G.; Abdelkhalik, O.; Korde, U.A.; Robinett, R.D. A comparison of control strategies for wave energy converterss. *Int. J. Mar. Energy* **2017**, *20*, 45–63. [CrossRef]
41. Falcão, A.F.d.O.; Henriques, J.C.C. Effect of non-ideal power take-off efficiency on performance of single- and two-body reactively controlled wave energy converters. *J. Ocean Eng. Mar. Energy* **2015**, *1*, 273–286. [CrossRef]
42. Strager, T.; Martin dit Neuville, A.; Fernandez Lopez, P.; Giorgio, G.; Muresan, T.; Andersen, P.; Nielsen, K.M.; Pedersen, T.S.; Vidal Sanchez, E. Optimising Reactive Control In Non-Ideal Efficiency Wave Eenergy Converters. In Proceedings of the ASME 2014 33rd International Conference on Ocean, Offshore and Arctic Engineering, San Francisco, CA, USA, 8–13 June 2014.

Journal of
Marine Science and Engineering

MDPI

Article

Optimization on Hydrodynamic Performance for First Level Energy-Capturing Enhancement of a Floating Wave Energy Converter System with Flapping-Panel-Slope

Tianyu Song [1], Ze Li [2,*], Honghao Zheng [3,*], Chujin Liang [2] and Zhanhong Wan [2]

1 Hangzhou Applied Acoustics Research Institute, Hangzhou 310023, China
2 Ocean College, Zhejiang University, Zhoushan 316021, China
3 School of Aerospace Engineering and Applied Mechanics, Tongji University, Shanghai 200092, China
* Correspondence: ocean-lize@zju.edu.cn (Z.L.); zhenghh@tongji.edu.cn (H.Z.)

Abstract: Based on the wave reflection principle, a floating flapping-panel wave energy converter was developed. The feasibility study and optimization study of the new WEC was carried out by laboratory research and computational fluid dynamics techniques. A numerical model was developed for an in-depth study to establish the relationship between slope tilt angle and power. The results for different wave periods show that the power take-off damping coefficient has a significant effect on the power. Meanwhile, the effects of flap length and wave height on converter resonance and power are investigated. Finally, a preliminary laboratory physical model test of the device is conducted. The flapping-panel-slope structure is very feasible and effective with good hydrodynamic performance.

Keywords: wave energy converter; floating flapping-panel; slope; hydrodynamic performance

check for updates

Citation: Song, T.; Li, Z.; Zheng, H.; Liang, C.; Wan, Z. Optimization on Hydrodynamic Performance for First Level Energy-Capturing Enhancement of a Floating Wave Energy Converter System with Flapping-Panel-Slope. *J. Mar. Sci. Eng.* **2023**, *11*, 345. https://doi.org/10.3390/jmse11020345

Academic Editor: Unai Fernandez-Gamiz

Received: 16 November 2022
Revised: 22 December 2022
Accepted: 30 December 2022
Published: 4 February 2023

1. Introduction

At present, there are various design models for wave energy conversion devices [1], which can be basically classified into offshore, nearshore and onshore devices, according to the installation location [2]. Offshore devices face huge survival pressure and high submarine cable laying costs, making commercial application very difficult. Nearshore and onshore devices are relatively simple and convenient in terms of survival and maintenance and may be a good option.

A number of wave energy converter (WEC) models with very good performance have emerged for both nearshore and onshore devices. The Oyster device developed by Queen's University in the UK, which has been updated for two generations, has performed well in all specifications and has successfully achieved long-term stable operation in real sea conditions [3]. Unlike the Oyster, the WaveRoller uses a fully submerged floating pendulum that can be formed into an array of power modules, with all floating pendulum motion occurring below the water surface, and full-size wave energy generator WaveRoller installed in Portugal [4]. Eco Wave Power, which fixes pendulum wave energy conversion devices on shore base or combined with breakwaters, has now deployed physical models in several locations around the world and has successfully captured wave energy for power generation [5]. The flexible combination of Wavestar and offshore wind energy capture devices enables multi-energy hybrid absorption and improves commercial competitiveness [6].

To test the feasibility of a WEC model, physical model tests can generally be taken to solve a series of problems by simulating the realization of the phenomena of the interaction between structures and waves in the laboratory. Bosma et al. obtained a series of hydrodynamic parameters by conducting wave pool experiments on FOSWEC for comparison with numerical simulations for validation [7]; Wei et al. successfully implemented the kinematic response of multi-pump multi-piston power take-off system in the laboratory using wave tank experiments [8]. Davey's team conducted long-term Round Robin tests of a Hinged

Raft Wave Energy Converter in several laboratories and found that wave tank depth had little effect on the results [9].Physical model experiments require the construction of an experimental platform, which is a very expensive and time-consuming process but is very meaningful for scientific research.

Another option is the computational fluid dynamics (CFD) technique and has been widely used as a cost-effective experimental tool in the optimal design studies of marine structures. A WEC-Sim code was developed at Sandia Labs, USA that can be used to calculate the motion response and power output of a WEC model in waves [10]; Poguluri et al. tested the WEC-rotor using CFD techniques that can be well-matched with physical experimental results [11]; Ransley et al. calculated the Wavestar through the open source program OPENFOAM software [12]; Tezdogan et al. conducted simulation experiments using Star-CCM+ on a full-size hull model in deep and shallow water and provided an in-depth analysis of the results in shallow water [13]. The development of the Smoothed Particle Hydrodynamics model for wave-structure interaction is more in line with engineering standards in terms of numerical calculations [14]. It is worth noting that different model scales in numerical simulations can cause different degrees of error in the experimental results [15].

Various experimental methods, which have played a great role and convenience in the calculation and testing of WEC, are becoming more and more mature. However, the commercialization and real use of WEC is still very slow; one of the biggest constraints is the hydrodynamic performance of WEC. Much of the current WEC optimization work is precisely around the hydrodynamic performance [16,17]; how to design a hydrodynamic performance of the WEC may be the key to solve the problem.

In conclusion, with the emphasis on renewable energy, the development of wave energy technology is also facing opportunities and challenges. Based on the novel wave energy conversion device, this paper will introduce the working principle of the novel WEC in Section 2, and the physical experimental method of the model and the experimental details of CFD technology will be developed in Section 3. Finally, the relevant experimental results will be obtained in Section 4, and relevant meaningful results will be derived around the experimental results, hoping that can drive the wave energy generation technology forward.

2. A Floating Flapping-Panel WEC

Wan et al. studied and designed a near-shore-based floating flapping-panel wave energy conversion device [18], as shown in Figure 1. In our previous work, our team focused on the overall system workflow of the novel wave energy conversion device and the significant improvement in wave energy capture efficiency by the slope structure [19]. In this paper, we focus on optimizing the hydrodynamic performance module and initially discuss the effect of slope angle on the capture efficiency of the device, as well as other factors, such as its own shape. This has a forward-looking effect on the local and overall optimization of the floating flapping-panel wave energy conversion device.

By arranging a freely adjustable slope structure under the floating flapping-panel, the wave energy from the lower layer of the free liquid surface is enhanced by reflecting it near the free liquid surface, which boosts the energy of the wave field near the floating flapping-panel; therefore, the overall wave energy absorption efficiency can be enhanced. The waves impact the floating flapping-panel, drive the hydraulic device to do work, output electrical energy and return to the original position by their own gravity, cycling the above process, continuously carrying out power output and merging electrical energy into the grid or storing it in the battery.

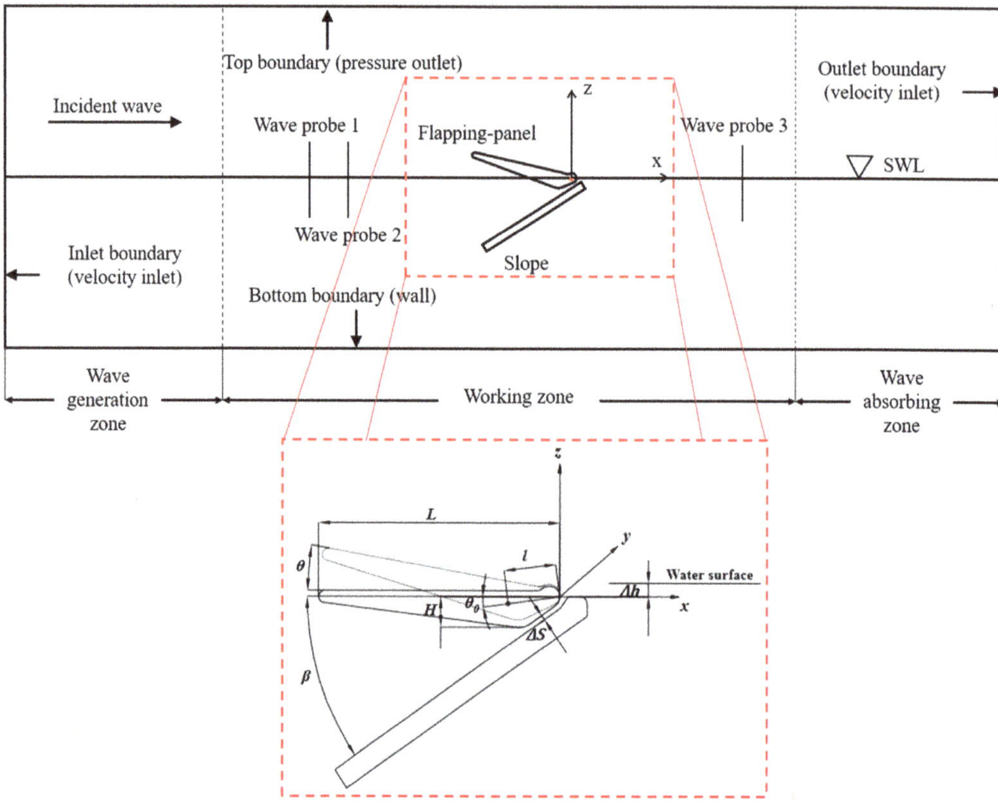

Figure 1. Schematic diagram of floating flapping-panel WEC.

The first described parameter was the floating swing force analysis under the wave force on the y-axis rotation. According to the momentum moment theorem, we have the dynamic equation:

$$J\ddot{\theta} = M_F + M_G + M_{PTO} \tag{1}$$

where $J\ddot{\theta}$ is the moment of inertia of floating flapping-panel around the shaft and $\ddot{\theta}$ represents the angular acceleration of the flapping-panel. M_F indicates the fluid moment expressed as:

$$M_F = \int_{\tau} r \times f_p d\tau \oint_{A} r \times P_n dA \tag{2}$$

where A is the floating force area; r is the distance from the cell area of the flapping-panel to the rotation; P_n is the tangential force of the buoyancy flapping-panel; f_p is the normal force of the flapping-panel; and M_G is the moment of gravity formulated as:

$$M_G = Mglcos(\theta - \theta_0) \tag{3}$$

where the variables are: θ_0¯the initial state of gravity axis connection and horizontal angle; θ—the rotation angle of the t moment;M¯mass of a buoyancy flapping-panel; g—gravity acceleration; and l—the distance from the center of gravity of the floating flapping-panel to the origin of coordinate. The power take-off (PTO) system torque moment is M_{PTO}. The linear moment used here is $M_{PTO} = -C\dot{\theta}$, where C is the coefficient of linear resistance moment, $\dot{\theta}$¯the flapping-panel angle acceleration and "$-$" means that the floating flapping-panel was performed outside. The nonlinear moment is a constant torque load, and the PTO system is instantaneous power: $P_{PTO} = M_{PTO}\dot{\theta}$.

The wave energy conversion device basically consists of a three-stage energy conversion process that converts wave energy into mechanical energy. The efficiency of a primary energy conversion depends on the hydrodynamic performance of the device [20]. Typically, the capture width ratio "c_f" is used to evaluate the hydrodynamic performance of the device. This capture width ratio is equal to the wave energy input by the wave energy device and the wave energy input within the device width. The buoyancy flapping-panel device can be calculated using (4)

$$c_f = \frac{P_{PTO}}{P_W \cdot w} \tag{4}$$

where w is the width of flapping-panel = 0.4 (two-dimensional case); P_{PTO} denotes conversion power; and P_W represents wave power.

After the wave energy device is captured, the first-order energy conversion is achieved by overcoming the PTO damping. The conversion power is based on using linear PTO damping [21] and the average power per unit W. The resistance moment T_{PTO} is proportional to the speed of the flapping-panel [22].

$$T_{PTO} = C\dot{\theta} \tag{5}$$

Furthermore, the conversion power of the device can be calculated by (6)

$$P_{PTO} = \frac{1}{T}\int_0^T C\dot{\theta}^2 dt = \frac{1}{2}\omega^2\theta_0^2 \tag{6}$$

By substituting (6), the expression of width capture ratio can be obtained.

$$c_f = \frac{8C\omega_0^2\theta_0^2}{\rho_0 g H_0^2\left(1 + \frac{2k_0 d_0}{sinh2k_0 d_0}\right)} \tag{7}$$

3. Model Test

This numerical simulation test uses Star-CCM+ software to establish a three-dimensional numerical wave flume, assuming that the slope structure is fixed and the floating pendulum does pitch motion near the still water surface and does not consider the motion of the water platform and the effect on the WEC, as shown in Figure 1. And three areas are set up, which are wave generation zone, working zone and wave absorbing zone.

Wan et al. built the computer-aided design software and imported it into Star-CCM+ [18]. The trimmed cell mesh and prismatic layer mesh were often used to generate the mesh in the Star-CCM+. The trimmed cell mesh was used to produce high-quality grids for complex grid generation problems. It can produce non-structural hexahedral meshes on complex geometric surfaces. The prismatic layer mesh can be used to control the boundary layer and make the Y+ out of the development zone.

In order to capture the free surface and the device, meshes should be refined according to The International Towing Tank Conference (ITTC), and there were at least 100 grids in the wavelength range on the free surface. In addition, a minimum of 20 grids was used in the wave height direction of the free surface. The mesh is shown in Figure 2A. In order to ensure more accurate calculations and not to waste computational resources, the meshes of the background calculation area, the motion overlap area and the meshes around the free surface regular waves are refined, as shown in Figure 2B.

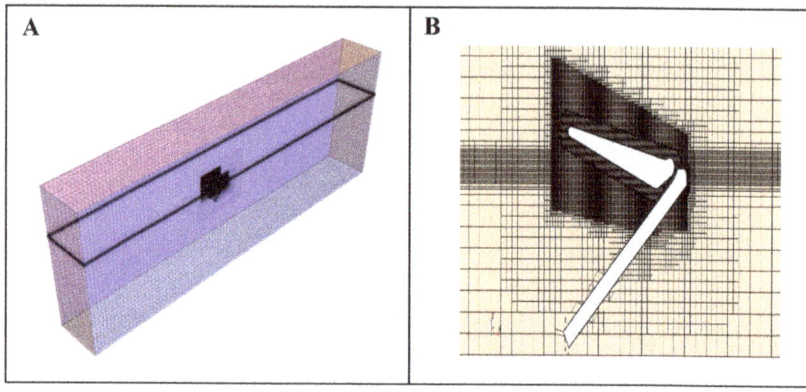

Figure 2. Mesh of the model (**A**), Mesh generation details of the WEC model (**B**).

It may be noted that the RANS solver [13], a segregated numerical flow model where the flow equations were solved in an uncoupled way, was adopted in all the simulations in this paper. Convection terms in the RANS formulae were discretized by employing second-order upwind schemes. The general solution was obtained based on a SIMPLE-type algorithm.

In order to predict realistic flapping-panel behavior, a Dynamic Fluid Body Interaction (DFBI) model was adopted. The DFBI model was integrated with the RANS solver to compute the exciting force and moments acting on the flapping-panel hull caused by waves and to solve the governing equations of rigid body motions to relocate the rigid body.

In order to reduce the computational complexity and requirements, only one-half of the floating flapping-panel device was calculated. The velocity inlet is in the negative × direction, where the incident regular wave was generated. The initial velocity at the inlet was set to the corresponding speed of the waves, as shown in Figure 1. Inside the tank, water and air are used as two-phase flow, so the pressure outlet is set at the top, which is consistent with the actual situation. The sliding wall was selected as the bottom boundary condition due to the large water depth. The symmetrical boundary in the positive y direction was used for the symmetry of the device, and the symmetrical boundary in the negative y direction was used to prevent the fluid from adhering to the wall.

4. Results & Discussion

4.1. Relationship between Slope Angle and Power

The bottom slope structure affects the reflection of waves from the underwater surface [23], and how to choose a reasonable slope angle β (the angle between the slope structure and the static water surface, as shown in Figure 1) is important to improve the efficiency of the wave energy conversion device. In the experiment, the angle of the slope is adjusted several times for different angles β, resulting in changes in the power and energy captured by the wave energy hydrodynamic device.

In this experiment, we take the following conditions as the basic initial conditions; the experimental process is always unchanged and all in the same sea state conditions. The flapping-panel length L was 8 m, and damping coefficient C was 3,500,000 $N{\cdot}m{\cdot}s{\cdot}rad^{-1}$.

Figures 3 and 4 show the power and captured energy at different flapping-panel-slope angles. It is observed that the optimum angle was 35 degrees. When the flapping-panel-slope angle increased above 35 degrees, the upward reflection of the slope on the fluid decreased so that the force of the fluid on the floating flapping-panel was reduced. When the slope angle was below 35 degrees, by decreasing the angle, there was a decrease in the incoming wave energy captured.

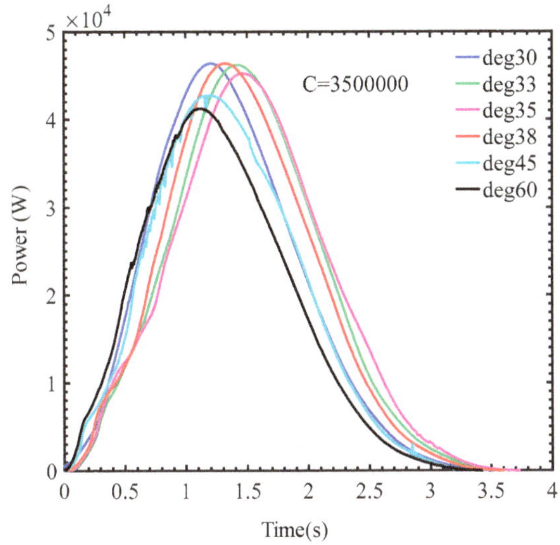

Figure 3. Power curves with different slope angles.

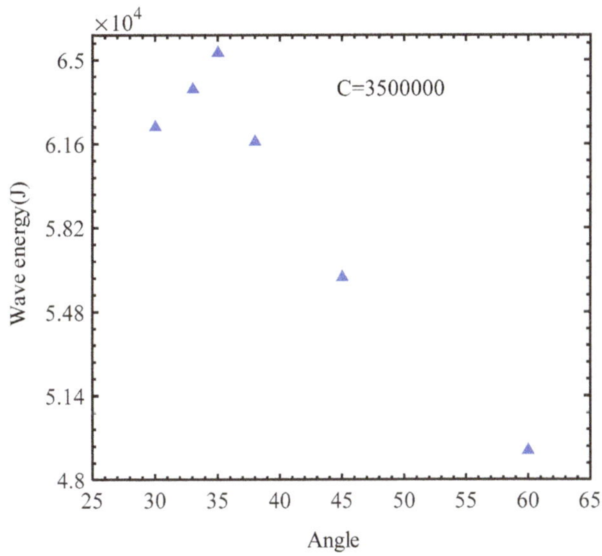

Figure 4. Captured wave energy with different slope angles.

Obviously, under the above conditions, the conclusions we obtained have some limitations, which are directly related to the sea conditions. For different wave heights and different wave periods, the optimal angle of the slope will change, and the next goal of the team will be to study the optimal angle of different slopes under multiple sea conditions to meet the needs of complex and variable sea conditions and to actively change the slope angle so that the floating flapping-panel is always in a large wave energy field, thus improving the first-level capture efficiency.

4.2. Relationship between PTO Damping Coefficients and Power

The PTO damping factor of the wave energy conversion device is a very important parameter in the process of absorbing wave energy, and many scholars have done research on this. There exists a proper PTO damping factor that can make the device resonate and significantly improve the effect of wave energy capture.

According to Zhao et al.'s study on bottom hinged pendulum wave energy device, the optimal PTO damping coefficient has the following relationship with radiation damping [24]:

$$C_{opt} = \sqrt{\left[\left(\frac{K}{\omega}\right) - \omega(I + I_a)\right]^2 + c_r^2} \tag{8}$$

In the above equation, K is the restoring force coefficient; I is the moment of inertia of the floating pendulum; I_a is the additional moment of inertia; and c_r is the radiation damping.

It is obtained from the above equation that the optimal PTO damping coefficient is equal to the radiation damping coefficient in order of magnitude, so in the selection of the test damping coefficient, then it is kept in the same order of magnitude as the radiation damping and increases in order from small.

The influence of different PTO damping coefficients on the floating flapping-panel efficiency is studied by performing numerical simulations of the floating flapping-panel at different damping coefficients C. When the interaction between the wave and the floating flapping-panel was stable, the selected power cycles are as shown in Figure 5 with the wave periods T = 5 s, T = 6 s and T = 7 s as well as the wave height of 1 m. It can be seen from Figure 5 that at different damping coefficients, the effective power was the smallest when the wave period was 5 s and the effective power was similar for T = 6 s and T = 7 s. The work time was the shortest when T = 5 s while the longest when T = 7 s. As the damping coefficient increased, the effective power of different wave periods increased at first and then decreased.

Therefore, there were optimal damping coefficients for different wave periods. For wave periods T = 5 s, T = 6 s and T = 7 s the effective power reached the maximum when the damping coefficient was 5,000,000 $N \cdot m \cdot s \cdot rad^{-1}$, 450,000 $N \cdot m \cdot s \cdot rad^{-1}$ and 5,500,000 $N \cdot m \cdot s \cdot rad^{-1}$, respectively, as shown in Figure 6. The maximum effective power values were observed to be 48.32 kW, 64.21 kW and 63.33 kW, respectively. This indicated that the floating flapping-panel efficiency depended on the PTO damping coefficient [25].

It is found that, when the wave period is T = 5 s, T = 6 s and T = 7 s, the energy increased at first and then decreased as the PTO damping coefficient increased. The power reached the maximum at the damping coefficient of 5,000,000 $N \cdot m \cdot s \cdot rad^{-1}$, 4,500,000 $N \cdot m \cdot s \cdot rad^{-1}$ and 5,500,000 $N \cdot m \cdot s \cdot rad^{-1}$. As shown in Table 1, longer wave periods led to lower wave energy conversion efficiency. It might be attributed to the effects of floating flapping-panel length L on the floating flapping-panel moving time and the wave energy conversion efficiency. Therefore, the value of L is changed to verify the above observation.

As shown in Table 2 and Figure 7, longer flapping-panel length became longer and improved the wave energy conversion efficiency under each wave period due to longer work time. However, as shown in Figure 8, the floating flapping-panel could not return to the origin to achieve resonance in the case when L_P = 10 m and T = 5 s, which was not conducive to structural stability. The floating flapping-panel worked well when $L_P = 10$ m and $T = 6$ s. When $L_P = 10$ m and $T = 7$ s, the floating flapping-panel returned to the origin and continued to move downward, which would result in a waste of wave energy. In summary, long flapping-panel caused a long time of wave action. However, since the floating flapping-panel contacted wave surface earlier when moving downward, it cannot return to the initial position. In the case of short flapping-panel length, the floating flapping-panel would return to the initial position as expected to achieve resonance.

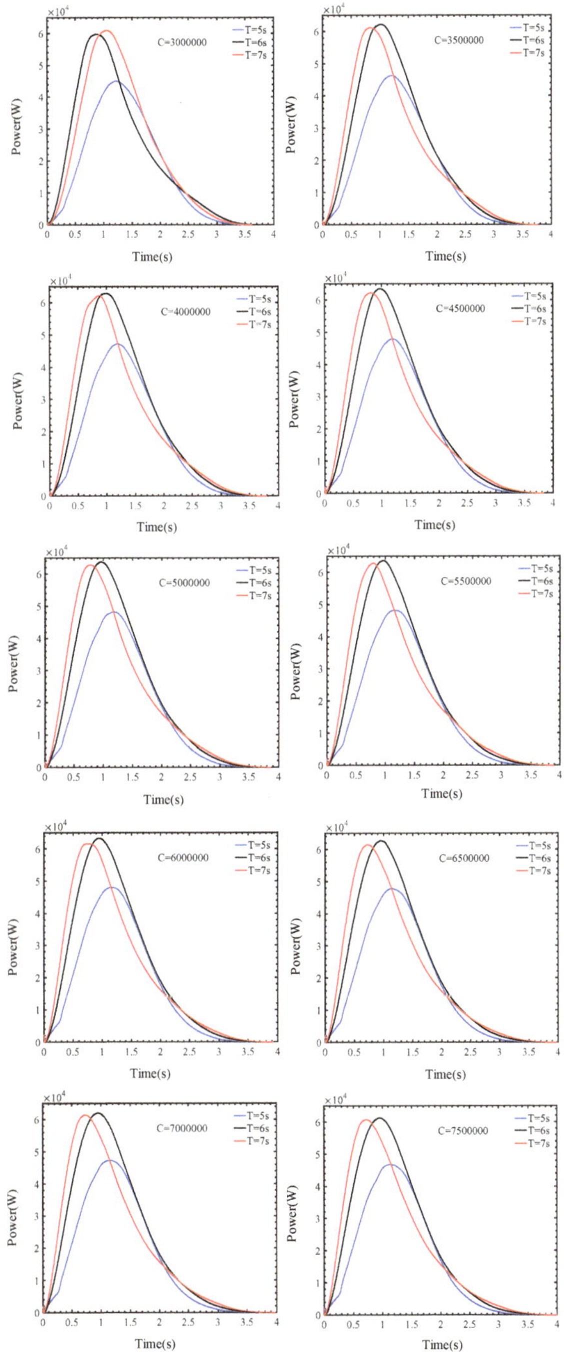

Figure 5. Power curves at different linear moment coefficient.

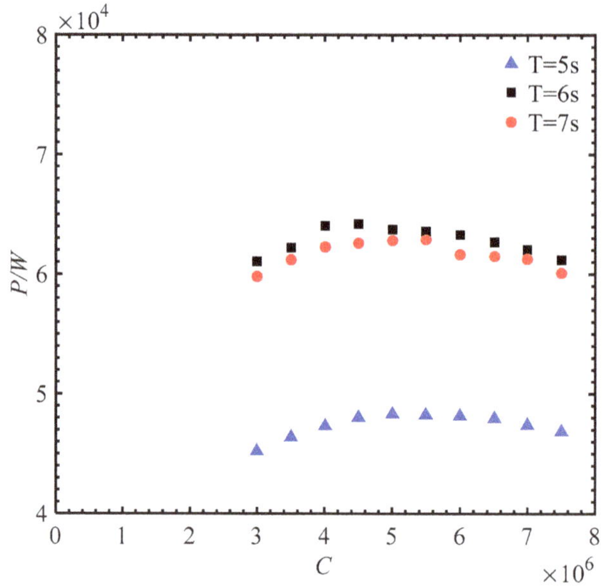

Figure 6. Different wave period power curves.

Table 1. Wave energy conversion efficiency.

T/s	E_m/kJ	W_{PTO}/kJ	P_{PTO}/KW	c_f/%
5	100	66.2	47.6	66.2
6	144	86.1	64.3	59.8
7	196	83.6	62.6	42.7

T—Wave period; E_m—The energy of a periodic wave acting on a floating flapping-panel generator; W_{PTO}—The capture energy of floating flapping-panel generator; c_f—Wave energy conversion efficiency.

Table 2. Different flapping-panel length efficiency comparison.

T/s	L_P/m	P_{PTO}/KW	W_{PTO}/kJ	c_f/%
5	8	47.3	65.8	65.8
	10	48.6	75.4	75.4
6	8	64	85.8	60
	10	63.7	94.9	66
7	8	62.3	83	42.3
	10	61.3	93	47.4

4.3. Physical Model Preliminary Test

The floating flapping-panel wave energy converter was fixed in the wave water tank, as shown in Figure 9. The water tank walls functioned as the walls of the floating flapping-panel wave energy converter. The horizontal position of the flapping-panel was 30 m from the wave-making area, long enough for waves to develop fully.

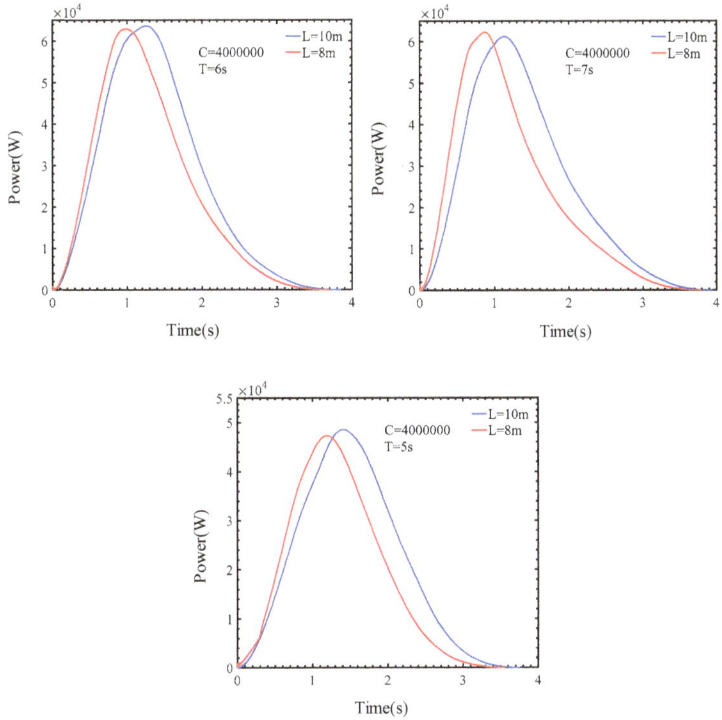

Figure 7. Power with different flapping-panel lengths.

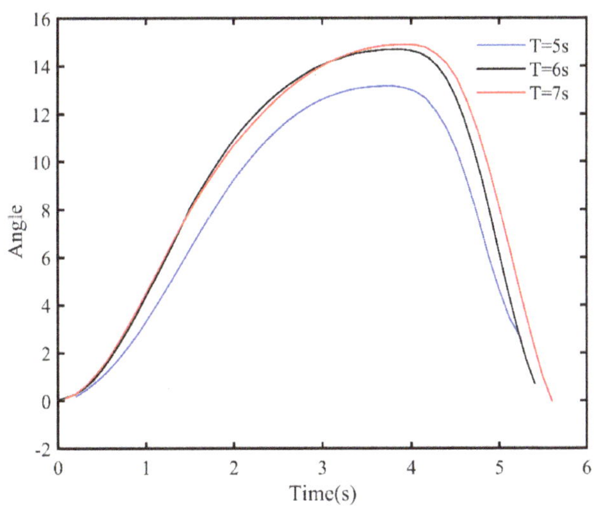

Figure 8. Floating swing angle over time.

Figure 9. Physical model preliminary test.

The flapping-panel movement was recorded by a camera and was post-processed to get the height displacement of the flapping-panel flapping end. During the measurement, the guide rope was loose when the flapping-panel flapped upward under the wave action; it became tight when the flapping-panel moved to the maximum height. It was smoothly controlled to slowly lay down the flapping-panel to avoid possible reflected waves which affect the incident wave.

To further study the feasibility of the above fundamental design, the whole system model was developed as presented in Figure 9, where the supporting frame, slope, flapping-panel, hydraulic rod pump, hydraulic motor and electric generator were included. The flapping-panel was manufactured by lightweight material, and the flapping-panel frame was hinged on the supporting frame. This system was installed in the wave tank according to the schematic design in Figure 1. During the experiment, mechanical energy held by the flapping-panel was converted to high-pressure hydraulic energy [26]. The hydraulic pump provided high-pressure hydraulic oil to drive the hydraulic motor, and the motor in turn powers the electric generator through speed accelerating the gearbox. After scale conversion, some base data and wave tank parameters are obtained, as shown in Table 3.

Table 3. Water tank test parameter table.

Wave tank width/m	0.5
Water depth/m	0.6
Water density/10^3 kg·m^{-3}	1.025
Gravity acceleration/m·s^{-2}	9.81
Flapping-panel width/m	0.4
Slope angle/°	30
Bracket quality/kg	0.81
Single weight quality/kg	1
Water period/s	1.2, 1.4, 1.6, 1.9,2.5,
Wave height/m	0.06

In the laboratory of Zhoushan Campus of Zhejiang University, the hydrodynamic feasibility study experiment of the floating flapping-panel WEC was carried out.

The laboratory physical model experimental results showed that this system can provide enough power to light the connected LED when the wave height H was 0.06 m and wave periods *T* were 1.2 s, 1.4 s, 1.6 s, 1.9 s as well as 2.5 s with a corresponding output pressure of hydraulic pump of 1 MPa. This performance confirmed the feasibility of the fundamental study and structure design in this paper. The test rocker hydraulic pump outputs high-pressure hydraulic oil, which directly drives the hydraulic motor. After the gearbox was accelerated, the generator was energized and fixed to the tank, as shown in Figure 9, the light "ZJU" character consists of 30 0.5 W LED lights, and the maximum instantaneous power is 15 W.

5. Conclusions & Future work

In this paper, the floating flapping-panel wave energy conversion device was tested by numerical simulation studies; its hydrodynamic performance was quantitatively analyzed; and finally, a laboratory physical model hydrodynamic study was conducted to initially verify the feasibility of the device. The following conclusions were obtained:

- The simulation test study showed that the angle between the slope and the horizontal plane significantly influenced the efficiency of the wave energy converter. The optimal slope angle is 35 degrees, and the wave energy captured in the experimental sea conditions is the largest.
- Meanwhile, from the output characteristics under different PTO damping coefficients for the linear damping, an optimal coefficient was obtained to give maximum captured energy. The wave energy converter became more efficient as PTO damping coefficients fell.
- The length of the flapping-panel will affect the efficiency and resonance of the wave energy converter. With the increase of incident wave height, the wave energy converter can capture more energy.
- In the follow-up work, we will continue to carry out in-depth research around it, optimize the design and achieve high-efficiency capture of wave energy.

Author Contributions: T.S., methodology, data curation, software; Z.L., data curation, analysis, writing—original draft, writing—review and editing; H.Z., software, data curation, analysis, writing—original draft; C.L., writing—review and editing, supervision; Z.W., conceptualization, methodology, formal analysis. All authors have read and agreed to the published version of the manuscript.

Funding: The support by National Natural Science Foundation of China (NOs. 11572283, 11602179), the Education Department Foundation of Liaoning Province (NO. LJKZ0303), the National Key Research & Development Plan of China (NOs. 2017YFC1403306), the Public Science and Technology Research Funds Projects of Ocean (201205015), the Program for Zhejiang Leading Team of S & T Innovation (NO. 2010R50036) and HPC Center OF ZJU (ZHOUSHAN CAMPUS) are all gratefully acknowledged.

Institutional Review Board Statement: Not applicable.

Informed Consent Statement: Not applicable.

Data Availability Statement: Data available on request due to restrictions of privacy.

Conflicts of Interest: The authors declare no conflict of interest.

References

1. Babarit, A.; Hals, J.; Muliawan, M.J.; Kurniawan, A.; Moan, T.; Krokstad, J. Numerical Benchmarking Study of a Selection of Wave Energy Converters. *Renew. Energy* **2012**, *41*, 44–63. [CrossRef]
2. López, I.; Andreu, J.; Ceballos, S.; Martínez de Alegría, I.; Kortabarria, I. Review of Wave Energy Technologies and the Necessary Power-Equipment. *Renew. Sustain. Energy Rev.* **2013**, *27*, 413–434. [CrossRef]
3. Whittaker, T.; Folley, M. Nearshore Oscillating Wave Surge Converters and the Development of Oyster. *Philos. Trans. R. Soc. Math. Phys. Eng. Sci.* **2012**, *370*, 345–364. [CrossRef] [PubMed]
4. AW-Energy. Waveroller. 2022. Available online: https://aw-energy.com/waveroller/ (accessed on 9 October 2022).
5. Tul Huda Ahmad, N.H.; Zamri Ibrahim, M.; Rahman, S.J.A.; Albani, A.; Mohad, S. The Development of Wave Energy Converter System Using Hydraulic Power Take Off at Terengganu Shoreline. In Proceedings of the 2018 International Conference and Utility Exhibition on Green Energy for Sustainable Development (ICUE), Phuket, Thailand, 24–26 October 2018; pp. 1–7.

6. Ghafari, H.R.; Ghassemi, H.; Neisi, A. Power Matrix and Dynamic Response of the Hybrid Wavestar-DeepCwind Platform under Different Diameters and Regular Wave Conditions. *Ocean Eng.* **2022**, *247*, 110734. [CrossRef]
7. Bosma, B.; Simmons, A.; Lomonaco, P.; Ruehl, K.; Gunawan, B. WEC-Sim Phase 1 Validation Testing: Experimental Setup and Initial Results. In Proceedings of the International Conference on Ocean, Offshore and Arctic Engineering, Busan, Republic of Korea, 19 June 2016; Volume 6, p. V006T09A025.
8. Wei, Y.; Barradas-Berglind, J.J.; Van Rooij, M.; Prins, W.A.; Jayawardhana, B.; Vakis, A.I. Investigating the Adaptability of the Multi-Pump Multi-Piston Power Take-off System for a Novel Wave Energy Converter. *Renew. Energy* **2017**, *111*, 598–610. [CrossRef]
9. Davey, T.; Sarmiento, J.; Ohana, J.; Thiebaut, F.; Haquin, S.; Weber, M.; Gueydon, S.; Judge, F.; Lyden, E.; O'Shea, M.; et al. Round Robin Testing: Exploring Experimental Uncertainties through a Multifacility Comparison of a Hinged Raft Wave Energy Converter. *J. Mar. Sci. Eng.* **2021**, *9*, 946. [CrossRef]
10. Ruehl, K.; Forbush, D.D.; Yu, Y.-H.; Tom, N. Experimental and Numerical Comparisons of a Dual-Flap Floating Oscillating Surge Wave Energy Converter in Regular Waves. *Ocean Eng.* **2020**, *196*, 106575. [CrossRef]
11. Poguluri, S.K.; Kim, D.; Bae, Y.H. Performance Assessment of Pitch-Type Wave Energy Converter in Irregular Wave Conditions on the Basis of Numerical Investigation. *Ocean Syst. Eng.* **2022**, *12*, 23–38.
12. Ransley, E.J.; Greaves, D.M.; Raby, A.; Simmonds, D.; Jakobsen, M.M.; Kramer, M. RANS-VOF Modelling of the Wavestar Point Absorber. *Renew. Energy* **2017**, *109*, 49–65. [CrossRef]
13. Tezdogan, T.; Incecik, A.; Turan, O. Full-Scale Unsteady RANS Simulations of Vertical Ship Motions in Shallow Water. *Ocean Eng.* **2016**, *123*, 131–145. [CrossRef]
14. Zhang, D.H.; Shi, Y.X.; Huang, C.; Si, Y.L.; Huang, B.; Li, W. SPH Method with Applications of Oscillating Wave Surge Converter. *Ocean Eng.* **2018**, *152*, 273–285. [CrossRef]
15. Pierart, F.G.; Fernandez, J.; Olivos, J.; Gabl, R.; Davey, T. Numerical Investigation of the Scaling Effects for a Point Absorber. *Water* **2022**, *14*, 2156. [CrossRef]
16. Guo, B.; Ringwood, J.V. Geometric Optimisation of Wave Energy Conversion Devices: A Survey. *Appl. Energy* **2021**, *297*, 117100. [CrossRef]
17. Zhang, H.; Zhou, B.; Vogel, C.; Willden, R.; Zang, J.; Geng, J. Hydrodynamic Performance of a Dual-Floater Hybrid System Combining a Floating Breakwater and an Oscillating-Buoy Type Wave Energy Converter. *Appl. Energy* **2020**, *259*, 114212. [CrossRef]
18. Wan, Z.; Zheng, H.; Sun, K.; Zhang, D.; Yao, Z.; Song, T. Simulation of Wave Energy Converter with Designed Pendulor-Slope Combination. *Energy Procedia* **2019**, *158*, 733–737. [CrossRef]
19. Wan, Z.; Li, Z.; Zhang, D.; Zheng, H. Design and Research of Slope-Pendulum Wave Energy Conversion Device. *J. Mar. Sci. Eng.* **2022**, *10*, 1572. [CrossRef]
20. Elwood, D.; Yim, S.C.; Prudell, J.; Stillinger, C.; Von Jouanne, A.; Brekken, T.; Brown, A.; Paasch, R. Design, Construction, and Ocean Testing of a Taut-Moored Dual-Body Wave Energy Converter with a Linear Generator Power Take-Off. *Renew. Energy* **2010**, *35*, 348–354. [CrossRef]
21. Johanning, L.; Smith, G.H.; Wolfram, J. Measurements of Static and Dynamic Mooring Line Damping and Their Importance for Floating WEC Devices. *Ocean Eng.* **2007**, *34*, 1918–1934. [CrossRef]
22. Zhao, H.; Sun, Z.; Hao, C.; Shen, J. Numerical Modeling on Hydrodynamic Performance of a Bottom-Hinged Flap Wave Energy Converter. *China Ocean Eng.* **2013**, *27*, 73–86. [CrossRef]
23. Midya, C.; Kanoria, M.; Mandal, B.N. Scattering of Water Waves by Inclined Thin Plate Submerged in Finite-Depth Water. *Arch. Appl. Mech.* **2001**, *71*, 827–840. [CrossRef]
24. Zhao, H.; Sun, Z.; Shen, J.; Ning, D.-Z.; Xu, X. Time-Domain Model on Hydrodynamic Performance of a Bottom-Hinged Flap Wave Energy Converter. *Chin. J. Hydrodyn.* **2013**, *28*, 151–158.
25. Falcão, A.F.D.O. Modelling and Control of Oscillating-Body Wave Energy Converters with Hydraulic Power Take-off and Gas Accumulator. *Ocean Eng.* **2007**, *34*, 2021–2032. [CrossRef]
26. Kofoed, J.P.; Frigaard, P.; Friis-Madsen, E.; Sørensen, H.C. Prototype Testing of the Wave Energy Converter Wave Dragon. *Renew. Energy* **2006**, *31*, 181–189. [CrossRef]

Journal of
Marine Science and Engineering

MDPI

Article

Experimental Modelling of Point-Absorber Wave Energy Converter Arrays: A Comprehensive Review, Identification of Research Gaps and Design of the WECfarm Setup

Timothy Vervaet [1,*], Vasiliki Stratigaki [1], Brecht De Backer [2], Kurt Stockman [2], Marc Vantorre [1] and Peter Troch [1]

[1] Department of Civil Engineering, Ghent University, Technologiepark 60, 9052 Zwijnaarde, Belgium; vicky.stratigaki@ugent.be (V.S.); marc.vantorre@ugent.be (M.V.); peter.troch@ugent.be (P.T.)
[2] Department of Industrial System and Product Design, Ghent University, Graaf Karel de Goedelaan 5, 8500 Kortrijk, Belgium; brecht.db@hotmail.com (B.D.B.); kurt.stockman@ugent.be (K.S.)
* Correspondence: timothy.vervaet@ugent.be; Tel.: +32-9-264-54-89

Abstract: Commercial wave energy exploitation will be realised by placing multiple wave energy converters (WECs) in an array configuration. A point-absorber WEC consists of a floating or submerged body to capture wave energy from different wave directions. This point-absorber WEC acts as an efficient wave absorber that is also an efficient wave generator. Optimising the WEC array layout to obtain constructive interference within the WEC array is theoretically beneficial, whereas for wind farms, it is only important to avoid destructive interference within an array of wind turbines due to wake effects. Moreover, the WEC array layout should be optimised simultaneously with the applied control strategy. This article provides a literature review on the state of the art in physical modelling of point-absorber WEC arrays and the identification of research gaps. To cover the scientific gap of experimental data necessary for the validation of recently developed (nonlinear) numerical models for WEC arrays, Ghent University has introduced the "WECfarm" project. The identified research gaps are translated into design requirements for the "WECfarm" WEC array setup and test matrix. This article presents the design of the "WECfarm" experimental setup, consisting of an array of five generic heaving point-absorber WECs. The WECs are equipped with a permanent magnet synchronous motor (PMSM), addressing the need for WEC array tests with an accurate and actively controllable power take-off (PTO). The WEC array control and data acquisition are realised with a Speedgoat Performance real-time target machine, offering the possibility to implement advanced WEC array control strategies in the MATLAB-Simulink model. Wave basin testing includes long- and short-crested waves and extreme wave conditions, representing real sea conditions. Within the "WECfarm" project, two experimental campaigns were performed at the Aalborg University wave basin: (a) a testing of the first WEC in April 2021 and (b) a testing of a two-WEC array in February 2022. An experimental campaign with a five-WEC array is under preparation at the moment of writing.

Keywords: wave energy converter (WEC); heaving point-absorber WEC; WEC array; WECfarm; physical modelling; real-time control; MATLAB-Simulink

Citation: Vervaet, T.; Stratigaki, V.; De Backer, B.; Stockman, K.; Vantorre M.; Troch, P. Experimental Modelling of Point-Absorber Wave Energy Converter Arrays: A Comprehensive Review, Identification of Research Gaps and Design of the WECfarm Setup. *J. Mar. Sci. Eng.* **2022**, *10*, 1062. https://doi.org/10.3390/jmse10081062

Academic Editor: Unai Fernandez-Gamiz

Received: 29 June 2022
Accepted: 28 July 2022
Published: 2 August 2022

1. Introduction

Wave energy can be extracted from ocean waves by using wave energy converters (WECs) and commercial wave energy will be realised by installing multiple WECs in an array to become economically competitive with other renewable energy resources. A point absorber is a WEC which consists of a floating or submerged body to capture energy from different wave directions. The point absorber's diameter should preferably be in the range of 5–10% of the prevailing wavelength [1]. Due to its ability to absorb energy from different directions, this WEC type is particularly suitable to put in arrays. In a WEC array, hydrodynamic interactions between the WECs occur through the absorption, radiation and diffraction of waves [2]. Both constructive and destructive interactions occur between individual WECs within a WEC array, called near-field interactions. The power absorption of the entire WEC array reduces the wave height behind the array, called far-field effects. Whether the averaged energy absorbed by each WEC in the array is higher or lower than the energy absorbed by an isolated WEC is defined by the q interaction factor [3]:

$$q = \frac{P}{n \cdot P_s} \tag{1}$$

where P is the total energy absorbed by the WEC array, P_s is the energy absorbed by a single isolated WEC and n is the number of WECs in the array. If wave interactions have a net constructive effect on the absorbed power, q will be larger than unity. If they have a net destructive effect, q will be lower than unity. Therefore, the q interaction factor is widely accepted to quantify near-field interactions (WEC–WEC interactions). This article focuses on the experimental identification of the WEC–WEC and wave–WEC interactions within an array of floating point-absorber WECs.

The layout of the WEC array should be optimised to maximise constructive interactions between the WECs and the power absorption should be improved by applying a control system tailored to the WEC array. In practice, the WEC natural frequency is typically much higher than the ocean wave frequency due to the WEC size constraint [4]. The majority of power produced by WECs occurs during resonant absorption, when the excitation force is in phase with the device velocity. Control strategies aiming to maximise power absorption generally attempt to alter the WEC system dynamics in order to achieve resonance [5]. The control is implemented to act on the WEC power take-off (PTO), the subsystem responsible for the energy transfer from the ocean waves. It is important that the WEC array layout and the control system are optimised simultaneously. A layout optimised without knowledge of the control system can be inferior to the extent of recovering 40% less energy than an array layout optimised with knowledge of the control system [6]. These optimized arrays of interacting WECs can be integrated in a WEC farm. The adopted definitions for a WEC array and a WEC farm in the presented article may differ from the ones used in other research works.

There is a need for available real-life data for the validation of WEC array modelling and optimisation [7]. This is a research gap for the full wave energy sector, not only in the optimisation of WEC arrays. As the computational power capabilities increase yearly, so do the numerical models capabilities, stressing the need for experimental data to validate the models. However, publicly available databases from WEC array experiments are scarce. Only a few physical experiments on arrays of point-absorber WECs have been carried out during the last years and some others were performed around a decade ago [8]. Limited experimental campaigns have been performed due to the high cost of constructing and testing in wave basin facilities, as well as due to the complexity of the experiments and related instrumentation [9].

In the framework of the Hydralab IV WECwakes project coordinated by Ghent University, the near-field interactions and far-field effects have been experimentally investigated for layouts of up to 25 heaving point-absorber type WECs [10–13]. The obtained unique database served for validation purposes of numerical models for studying wave–WEC interactions (near-field interactions) and wave propagation through WEC arrays (far-field effects) [14–17]. Many numerical models have progressively advanced since the completion of the WECwakes project. Therefore, the new experimental "WECfarm" project, introduced by Ghent University and its partners (Queens University Belfast, Aalborg University and University of Vigo) in 2018, aims to deliver a dataset to cover the research gap on the need for publicly available real-life and reliable data to validate these new advanced numerical models. Within the "WECfarm" project, a high-accuracy data acquisition, an actively controllable PTO system and a control platform to implement advanced WEC array control strategies are prioritised. Given the available funding and the complexity of the latter, the number of WECs is limited to five. The WECfarm WEC is intended to be used only for research purposes and is thus not intended for full-scale deployment. Within the "WECfarm" project, wave basin testing with WEC arrays of up to five real-time controllable heaving point absorbers is planned.

The purpose of this article is threefold. Firstly, Section 2 provides a comprehensive literature review on experimental modelling of point-absorber WEC arrays. Secondly, Section 3 provides the research gaps, resulting from the literature review in Section 2. This Section addresses the motivation for the "WECfarm" project. Thirdly, Section 4 discusses how the design of the "WECfarm" project experimental setup aims to address these research gaps. Therefore, the "WECfarm" WEC is decomposed in its respective subsystems: hydrodynamic, electromechanical, data acquisition system (DAQ) and control platform. This Section addresses the objectives and methodology of the "WECfarm" project. Section 5 provides a discussion and conclusions.

2. Literature Review on Experimental Modelling of Point-Absorber WEC Arrays

The presented literature review considers point-absorber WECs, consisting of a floating or submerged body, named floater or buoy hereafter. For the theoretical background on hydrodynamics of floating structures and on wave energy extraction by point-absorber WEC arrays, the reader is referred to [18,19] and to [20,21], respectively. A theoretical analysis on physical modelling of WECs is presented in [22]. The presented review does not consider point-absorber oscillating water columns (OWC), since their operating principle with a water column oscillating in a plenum and a turbine PTO system can be categorised separately. Gaebele et al. discussed the hydrodynamic interaction within an array of OWCs [23]. Kelly et al. performed experiments with an array of 32 OWCs [24]. Lamont-Kane performed experiments with a single OWC, an array of four OWCs and an array of five OWCs [25]. Doyle et al. performed experiments with a single OWC, an array of three OWCs and a three-chamber OWC [26,27]. A minimum of two WECs is needed to have an array with corresponding WEC–WEC interactions. Within a WEC array, a WEC row refers to a number of WECs placed equidistant to the wavemaker, while a WEC column refers to a number of WECs aligned with the wave propagation direction. For a $(j \times k)$-WEC array, j is the number of columns, while k is the number of rows. As a result, a $(1 \times k)$-WEC array is a column and a $(j \times 1)$-WEC array is a row. The presented review only covers experimental campaigns which have been publicly documented, hence excluding WEC array experiments not disclosed by research institutes or companies. The review is in chronological order, ending with the most recent projects. The focus is on the methodology and research objectives of the performed experiments, rather than on the obtained results. Additionally, the wave testing facility, the design of a single WEC, the tested wave conditions, the number of WECs, the tested WEC separation distances, the WEC array layouts and the target measurements are discussed. In total, 17 WEC array experimental campaigns were identified. Since some campaigns make use of the same WEC model, the number of research projects was reduced to 12. This literature review allows us to identify research gaps in Section 3, which result in unique research objectives and design features of the "WECfarm" project in Section 4.

Pioneering research with WEC arrays was performed by Budal et al. in 1979 [28] and by Count et al. in 1980 [29]. Budal et al. performed experiments on a row of heaving buoys optimised for maximum wave power absorption [28]. The optimum phase was obtained by resonance tuning or by locking the heave motion during controlled intervals of each wave cycle [28]. At a similar time, tests were performed by Count et al. at the University of Edinburgh to measure the q interaction factor of a linear array of both two and ten WECs at different spacings [29]. After these pioneering experimental campaigns, it took some decades until new experiments were set up with point-absorber WEC arrays. The work by Budal et al. and Count et al. is not included in the overview Tables 1–4 due to limited available information on the details of the experiments [28,29].

2.1. Manchester Bobber WEC Array

The first WEC array experimental setup discussed is the Manchester Bobber WEC array. The project was initiated with the purpose to deploy a full-scale WEC array in an ocean environment for power production. Figure 1a shows the employed laboratory-scale model. The only component in contact with the water is a partially immersed axisymmetric floater with a radius a of 0.075 m. The considered scale factor is 1:67. The floater mass m_f is connected to a cable, which is supported by two pulleys. A counterweight of mass m_c applies an opposing force, holding the cable taut and the floater at an equilibrium draft z_{eq} in still water. Both the floater and the counterweight masses can be altered to adjust the draft, the heave and the surge natural frequencies and the hydrodynamic characteristics. In Stallard et al., Thomas et al. and Alexandre et al., a hemisphere-cylinder was considered, while in Weller et al. a flat-bottomed hemisphere-cylinder was considered [30–33]. The pulley nearest to the floater is joined to one side of a freewheel clutch by a shaft. On the other side of the clutch, there is a flywheel providing inertia to the rotating system. The flywheel is connected to a 12 V permanent magnet direct current (PMDC) motor via a flexible coupling. A mechanical torque resists the rotation of the flywheel, thereby applying damping on the system and extracting energy. A motor encoder and code wheel are used to read angular displacements of the motor shaft and pulley, respectively. Changes in the water surface elevation cause the floater to move from its equilibrium position, in a motion similar to water particle orbits. Tethers should ensure only heave motion; however, in order to avoid large horizontal loads on the structure, surge motion has also been allowed for specific extreme wave conditions [34]. Figure 1b shows the experimental setup in the wave flume of Manchester University (5.0 m wide, 18.5 m long and 0.45 m deep).

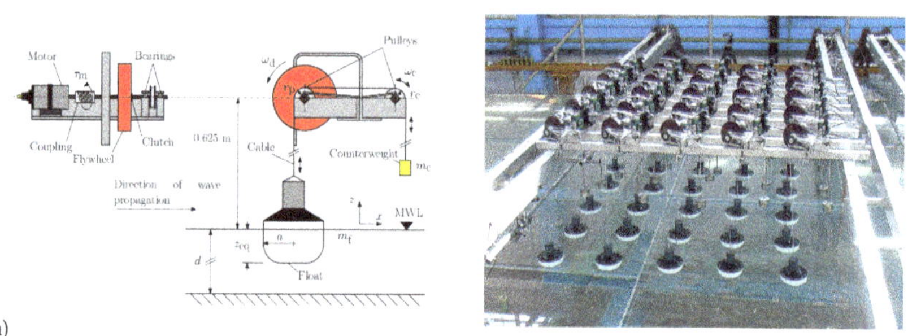

(a)

(b)

Figure 1. Experimental setup of the Manchester Bobber: (**a**) drawing of the WEC model, reprinted with permission from Ref. [34], 2010, Weller and (**b**) picture of the 5 × 5 WEC array in the wave flume of Manchester University, reprinted with permission from Ref. [35], 2008, Manchester University.

Next, the research performed on this setup by Stallard et al., Thomas et al., Alexandre et al. and Weller et al. is discussed [30–33]. Although each time a different number of WECs was considered, the close centre-to-centre spacing equal to four times the WEC radius was retained. Therefore, this array is a closely spaced WEC array, with higher WEC–WEC interactions compared to other research projects discussed later.

Stallard et al. studied the effect of both WEC array size and configuration on the power output and response of individual WECs in the array compared to an isolated WEC [30]. Three configurations were tested: a 1 × 3 column, a 3 × 3 array and a 3 × 4 array. Figure 2a shows the (1 × 3)- and the (3 × 3)-WEC array. Regular waves with an amplitude of 0.013 m and varying frequency were used [30].

Thomas et al. measured values of the WEC free response amplitude for rectilinear (1 × 5)- and (5 × 1)-WEC arrays [31]. Figure 2b shows the tested 1 × 5 column layout and the 5 × 1 row layout. Small scale monochromatic waves with a wave height of 0.026 m and wave periods varying from 0.57 to 1.33 s were applied [31].

Alexandre et al. experimentally measured the transformation of an irregular wave field in the vicinity of a WEC array [32]. Two WEC array configurations were investigated: a single row of five WECs (5 × 1) and a double row of five WECs (5 × 2). Figure 3a shows the tested 5 × 1 row layout. Irregular long-crested waves with a significant wave height of 0.040 m and a peak wave period of 1.31 s were applied. Water surface elevations were measured both before the deployment of the WEC arrays (in an empty wave flume) and with the WEC arrays in place, using a network of three wave gauges installed up- and downwave of the WEC array [32]. The Bretschneider spectrum was constructed with the wave gauge data and used to characterise the influence of the WEC array on the surface elevation.

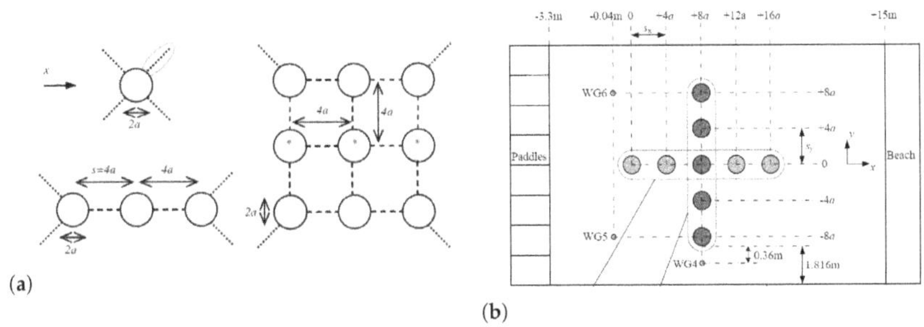

Figure 2. Wave flume of Manchester University's WEC array layout (**a**) used by Stallard et al., adapted with permission from Ref. [30], 2008, Stallard et al. and (**b**) used by Thomas et al., adapted with permission from Ref. [31], 2008, Thomas et al.

Weller et al. experimentally measured the power absorbed by a two-dimensional rectilinear (3 × 4)-WEC array [33]. Figure 3b shows the tested 3 × 4 layout. Regular and irregular long-crested waves with (significant) wave heights varying from 0.015 m to 0.064 m and with (peak) wave periods ranging from 0.61 s to 2.00 s were generated. The capture width, defined as the width of the wave front containing the same available power as the useful power captured by the WEC in the same wave climate, was reported [33].

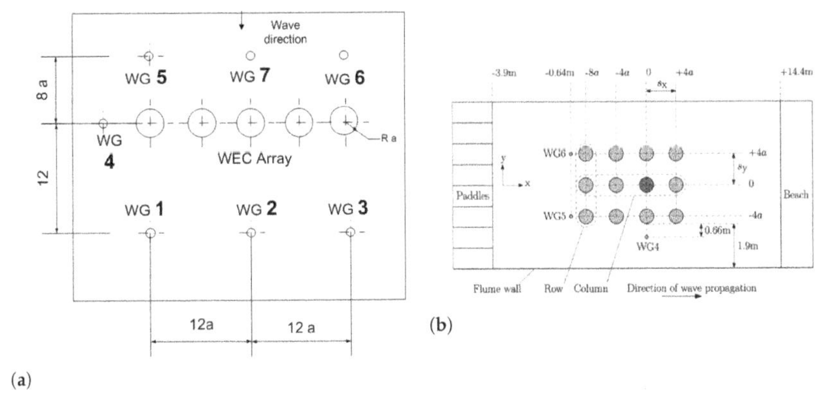

Figure 3. Wave flume of Manchester University' WEC array layout (**a**) used by Alexandre et al., reprinted with permission from Ref. [32], 2009, Alexandre et al. and (**b**) used by Weller et al., reprinted with permission from Ref. [33], 2010, Weller et al.

2.2. *PerAWaT Project*

The second experimental setup of WEC arrays discussed is the "Performance Assessment of Wave and Tidal array systems" (PerAWaT) project. Child et al. and Folley et al. experimentally tested WEC arrays with 22–24 WECs and Lamont-Kane et al. experimentally tested WEC arrays with four WECs at the Queen's University Belfast (QUB)'s wave basin in Portaferry, Northern Ireland [36–38]. The wave basin is 18 m long and 16 m wide with a water depth of 0.625 m.

The WEC models had a cylindrical part with 0.25 m diameter, with a hemispherical end and a draft of 0.25 m in still water. A Froude scale factor of 1:80 was applied. The WECs were constrained to move in heave by a pressurised rectilinear air bushing. The heave motion of the WEC was measured using a potentiometer. Power was extracted using a coulomb friction brake, which provides an approximation to a constant pressure hydraulic power take-off. In the physical model, the braking force was adjusted by varying the voltage to a solenoid. The brake force was measured using a cantilever load cell, which measured the combined force due to the clutch and potentiometer [36,37]. Figure 4a shows a drawing of the PerAWaT WEC.

Lamont-Kane et al. experimentally tested WEC arrays with four WECs to assess the primary uncertainties encountered in physical array tests [38]. The sources of uncertainty were described, along with statistical measures used to assess their impact and ensure that the model data collected were of sufficient quality to allow numerical validation [38]. Figure 4b shows the general layout of the setup in the wave basin and Figure 4c shows a picture of the setup.

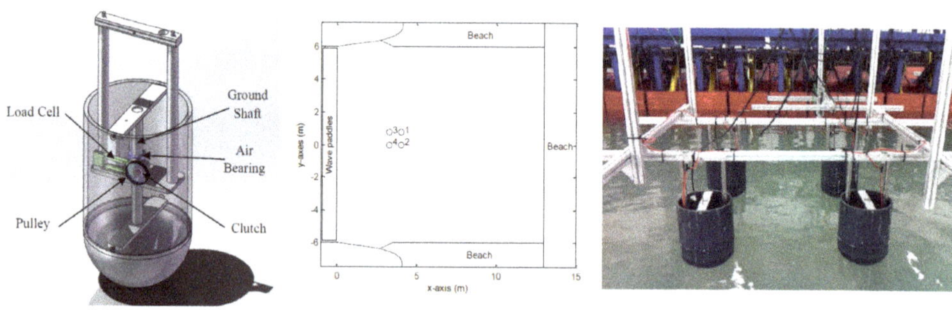

(a) (b) (c)

Figure 4. Experimental setup used by Lamont-Kane et al., reprinted with permission from Ref. [38], 2013, Lamont-Kane et al.: (**a**) WEC model; (**b**) QUB wave basin WEC array layout; (**c**) QUB wave basin setup.

Child et al. used a modified third-generation spectral wave solver (TOMAWAC) to model the power absorption and evolution of wave energy spectra across the WEC array [36]. Folley et al. compared physical model array interactions to those predicted by a spectral-domain numerical model of the WEC array and the suitability of the numerical and physical models was analysed [37]. Both long-crested and short-crested irregular sea states were tested to provide validation for numerical models in real sea conditions. Figure 5a shows the three tested configurations and Figure 5b shows the layout of the second configuration in the wave basin. Figure 5c shows a picture of the setup.

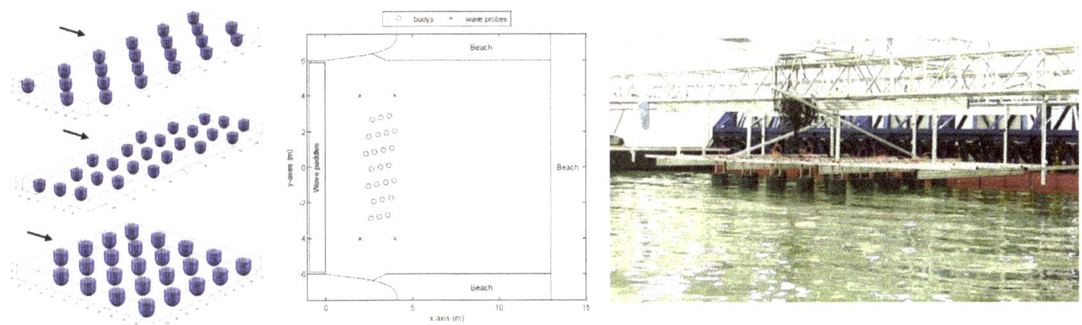

(a) (b) (c)

Figure 5. Experimental setup used by Child et al. and Folley et al., adapted with permission from Ref. [36], 2013, Child et al.: (**a**) three tested configurations; (**b**) QUB wave basin WEC array layout; (**c**) QUB wave basin setup picture.

2.3. *Wavebob WEC Array*

Mackay et al. compared tests with an array of four Wavebob point-absorber-type WECs on a scale of 1:19 to numerical predictions from a time-domain modelling tool for WECs, called WaveDyn [39]. The tests were executed in the Seakeeping and Manoeuvring basin at MARIN (Wageningen, The Netherlands), which is 170 × 40 m in area and 5 m deep. Similar to the Manchester Bobber and the PerAWaT project, this project was initiated with the aim to deploy a full-scale WEC array in an ocean environment [39].

The Wavebob WEC consists of two bodies: a torus and a central spar, known as the float-neck-tank (FNT). The torus is essentially a wave follower over the range of the wave periods of interest for power conversion, whereas the FNT has a much lower natural period and acts as a source of reference for the motion of the torus. Figure 6a shows four Wavebob WECs on a scale of 1:19. The torus and the FNT are linked via a PTO system, which is modelled using a linear servomotor. At full-scale, the torus has a diameter of 17.60 m, a draft of 4.86 m and a freeboard of 3.00 m. The FNT has a draft of 57.00 m. The PTO is connected to a six-DOF force frame to measure the forces and rotational moments between the torus and FNT. Measurements from the six-DOF frame are used in the control algorithm in order to compensate for friction and stiction in the system. Using this feedback, it was possible to achieve an accurate match to the target PTO profile. The model is moored using three lines spaced at 120° around the model. Each line consists of an anchor line connected to a submerged float which is then connected to a horizontal line section and subsequently to a bridle line arrangement, and finally, to the FNT at two points [39].

The test program comprised radiation tests, where the PTO of one WEC was excited in initially calm water and the effects of the radiated waves on the other WECs were measured, and also tests in regular and irregular long-crested waves. Figure 6b shows the tested diamond layout, with a spacing of 149 m (full-scale) between the model centres, equivalent to 8.5 diameters. Due to the large spacing, this array is not considered a closely spaced WEC array. The waves from 0° correspond to the long-side wave makers and the waves from 90° correspond to the short-side wave makers. The array layout was selected so that adjacent WECs shared anchor points, whilst maintaining a realistic separation distance. Figure 6c shows two Wavebob WECs during a radiation test [39].

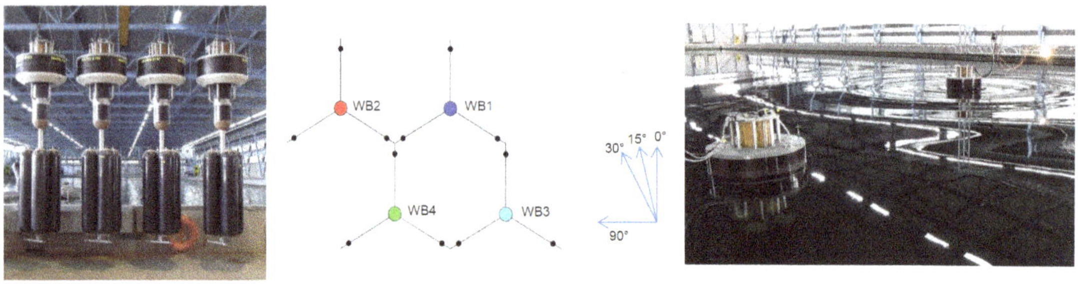

(a) (b) (c)

Figure 6. Experimental setup used by Mackay et al., reprinted with permission from Ref. [39], 2013, Mackay et al.: (**a**) WEC model; (**b**) WEC array layout in the MARIN wave basin; (**c**) MARIN wave basin two-WEC array setup picture.

2.4. *WECwakes Project*

In the framework of the Hydralab IV WECwakes project, Stratigaki et al. performed experiments on large arrays of up to 25 heaving point-absorber WECs for different geometric rectilinear and staggered WEC array configurations [10–12]. The experiments were performed in the Shallow Water Wave Basin of DHI (Hørsholm, Denmark), which is 35.0 m long and 25.0 m wide, with an overall depth of 0.8 m. These tests were executed with the purpose to investigate near-field interactions and far-field effects and obtain a dataset for numerical modelling validation. Therefore, this project considered a generic WEC concept which was not intended to become commercially deployed in ocean conditions.

Each WEC consisted of a hemisphere-cylinder buoy with a diameter and draft of 0.315 m. The PTO system consisted of polytetrafluoroethylene (PTFE) blocks pressed by springs against a fixed steel shaft, realising damping of the WEC's motion through friction based energy dissipation. Adjusting the length of these springs allows one to adjust the passive damping coefficient, as illustrated on the sketch in Figure 7a. Figure 7b shows a picture of the WECwakes's WEC model. Figure 7c shows a picture of the 5 × 5 rectilinear array setup.

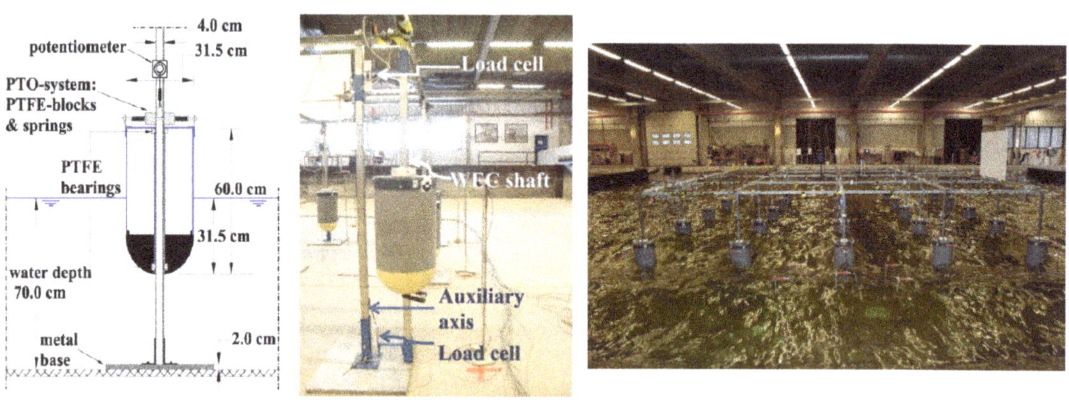

(a) (b) (c)

Figure 7. Experimental setup used by Stratigaki et al., reprinted with permission from Ref. [10], 2014, Stratigaki et al.: (**a**) WEC model; (**b**) WEC model; (**c**) DHI's Shallow Water Wave Basin setup.

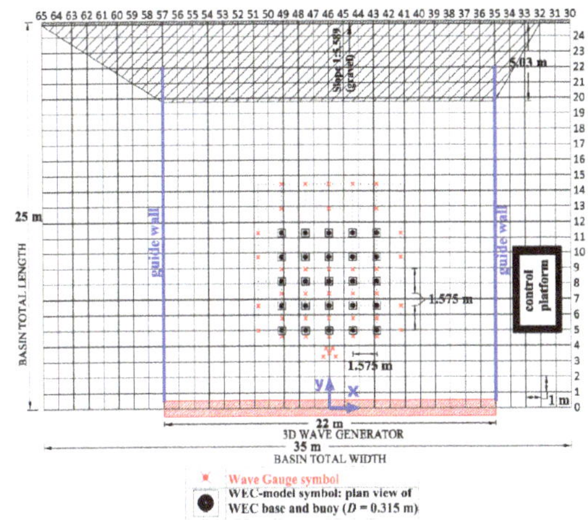

(a)

Configuration	Layout sketch	Configuration	Layout sketch
Waves (no WECs)	N/A	10-WEC column	
Individual WEC		5x5-WEC aligned	
2-WEC column		5x5-WEC staggered	
2-WEC row		3x3-WEC aligned 10D	
5-WEC column		3x3-WEC aligned 5D	
5-WEC row		13-WEC staggered	

(b)

Figure 8. Experimental setup used by Stratigaki et al., adapted with permission from Ref. [11], 2015, Stratigaki et al.: (**a**) DHI's wave basin 5 × 5 rectilinear WEC array layout; (**b**) sketches of tested WEC (array) configurations at the DHI's wave basin.

A rectilinear arrangement of WEC support structures was employed such that several WEC array configurations could be studied. The configurations tested were: individual WEC, 2-WEC column, 2-WEC row, 5-WEC column, 5-WEC row, 10-WEC array with 2 columns, (5 × 5)-WEC rectilinear array, (5 × 5)-WEC staggered array, (3 × 3)-WEC rectilinear array with a WEC interdistance of 5 D and 10 D and 13-WEC staggered array. Due to the large spacings, these arrays are not considered closely spaced WEC arrays. Figure 8a shows the 5 × 5 rectilinear array layout in the DHI's wave basin and Figure 8b shows the tested configurations. Power absorption tests with regular, panchromatic, long- and short-crested irregular waves were executed. Furthermore, decay tests and wave diffraction tests were executed. The WEC response and modifications of the wave field were measured to provide data for understanding WEC array interactions and to evaluate numerical models of array interaction. Wave gauges were located within and around the WEC array. By extracting spectra at different locations within and around the array, the wave field modifications were studied.

2.5. Wavestar WEC Array

Ruíz et al. performed experiments with a staggered array of five WECs [40] in the deep-water wave basin at Aalborg University (Denmark), which is 15.7 m long, 8.5 m wide, and has a water depth of 0.6 m. The study used experimental data for the validation and uncertainty assessment of the WEC array hydrodynamics tool implemented in DTOcean open-source software, relying on the linearised potential flow theory [40].

The WEC considered for the experiments was a small-scale version of one of the floaters of the Wavestar WEC, reproduced at a scale of 1:20 according to the prototype version of the Wavestar WEC installed in Hanstholm (Denmark) [41]. Figure 9a shows the WEC used in the experiments, consisting of a semisubmerged buoy of 0.254 m diameter and 0.104 m draught connected to a 0.680 m long lever arm.

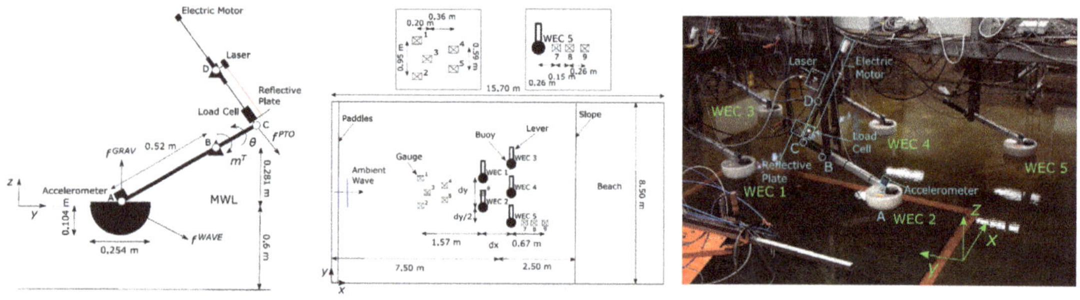

(a) (b) (c)

Figure 9. Experimental setup used by Ruíz et al., adapted with permission from Ref. [40], 2017, Ruíz et al.: (**a**) WEC model; (**b**) Aalborg University wave basin WEC array layout; (**c**) Aalborg University's wave basin setup.

The lever arm was hinged at a height of 0.281 m from the mean water level (MWL) allowing a rotation θ around the x-axis. The WECs were equipped with a linear motor capable of accurately generating linear loads (force directly proportional to velocity), reducing the uncertainty in the physical model [40]. Figure 9b shows the general layout of the setup in the wave basin and Figure 9c shows a picture of the setup. During the experimental campaign only a staggered WEC array was considered with dx = 1.28 m, dy = 1.28 m and a position shift along the y-axis of dy/2 between rows. The WEC array consisted of five WECs split into two rows consisting of two and three WECs, respectively. Diffraction tests with regular waves, radiation tests and power absorption tests with regular and irregular long-crested waves were executed in order to validate the tool. Experimental measurements were then compared with numerical predictions in order to estimate the uncertainty in both wave forces and power absorption [40].

2.6. Australian Maritime College WEC Array

Nader et al. performed array experiments with up to six fully submerged point-absorber WECs moving in six DOFs [9]. The Australian Maritime College's (University of Tasmania) Model Test Basin was used to perform the experiments. The basin is 35 m long, 12 m wide and the water depth was kept constant at 0.6 m. A pit was used to house the linear motor and frame for the active model, such that this equipment was located in the basin floor. The study experimentally investigated the hydrodynamic aspects of an array of generic WECs by separating the problem into its diffraction and radiation problems, removing the need for PTO modelling and control. The experiments measured the interaction factor for two WECs moving in heave and surge. The stereovideogrammetry method was used to measure the wave field around and in the lee of the array [9].

The WECs considered had a generic spherical floater shape. One active model was used for the radiation problem and additional passive models were installed to measure the WEC array interactions when separated at various distances. The passive models were composed of three main components: a vertical post, a hollow sphere and a six-DOF load cell that was fitted between the post and sphere. These custom-built six-DOF load cells measured the hydrodynamic forces and moments experienced by the models during the experiments. The general characteristics of the active model were the same as those of the passive models. However, for the radiation problem, the active model was forced to heave or surge. This behaviour was ruled by an electromagnetic linear motor [9]. Figure 10a shows the active model with the linear motor heave and surge motion configurations.

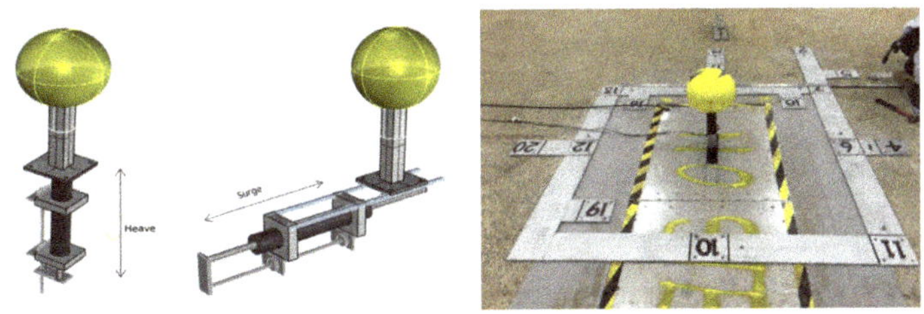

(**a**) (**b**)

Figure 10. Experimental setup used by Nader et al., reprinted with permission from Ref. [9], 2017, Nadar et al.: (**a**) active WEC model; (**b**) frame and active WEC.

In order to install and facilitate changes between the different WEC array configurations, a steel positioning frame marking 20 different locations was fixed on the floor of the basin. The frame was positioned in the middle of the tank surrounding the pit where the active model system was installed. The passive models were then fixed at the desired locations for the required configuration. Figure 10b shows this frame and Figure 11a shows the relative locations of the active model and various passive models in the WEC array. Figure 11b shows the various array configurations investigated experimentally. A total of 1447 experimental runs were completed during the experimental campaign. Regular waves were used with wave amplitudes of 0.015 m and 0.030 m and wave frequencies from 0.5 to 1.7 Hz in steps of 0.1 Hz.

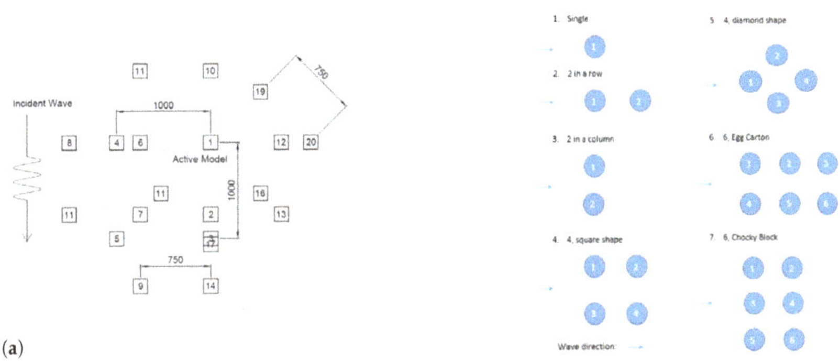

(**a**) (**b**)

Figure 11. Experimental setup used by Nader et al., reprinted with permission from Ref. [9], 2017, Nadar et al.: (**a**) WEC array layout; (**b**) sketches of tested WEC array configurations.

2.7. *Tu Delft WEC Array*

Boere et al. studied experimentally the hydrodynamic interaction of two cylindrical point-absorber WECs in close vicinity [42]. The experiments were performed in the towing tank of TU Delft (The Netherlands), which is 85 m long, 2.75 m wide and 1.20 m deep. The study applied the boundary element method (BEM) model Nemoh and a time-domain WEC Simulator (WEC-Sim) to compute the power output of two WECs while varying their relative position. The trends found in the simulation were experimentally validated by testing scaled versions of two-WEC configurations in the wave tank [42].

To perform the experiments, a generic WEC was identified, namely, an axisymmetric semisubmerged cylinder with a draft of 2.0 m and a diameter of 10.0 m, scaled by a factor of 1:40. The required one-DOF vertical movement of the WECs was secured by mounting the WEC buoy to a rod hanging in linear sliding bearings. The damping of the PTO system was realised by a viscous damper, with a designed damping coefficient of 80 Ns/m. An elevation sensor was attached to the WEC buoy to accurately measure its vertical position. A load cell was used to connect the piston of the damper to the WEC in order to measure the forces applied by the damper. The combined data of the elevation sensor and the load cell gave insight on both the damper characteristics and the power output of the entire system. Figure 12a shows a rendering of the WEC used in the experiments, with indication of the subsystems.

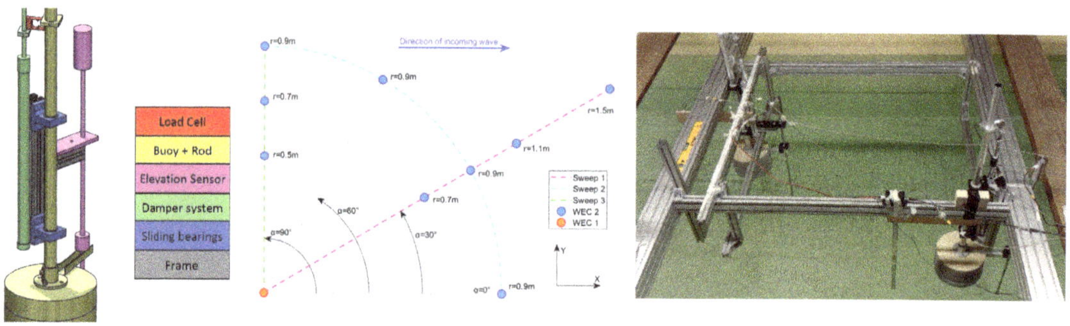

(b) (c)

(a)

Figure 12. Experimental setup used by Boere et al., reprinted with permission from Ref. [42], 2018, Boere et al.: (**a**) WEC model; (**b**) tested WEC array configurations; (**c**) TU Delft's towing tank setup picture.

The two WECs were mounted in a frame of sliding aluminium beams, making it easy to adjust the positions of the two WECs in the array. Figure 12c shows the setup of the two-WEC array, where waves are coming from the right. Four regular wave conditions were chosen and correspondingly labelled as weak, fair, moderate and rough. Each wave condition was tested for each of the ten WEC array configurations, resulting in 40 tests. The ten different WEC array configurations consisted of three sweeps as shown in Figure 12b: Sweep 1 with a constant angle of 30° and four different radii, Sweep 2 with a constant radius and four different angles and Sweep 3 with a constant angle of 90° and three different radii. The power output was measured in each of the performed tests [42].

2.8. Multipoint-Absorber WEC

Do et al. experimentally tested a multipoint-absorber WEC (MPAWEC) consisting of ten hemispherical–cylindrical heaving–surging buoys [43]. The research considered a hydrostatic transmission PTO concept with a focus on increasing the absorbed power by smoothing the output power, utilising a common pressure rail connected to a high-pressure accumulator. All of the components of the MPAWEC were mathematically modelled, and simulation results were compared with the experimental results. The test were performed in an ocean wave simulator tank at the Research Institute of Small and Medium Shipbuilding (RIMS), Busan, Korea [43,44]. This project was initiated with the purpose to deploy a full-scale WEC array in an ocean environment for power production.

Figure 13a shows a 3D rendering of the MPAWEC. Ten floating buoys are connected to a moving shaft. A cylinder functioning as a hydraulic pump is attached to each shaft. One end of the cylinder is joined to the moving shaft, while the other is joined to the frame. The frame and all of its attached components are supported by columns, which are fixed to the bottom. The wave excitation force is decomposed into a vertical component and a horizontal component. The resultant force has the same direction as the moving direction of the floating buoy with shaft, or differs from it by a small angle. Therefore, friction between moving parts is reduced, and more power is absorbed from the waves. The piston of the cylinder moves up and down, thereby generating high-pressure fluid at the output port of the cylinder. The sliding angle of a floating buoy can be adjusted by an actuator, as shown in Figure 13b. The power output was measured during tests with regular waves [43,44].

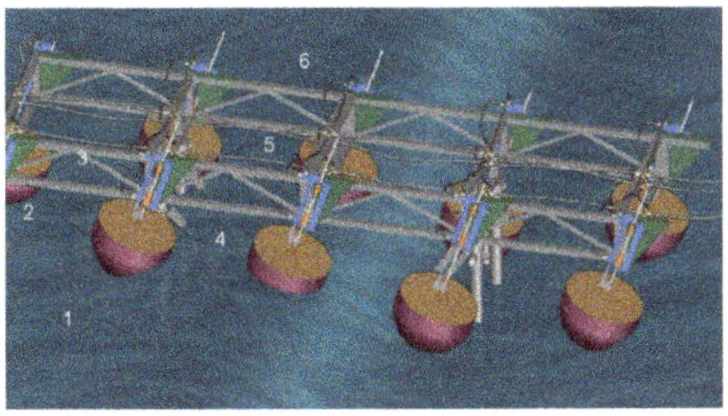

(a)

(b)

Figure 13. Experimental setup used by Do et al., adapted with permission from Ref. [43], 2018, Do et al.: (**a**) 3D rendering of the WEC array layout; (**b**) RIMS's wave basin setup picture.

2.9. *Uppsala University's WEC Array*

Giassi et al. performed an experimental campaign of arrays with six direct-driven point-absorber WECs. The experimental campaign was carried out in the COAST Lab at the University of Plymouth, UK. The ocean basin is 35 m long and 15.5 m wide, with a moveable floor set to operate at a depth of 2.5 m. The research studied and compared the performances of three different array layouts under several regular and irregular long-crested waves based on the absorbed power. Figure 14b shows the wave basin's WEC array layout of these three configurations.

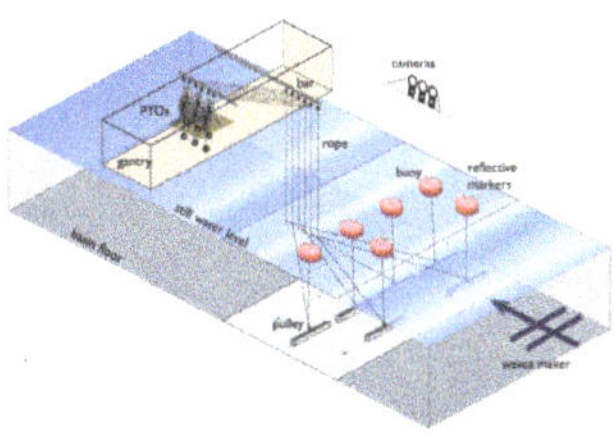

(a)

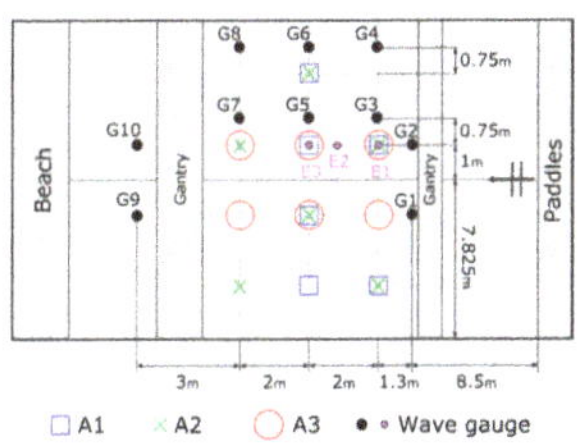

(b)

(c)

Figure 14. Experimental setup used by Giassi et al.: (**a**) sketch of the setup's components, reprinted with permission from Ref. [45], 2020, Giassi et al.; (**b**) wave basin WEC array layout, reprinted with permission from Ref. [8], 2019, Giassi et al.; (**c**) the University of Plymouth's COAST Lab setup, reprinted with permission from Ref. [8], 2019, Giassi et al.

It was determined whether the numerical predictions of the best performing array layouts were confirmed by experimental data. The simulations were executed with a frequency-domain model restricted to heave, which was a computationally fast approach that was merged into a genetic algorithm optimisation routine and used to find optimal array configurations [8,45]. The WEC model tested was a 1:10 scaled prototype based on the point-absorber WEC concept developed at Uppsala University (Sweden). Six identical ellipsoidal floating buoys with a diameter of 0.488 m and a height of 0.280 m moving in six DOFs were connected via a system of highly stiff ropes and pulleys to the PTO systems located on the main gantry. Each WEC had a rotating PTO system, and the damping of the PTO was changed by changing the air gap between magnets and an aluminium disc, changing the magnetic flux inside the disc. Additional inertia allowed them to change the WEC natural frequency, allowing to tune the WEC for a specific sea state [46]. Each WEC buoy was equipped with five reflective markers for motion capture, which were tracked by a set of eight Qualisys six-DOF cameras. Figure 14a shows a sketch of the experimental setup components and Figure 14c shows a picture of the setup at the COAST lab [8,45].

2.10. M4 WEC

The M4 WEC originally consisted of three in-line floaters increasing in diameter (and draft) from bow to stern so that drift forces increased with distance from the floater, causing the WEC to head naturally in the wave direction [47]. Different sizes of floater gave a range of natural periods to enable power capture across a range of wave periods for a given offshore site. Although the M4 WEC was conceived as one single WEC due to the connection between the floaters, the different floaters acted as a WEC array. The floaters cannot be called WECs, since not every floater was equipped with a PTO system. The M4 project was initiated with the purpose of a full-scale ocean deployment for power production. Stansby et al. performed experiments with this M4 WEC with three floaters in the Ocean Wave Basin of the Plymouth COAST laboratory [48,49]. Santo et al. assessed the performance of the three-floater M4 wave energy converter in Albany (on the south coast of western Australia) compared to EMEC in Orkney (UK) based on the experimental data [50]. Moreno et al. conducted an experimental campaign with a six-floater M4 WEC at a scale of 1:50 at the Lir's Ocean Basin at the Beaufort Centre, University College Cork (UCC), Ireland and presented cost estimates for several offshore sites [47]. The basin is 25.0 m by 17.0 m and 1.0 m deep. Liao et al. theoretically analysed the linear hydrodynamic model of the M4 WEC with eight floaters and four PTO units in a state-space form to make it possible to implement advanced control algorithms in real time [51]. Since testing of the M4 WEC resulted in gradual model improvements, only the latest testing by Moreno et al. is included in the presented literature overview [47]. Figure 15a shows a sectional drawing of the six-floater M4 WEC with dimensions and indication of the PTO system with two linear actuators and the mooring line connection.

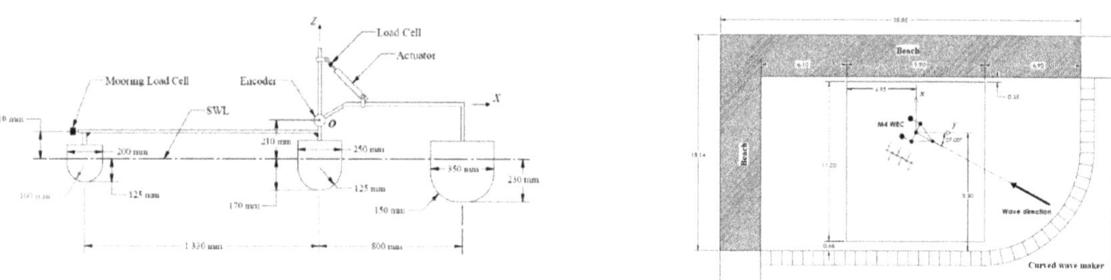

(a) (b)

Figure 15. Experimental setup used by Moreno et al., reprinted with permission from Ref. [47], 2019, Moreno et al.: (**a**) six-floater M4 WEC model with dimensions; (**b**) Lir's Ocean Basin WEC array layout.

The bow and mid floaters were rigidly connected by a beam, effectively forming a single body, and a beam from the stern floater was connected to the hinge point above the mid floater to work as PTO. The PTO damping coefficient was not controlled. Figure 15b shows the layout of the experimental setup in the Lir's Ocean Basin [47]. Stansby et al. used the six-floater M4 configuration with two PTOs to assess the results of the time-domain linear diffraction model for surge, pitch and heave floater motions in long-crested waves for multifloater configurations [52]. The angular motion at the PTOs and mooring forces were measured during different power absorptions tests. Irregular long-crested waves with JONSWAP spectra and irregular short-crested waves were applied. A test without PTO to give a worst-case response were performed as well [47].

2.11. Floating Offshore Platform with 12 WECs

The last two experimental campaigns discussed consider the integration of point-absorber WECs in a floating platform for offshore wind turbines. These technologies are in line with a recent trend to develop hybrid/synergy solutions for renewable offshore energy applications. Kamarlouei et al. studied experimentally 12 WECs concentrically arranged and attached on a floating offshore platform model [53]. Figure 16a shows the conceptual design rendering.

The 1:27 scale model was designed, built and tested in two main situations, without and with 12 cone-shaped WECs. Figure 16c shows a picture of the scale model. Besides the potential power output of the WECs, it was meant to obtain interaction between the WEC buoys and the platform to reduce the heave and pitch motions of the platform. The test was performed in the Lir NOTF medium size ocean basin at the MaREI centre of UCC, Ireland. The basin is 25 m by 17 m and 1 m deep with a moveable floor, which increases the depth up to 2.5m [53].

A wind turbine mast and nozzle were installed as a central column on the deck, thus the heeling moment caused by the weight of the wind turbine was fairly represented. The hexagonal steel frame used as the deck of the platform included connections for 12 point-absorber WECs via aluminium arms. Moreover, a set of 12 rotational friction dampers were installed in the hinge connection of the arm to the hexagonal deck. Figure 16b shows the model installation in the wave basin in detail. The model was set in the middle of the basin with a catenary mooring system with three lines. Experiments were performed for regular and irregular long-crested waves. The regular waves were simulated with 0.01, 0.02 and 0.04 m wave heights and various wave periods from 0.6 to 4.0 s. The irregular waves were simulated in seven different sea states. The floating platform was tested with and without WECs. For the former, tests were executed with and without the dampers [53].

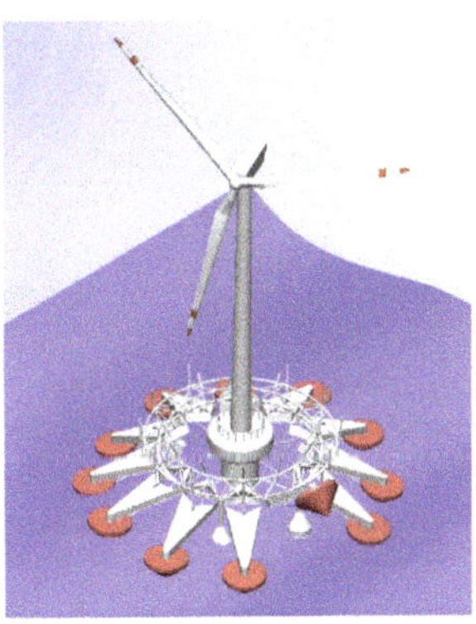

(a)

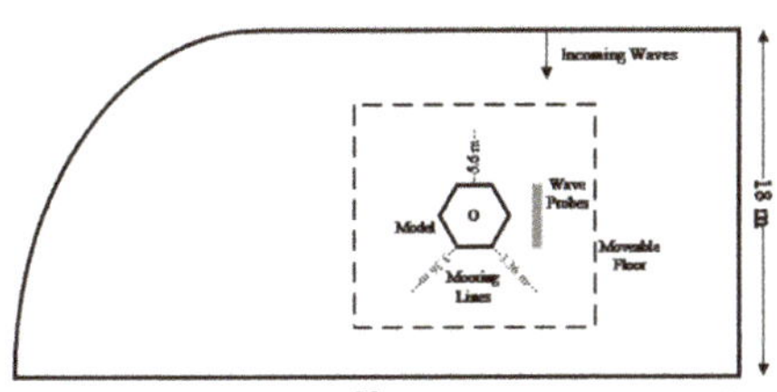

(b)

(c)

Figure 16. Experimental setup used by Kamarlouei et al., reprinted with permission from Ref. [53], 2020, Kamarlouei et al.: (**a**) conceptual design; (**b**) Lir's Ocean Basin WEC array layout; (**c**) model of the concentric WEC array installed on a floating offshore platform.

2.12. Three-Pontoon Semisubmersible Platform with Six WECs

Sun et al. performed experiments with a three-pontoon semisubmersible platform with six point-absorber WECs [54]. The platform was constructed on a 1:40 scale with a simplified hydraulic PTO system and a mooring system to restrict the platform movement. The prototype of the semisubmersible platform was WindFloat [55]. Tests were conducted in the towing tank of the National Ocean Technology Center, Tianjin, China. The towing tank is 130 m by 18 m and a water depth of 4.5 m was used. The study aimed to clarify the interaction between the floating platform and the WEC array, without taking the influence of the wind turbine into account. The effect of the floating platform on the energy conversion of the WECs and the effect of the WECs on the hydrodynamic performance of the platform were investigated [54].

Figure 17a shows a 3D rendering of the cone bottom floater with dimensions. The PTO system is mainly made up of two hydraulic accumulators and an energy storage cylinder. Figure 17b shows the three different wave headings, their mooring system and floaters placement considered in the tests. Since the hydrodynamic performance of the floaters is different for changing wave direction, these floaters are numbered. Figure 17c shows the platform consisting of three pontoons with damping plates.

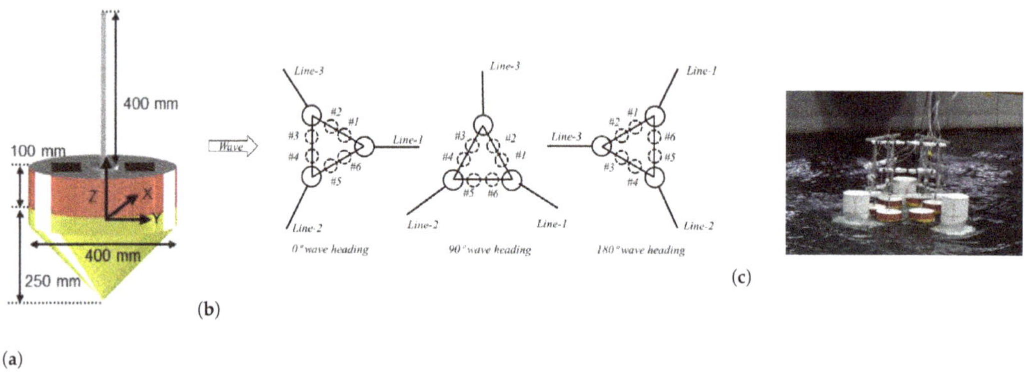

Figure 17. Experimental setup used by Sun et al., reprinted with permission from Ref. [54], 2021, Sun et al.: (**a**) 3D model of the WEC buoy with dimensions; (**b**) towing tank WEC array layout; (**c**) National Ocean Technology Center's towing tank setup.

Six floaters were fixed on the upper beam, whereas every two floaters were evenly arranged between every two pontoons. The platform's motion was measured by a six-DOF measurement system. Regular waves with wave height of 0.08 m and period of 1.0–2.2 s and irregular long-crested waves with significant wave height of 0.0975 m and peak period of 1.5 s were tested. The heave, pitch and surge response amplitude operator (RAO) of the platform were presented, for tests with and without the WECs. The RAO was defined by the ratio of the heave amplitude z and the wave amplitude η. The hydraulic power output for each of the six WECs was presented for the configurations displayed in Figure 17b [54].

3. Research Gaps

3.1. Wec Shape

Based on the literature review presented in Section 2, research gaps, knowledge gaps and opportunities within experimental modelling of point-absorber WEC arrays can be identified. Table 1 gives an overview of the point-absorber WEC array experimental campaigns available in the literature, as discussed in Section 2. For each experimental campaign, labelled with a campaign ID, the authors, year, wave testing facility and DOFs for a single WEC are summarised. The identified experimental campaigns have been carried out over the last 15 years, mainly in Europe. Besides the heave DOF, some WECs have additional DOFs. Only the experimental campaigns performed by Stratigaki et al., Nadar et al., Boere et al. and Sun et al. considered a generic WEC concept [9–12,42,54]. The other campaigns considered a scaled model of a WEC concept which was targeted for a full-scale ocean deployment in a WEC array. To the knowledge of the authors, only the Hydralab IV WECwakes project database (campaign ID 9) is made publicly available [10–13]. Göteman et al. stressed the need for available real-life data for the validation of WEC array modelling and optimisation [7]. Consequently, there is a research gap of publicly accessible experimental datasets. Preferably, a generic WEC concept should be considered, resulting in data useful for a wide range of point-absorber WEC concepts. Tables 2–4 allow us to identify the focus and needs within these WEC array experiments.

3.2. PTO Design and Control Platform

Table 2 summarises for each experimental campaign the WEC shape, WEC diameter, PTO system and applied control strategy. Most test campaigns considered a hemisphere-cylinder for the WEC floater shape. The diameter was limited, resulting in small scale models. To limit scale effects, it is preferable to increase the WEC diameter. The WEC diameter should comply with the wave generation possibilities and dimensions of the wave testing facility, since the resonance frequency of the WEC and the WEC spacing within a WEC array depend on the diameter.

Table 1. Overview of WEC array experimental campaigns with point-absorber WECs available in the literature.

Campaign ID	Authors	Year	Wave Testing Facility	DOFs
1	Stallard et al.	2008	Wave flume, Manchester Univ.	Heave, surge
2	Thomas et al.	2008	Wave flume, Manchester Univ.	Heave, surge
3	Alexandre et al.	2009	Wave flume, Manchester Univ.	Heave, surge
4	Weller et al.	2010	Wave flume, Manchester Univ.	Heave, surge
5	Child et al.	2013	Wave basin, QUB	Heave
6	Folley et al.	2013	Wave basin, QUB	Heave
7	Lamont-Kane et al.	2013	Wave basin, QUB	Heave
8	Mackay et al.	2013	Wave basin, MARIN	6-DOF
9	Stratigaki et al.	2014	Shallow Water Basin, DHI	Heave
10	Ruiz et al.	2017	Wave basin, Aalborg University	Heave
11	Nadar et al.	2017	Wave basin, University of Tasmania	Heave, surge
12	Boere et al.	2018	Towing tank, TU Delft	Heave
13	Do et al.	2018	Wave basin, RIMS Korea	Heave, surge
14	Giassi et al.	2019	Wave basin, Univ. of Plymouth	6-DOF
15	Moreno et al.	2019	Lir wave basin, UCC	Heave, pitch, surge
16	Kamarlouei et al.	2020	Lir wave basin, UCC	Heave, pitch
17	Sun et al.	2021	Towing tank, Tianjin	Heave

It is well known that the physical design and operational quality of the PTO system used on small scale WEC models can have vast effects on the tank testing results [56]. Table 2 shows that most PTO systems are modelled as mechanical elements, such as friction brakes (campaign IDs 5, 6, 7, 9 and 16) or hydraulic dampers (campaign IDs 12, 15 and 17). These simplified PTO systems show nonlinear behaviours (e.g., static friction, temperature dependency and backlash), of which the effects propagate into the WEC power production data. The uncertainties introduced by these simplified PTO systems are seen as a major limitation for the data to be used for numerical validation and the extrapolation of the tank test results to a meaningful full ocean scale. A solution to this problem is to use actively controlled actuators for the PTO simulation by using force- (or torque)-controlled feedback systems with suitable instrumentation, enabling the PTO system to exert any desired time- and/or state-dependent reaction force [56].

Table 2 shows that most PTO systems produce a constant Coulomb damping force or resistive control, which is a simple proportional feedback with a PTO force equal to the WEC velocity times a damping factor. To implement more advanced control strategies, a highly accurate real-time controllable PTO system with bidirectional power flow is required. Test campaign 14 considered a rotational direct-driven PTO system with constant damping, showing a simple and accurate way to get a low friction PTO system with a nearly ideal velocity dependent force [8,45,46]. Only the experiments with the Manchester Bobbers (campaign IDs 1–4), the experiments with the four Wavebob WEC models (campaign ID 8), the experiments with an array of five Wavestar WECs (campaign ID 10) and the experiments by Nadar et al. (campaign ID 11) modelled the PTO system with a motor capable of providing a bidirectional power flow (motor and generator functionality) and capable of accurately generating linear loads (force directly proportional to the velocity) [9,30–34,39,40]. Although these motors are capable of imposing advanced control strategies, advanced control was not implemented in the respective experimental campaigns. In CFD-based numerical wave tanks (CNWT), the PTO system is mostly modelled as a linear spring–damper system, not representing realistic PTO dynamics and inefficiencies and undermining the overall model fidelity [57]. To validate CNWT considering WEC arrays, it is desirable to incorporate a realistic, nonlinear PTO model. Therefore, there is a research gap of WEC array experimental data considering WECs equipped with a highly accurate real-time controllable PTO with a bidirectional power flow.

Table 2. Overview of the WEC design characteristics of the WEC array experimental campaigns of Table 1.

Campaign ID	WEC Floater Shape	WEC Diameter (m)	PTO System	Control Strategy
1	Hemisphere-cylinder	0.15	PMDC motor	Resistive control
2	Hemisphere-cylinder	0.15	PMDC motor	Resistive control
3	Hemisphere-cylinder	0.15	PMDC motor	Resistive control
4	Hemisphere-cylinder	0.15	PMDC motor	Resistive control
5	Hemisphere-cylinder	0.25	Friction brake	Coulomb damping
6	Hemisphere-cylinder	0.25	Friction brake	Coulomb damping
7	Hemisphere-cylinder	0.25	Friction brake	Coulomb damping
8	FNT with torus	0.93	Linear motor	Resistive control
9	Hemisphere-cylinder	0.315	Friction brake	Resistive control
10	Hemisphere	0.254	Linear motor	Resistive control
11	Sphere	0.25	Linear motor	Resistive control
12	Cylinder	0.25	Linear hydraulic damper	Resistive control
13	Hemisphere-cylinder	1.20	Hydraulic motor	PID velocity control
14	Ellipsoidal	0.488	Rotational direct-driven generator	Resistive control
15	Hemisphere-cylinder	0.20–0.35	Linear hydraulic damper	Resistive control
16	Cone	n.a.	Rotational friction damper	Resistive control
17	Cone	0.40	Linear hydraulic damper	Resistive control

A model predictive control (MPC) and a linear time-invariant (LTI) energy-maximising control strategy were implemented and tested for a single Wavestar WEC [56,58]. However, extending these advanced control strategies from a single WEC to an array of multiple WECs has not yet been addressed in an experimental campaign. Bacelli et al. compared two model-based WEC array control strategies, namely, global control (GC) and independent control (IC) [59]. GC is based on a centralised control algorithm which uses the complete hydrodynamic model of the WEC array whereas, with IC, each WEC is controlled independently using the hydrodynamic model of a single isolated WEC. The theoretical study suggested that a significant performance improvement (up to 10%, or more) could be obtained using GC of arrays, compared to IC [59]. Similar to the requirement of realistic PTO models within the CNWT, the incorporation of advanced control strategies into the WEC model is desirable. The nonlinear hydrodynamic environment of the CNWT allows for a more realistic evaluation of control strategies, thereby accelerating the development and implementation of energy-maximising control for WECs [57]. Therefore, there is a research gap of WEC array experimental data considering GC and advanced control on WEC array level.

3.3. Wave Conditions

Table 3 summarises for each experimental campaign the tested wave conditions. Most experimental campaigns considered both regular and irregular long-crested waves with a range of different (significant) wave heights and (peak) wave periods. In this case, the table does not make a distinction between the wave height and wave period for the regular waves and the significant wave height and peak wave period for the irregular waves, as the main purpose is to indicate the order of magnitude of the tested wave conditions.

Table 3. Overview of the tested wave conditions of the WEC array experimental campaigns of Table 1.

Campaign ID	Wave Conditions	(Significant) Wave Height (m)	(Peak) Wave Period (s)
1	Regular	0.026	0.50–1.55
2	Regular	0.026	0.57–1.33
3	Irregular (Bretschneider spectrum)	0.040	1.31
4	Regular and irregular long-crested	0.015–0.064	0.61–2.00
5	Irregular (JONSWAP spectrum)	0.025–0.038	0.89–1.26
6	Irregular short- and long-crested (JONSWAP spectrum)	0.025–0.050	0.70–1.26
7	Regular and irregular long-crested	0.014–0.075	0.67–1.26
8	Regular and irregular (JONSWAP spectrum)	0.095–0.191	1.42–2.48
9	Regular and irregular long- and short-crested	0.024–0.104	0.87–1.51
10	Regular and irregular long-crested	0.045–0.060	0.76–2.00
11	Regular	0.030–0.060	0.59–2.00
12	Regular	0.063–0.113	1.11–1.58
13	Regular	0.26	3.79
14	Regular and irregular long-crested	0.124–0.175	1.11–2.37
15	Regular and irregular long- and short-crested	0.04–0.06	0.70–2.0
16	Regular and irregular long-crested	0.010–0.139	0.60–4.0
17	Regular and irregular long-crested	0.08–0.0975	1.0–2.2

When taking the WEC diameter into account, these wave heights are mostly small, complying with the assumptions of linear potential flow theory. Only a limited number of experimental campaigns (campaign IDs 6, 9 and 15) considered irregular short-crested waves [10–12,37,47]. Focused waves are a practical laboratory method for reproducing extreme waves. From a CFD numerical modelling point of view these waves are an interesting case for validation purposes due to their peak loading and short duration time, limiting the computational cost. Therefore, there is a research gap of experimental data of WEC array tests covering extreme wave conditions, short-crested waves and focused waves, to validate recently developed nonlinear numerical models considering nonlinear effects as impact loading, friction, viscous drag, bottom slamming and vortex shedding for real sea conditions [60,61].

3.4. Number of WECs and WEC Array Layouts

Table 4 summarises for each experimental campaign the number of WECs in an array, WEC axis-to-axis separation distance, WEC array layouts, target measurements and literature references.

The number of WECs within the array ranged from 2 to 25. The axis-to-axis separation distance expressed in WEC diameters (D) ranged from 1.25 D to 20 D. The WEC array can be called closely spaced when hydrodynamic WEC–WEC interactions are significant, which also depends on the WEC floater shape besides the separation distance. The WEC array layouts tested were mainly rectilinear and staggered configurations. The target measurements to quantify near-field interactions were the absorbed power, used to calculate the capture width and the (q) interaction factor. The target measurements to quantify far-field effects were the wave field measurements. Windt et al. reported that only a few studies modelled WEC arrays using a computational fluid dynamics (CFD) approach due to the increased computational burden when modelling WEC arrays [57]. To address the need for experimental data for the validation of CFD-based numerical models for WEC arrays, experiments should consider closely spaced WEC arrays to maximise hydrodynamic interaction. It is also preferable to increase the number of WECs in the experiments starting from two, to limit the computational cost of CNWT simulations. Therefore, there is a research gap of experimental data of WEC array tests considering a limited number of closely spaced WECs. The above-mentioned research and knowledge gaps were translated into the "WECfarm" project objectives and design requirements, summarised in Section 4.1.

Table 4. Overview of the WEC array characteristics of the WEC array experimental campaigns of Table 1.

Campaign ID	Number of WECs in an Array	Separation Distance	WEC Array Layouts	Target Measurements	Literature References
1	1, 3, 9, 12	2 D	Figure 2a	Capture width, (q) factor	[30]
2	5	2 D	Figure 2b	Heave amplitude	[31]
3	5, 10	2 D	Figure 3a	Wave field	[32]
4	12	2 D	Figure 3b	Capture width, (q) factor	[33–35]
5	22, 24	3 D	Figure 5a	Absorbed power, (q) factor	[36,62]
6	22, 24	3 D	Figure 5a	(q) factor	[37]
7	4	3 D	Figure 4b	Capture width, wave field	[38]
8	4	8.5 D	Figure 6b	Absorbed power	[39]
9	1–25	5–10–20 D	Figure 8b	(q) factor, wave field	[10–12]
10	5	5 D	Figure 9b	Absorbed power	[40,58]
11	1, 2, 4, 6	6 D; 4 D	Figure 11b	(q) factor	[9]
12	2	2 D–6 D	Figure 12b	Absorbed power	[42]
13	1, 10	>3 D	Figure 13a	q factor	[43,44]
14	6	4 D	Figure 14a	Absorbed power, (q) factor	[8,45,46]
15	3, 6, 8	2 D–7 D	Figure 15a	Absorbed power, capture width	[47]
16	12	<2 D	Figure 16b	Rotational velocity	[53]
17	6	1.25 D	Figure 17b	Absorbed power	[54]

4. Design of the WECfarm Project Setup

4.1. Research Objectives and Design Requirements

The experimental "WECfarm" project was initiated by Ghent University in 2018 targeting to deliver an experimental dataset on wave–WEC and WEC–WEC interactions for point-absorber WEC array tests. This dataset aims to cover the research gap on the need for publicly available real-life and reliable data for the validation of WEC array modelling and WEC array optimisation. The dataset will contain measurement time series at a sufficient high sample rate of the parameters fully characterising the experiments, aiming to minimise measurement uncertainties. The objective of the "WECfarm" project is to design, construct and test a point-absorber WEC array which addresses the research and knowledge gaps discussed in Section 3. Summarised, the WECfarm project objective is to address the need for WEC array tests with:

1. A generic WEC concept, resulting in data useful for a wide range of point-absorber concepts.
2. WECs equipped with a highly accurate, real-time actively controllable PTO system with bidirectional power flow.
3. A control platform to implement GC and advanced control on WEC array level.
4. A wide range of wave conditions. These should include regular waves, irregular long- and short-crested waves, focused waves and extreme wave conditions.
5. A limited number of closely spaced WECs.

Section 4.2 discusses how the "WECfarm" WEC hydrodynamic design addresses requirements 1 and 5, Section 4.3 discusses how the "WECfarm" WEC electromechanical design addresses requirement 2 and Section 4.4 discusses how the "WECfarm" WEC array DAQ and control platform addresses requirement 3. Section 4.5 discusses the methodology within the "WECfarm" project, mentioning the wave basin testing performed so far and the scheduled testing, addressing requirement 4.

4.2. Hydrodynamic Design

To simplify the physical and numerical modelling of the WEC and the WEC arrays, the DOFs for the "WECfarm" WEC were restricted to heave. Shadman et al. performed a numerical geometrical optimisation for heaving point-absorber WECs [63]. The study evaluated cylindrical buoys for different diameter-to-draft ratios based on the maximum power absorption, the natural period and the resonance bandwidth, defined as the frequency interval in which the absorbed power is more than half of its maximum value. These metrics can be derived from the RAO. The study found that cylindrical buoys with larger diameters and relatively small drafts provided a good resonance bandwidth, while the maximum energy absorption was achievable in the same diameter range with larger drafts. In general, a larger diameter improved the buoy performance by increasing the maximum power, natural period and resonance bandwidth. Increasing the draft led to a higher maximum power and natural period with a lower resonance bandwidth. Therefore, modifying the draft could significantly influence the point absorber performance. A numerical optimiser defined the best set of geometrical parameters which jointly maximised the maximum power and resonance bandwidth for the design sea state, resulting in a diameter-to-draft ratio of 4.5 [63].

Bacelli et al. designed and tested a generic six-DOFs WEC on scale of 1:17 [64]. The study found that a good point absorber was characterised by a flat response, meaning that it was able to extract power from waves without requiring reactive power from the PTO system over a broad frequency range. Therefore, a slender cylinder was not a good absorber since it had a high sensitivity of the absorbed power with respect to the wave frequency. An additional consideration was that good absorbers were also characterised by a relatively small oscillation amplitude, resulting in a smaller PTO stroke. The designed WEC buoy had a cylindrical shape with a diameter-to-draft ratio of 3.3 and a 34.9° truncation of the bottom extending over 7/10 of the draft [64].

Based on the findings and insights of Shadman et al. [63] and Bacelli et al. [64], an iterative numerical optimisation study was performed on the geometry of the "WECfarm" WEC. The scale of the "WECfarm" WEC was chosen based on a combination of the Coastal and Ocean Basin's (COB) (Ostend, Belgium) characteristics [65] and on a workable and fundable model scale. Given the available funding for the "WECfarm" project experimental setup, the complexity related to requirements 2 and 3 in Section 4.1, the number of WECs was set equal to five, complying with requirement 5 in Section 4.1. Linear potential flow simulations with WEC's axis-to-axis spacing of 2 D confirmed that this number was sufficient to obtain significant WEC–WEC interactions and to study the wave–WEC interactions for WEC arrays with two to five WECs.

To reduce the surge force loading and increase the radiation capacity to enhance near-field interactions, the WEC buoy was designed with a diameter-to-draft ratio of 3.75. The WEC buoy bottom was truncated under an angle of 45° over a height of 0.10 m. Therefore, the radius of the WEC buoy at the truncation part ranged from 0.20 m to 0.30 m. The WEC buoy was 0.32 m high and designed with a draft of 0.16 m. This draft corresponded with a submerged volume of 0.03683 m^3. Therefore, the mass of the WEC buoy and heaving parts on top of it was 36.83 kg. The angle of 3° on the cylindrical part was required for the thermoforming construction process to allow the removal of the acrylonitrile butadiene styrene (ABS)-truncated cylindrical WEC buoy from the wooden mould. Figure 18a shows a 2D rendering of the final geometry of the "WECfarm" WEC buoy with its dimensions and the adopted coordinate system. The WEC buoy is axisymmetric around the z-axis.

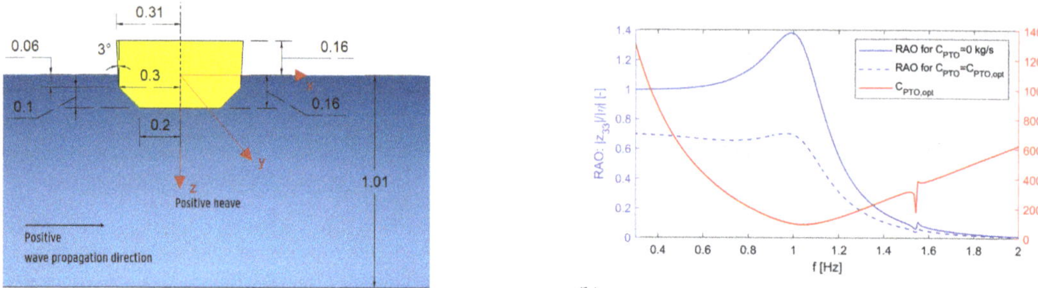

(a) (b)

Figure 18. WECfarm's WEC's buoy: (**a**) 2D rendering with dimensions in m; (**b**) RAO and $C_{PTO,opt}$.

Figure 18b shows the RAO for the free response case, equivalent to a PTO damping of 0 kg/s and the RAO for the case with the optimal PTO damping for the resistive control strategy, $C_{PTO,opt}$, according to:

$$C_{PTO,opt}(\omega) = \sqrt{B^2(\omega) + \left(\omega(M + A(\omega)) - \frac{\rho g S}{\omega}\right)^2} \quad (2)$$

in which $B(\omega)$ corresponds to the hydrodynamic damping and $A(\omega)$ to the added mass [66]. The mass (M) was equal to 36.83 kg, ρ is the water density equal to 1000 kg/m^3, (g) is the gravitational acceleration of 9.81 m/s^2 and S is the cross-sectional area of the buoy at the unperturbed SWL. Figure 18b also displays $C_{PTO,opt}$. The heave wave excitation coefficients, $B(\omega)$ and $A(\omega)$ were obtained with the open-source software package openWEC [67], with the integration of the BEM code Nemoh. As recommended by Shadman et al. [63] and Bacelli et al. [64], the RAO had a flat response, resulting in a high resonance bandwidth. The numerically obtained natural period was slightly higher than 1.0 s. This relatively small natural period had the advantage to avoid a sharp power absorption decrease for lower wave periods, since the WEC buoy was designed with dimensions to operate in wave conditions with higher wave periods.

Since the "WECfarm" project targets experimental data necessary for the validation of nonlinear numerical models for WEC arrays, viscous drag effects are important [61]. Moreover, the limited draft in combination with the flat bottom will result in important bottom slamming effects for tests where the WEC buoy re-enters the water after being lifted out [60].

4.3. Electromechanical Design

Figure 19 shows a 3D rendering of the final design of the "WECfarm" WEC. The used right-handed coordinate system has its origin at the intersection of the still water level (SWL) with the vertical axis through the centre of the WEC buoy. This allows one to express displacements of the WEC buoy relative to the SWL. The x-axis corresponds with the positive wave propagation direction. The y-axis follows from the motor sign convention: a positive torque results in a downward motion of the WEC buoy. Therefore, the z-axis is pointed downwards to define positive forces, displacements, velocities and accelerations. Figure 19 also indicates the three translations, surge (1), sway (2) and heave (3), and the three rotations, roll (4), pitch (5) and yaw (6).

To exclude friction in the linear guiding, air bushings were used. The surge excitation force, acting in the centre of buoyancy, results in a moment on the linear guiding system. This moment can be decoupled in normal forces by a configuration of three air bushings. The air bushings are characterised by a load versus pressure curve, where one 40 mm OAV (OAV Air Bearings, Princeton, NJ, USA) air bushing can cope with a maximum radial load of 720 N, for a nominal pressure of 5.5 bar [68]. A configuration of three OAV 40 mm air bushings guarantees a permanent layer of air between the guide shafts and the bushings for the most extreme wave conditions, resulting in zero-friction linear guiding on the condition of a proper alignment. The air bushings coefficient of friction is a function of air shear from motion, not from surface contact. Therefore, at zero velocity, there would be zero friction and making infinite motion resolution theoretically possible. A friction coefficient of 0.00008 in log-scale due to the contribution from air molecules and gravitation can be taken into account [68]. To obtain this air gap, a compressor with a two-stage air filter supplies these bushings with clean and dry air under a nominal pressure of 5.5 bar. The OAV 40 mm air bushings require a shaft's outside diameter of 40.00 mm +0.00/−0.02 mm (tolerance class g6). Therefore, MiSUMi (MiSUMi Europe, Frankfurt am Main, Germany) shafts with tolerance class g6 were custom configured.

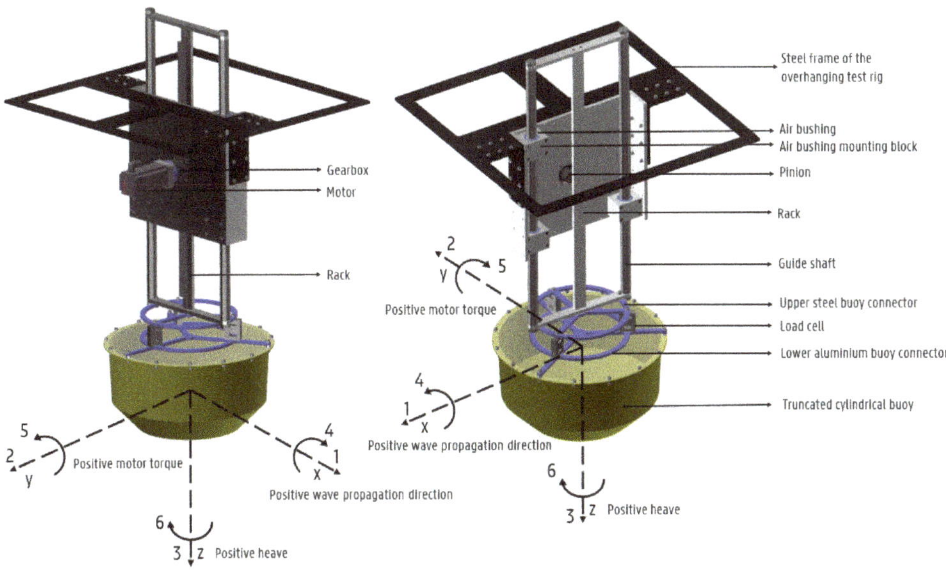

Figure 19. Rendering of the "WECfarm" WEC, made with Autodesk Inventor (Autodesk, San Rafael, CA, USA).

The WEC was mounted on a steel frame structure, designed to be rigid enough to avoid resonance at low frequencies. The total structure was designed for flexible (dis)assembly and installation in any facility. The steel frame was provided with M8 connection holes to allow mounting on MiniTec (MiniTec, Schönenberg-Kübelberg, Germany) profiles, or equivalent. This frame can be flipped 180° to install the WEC on the side walls of a wave flume or water basin.

The PTO system of the WEC was designed as a permanent magnet synchronous motor (PMSM) connected to a gearbox powering a rack and pinion system. A gearbox can offer a solution to deal with high forces at low velocities in wave energy applications. Bacelli et al. described that the Coulomb friction in the drivetrain attributed to the gearbox significantly degraded the quality of open-loop force control, necessitating the use of closed-loop control [59]. Therefore, a single-stage low-ratio transmission was desired. A Wittenstein (Wittenstein, Igersheim, Germany) single-stage gearbox "NPR 025S-MF1-4-2E1-1S" with a ratio i = 1:4 connects the pinion with the motor. The gearbox has a slotted-hole connection, allowing to slide the pinion in the rack to minimise backlash in the transmission. The pinion has 20 teeth and a pitch circle radius R_{pinion} of 0.0212205 m. Given the pinion helix angle of 19.5283°, the backlash can be at a maximum +−0.3 mm. The velocity on the pinion will be four times less than on the motor shaft, while the torque on the pinion will be four times more than on the motor shaft.

A PMSM was chosen since it can provide the bidirectional power flow necessary for reactive control, high-bandwidth open-loop force control and high-accuracy control and feedback (position, velocity and torque). The motor's torque–speed curves for a range of target test wave conditions (regular and irregular long-crested) were calculated with a MATLAB Simscape electromechanical system model coupled with the open-source code WEC-Sim for simulating WECs. This design procedure resulted in the selection of a Beckhoff (Beckhoff Automation, Verl, Germany) PMSM "AM8041-2D11-0000" with an inertia of 1.73 kg cm^2, a rated torque of 2.33 Nm and a rated speed of 1500 RPM, for a 230 V AC power supply [69]. The motor is equipped with a holding brake, mechanically operated by a solenoid, to make it possible to keep the WEC buoy lifted out of the water. When the motor is not operational, the mechanical brake is activated. The Beckhoff PMSM is powered and controlled by a Beckhoff motor drive. For the setup of five WECs, two Beckhoff motor drives type "AX5203-0000-0213" power each two motors and a third Beckhoff motor drive type "AX5103-0000-0212" powers the fifth WEC. The three drives will be daisy-chained with EtherCAT cables. The drive consumes the power generated by the WECs in the intermediate circuit. Therefore, no ballast resistor is needed.

4.4. DAQ and Control Platform for the Five-WEC Array

Figure 20 shows a scheme of the data acquisition and control flow for the "WECfarm" setup with five WECs, with a legend indicating the signal type. The MATLAB-Simulink real-time control model was built on the host PC and loaded on the Speedgoat (Speedgoat, Köniz, Switzerland) Performance real-time target machine via Ethernet communication. This target machine ran the Simulink model and processed the input/output (I/O) at a sample frequency of 1000 Hz. In this context, the high sample frequency corresponded with the defined "real-time" terminology. The target screen displayed diagnostic info of the execution process steps of the Simulink model. The sensor input for the Speedgoat target machine was realized with a Speedgoat IO133-Performance analogue and digital I/O module, which is a fast, simultaneous-sampling, 16-bit analogue input and output module with Simulink driver blocks. A scheme on the bottom of Figure 20 shows the sensor input for the Speedgoat IO133 terminal board for each of the five WECs. The accelerometer ADXL335 (Analog Devices, Norwood, MA, USA) was used to measure the acceleration of the WEC buoy in the heave direction and was attached on top of the rack, the furthest position on the WEC from the water. Three Althen (Althen Sensors & Controls BV, Rijswijk, The Netherlands) FLH3 50 kg load cells were placed between the hydrodynamic part (the buoy) and the electromechanical part (the motor) to measure the actual applied forces. These three load cells allowed the possibility to close the loop around the force. However, Bacelli et al. stressed that closing the loop around a force sensor may induce negative consequences for the design of higher-level control loops [59]. A configuration of at least three load cells was required to avoid torsion and bending influencing the measurements. A TLE analogue weight transmitter (Laumas Elettronica, Montechiarugolo, Italy) was used to amplify these three analogue signals and to sum them to one analogue signal. In case the WEC is locked, the wave heave excitation force can be measured. In case the motor is active, the load cells measure the PTO force. The upper microswitch and the lower microswitch were used as safety limit switches. It is necessary to limit the amplitude of the WEC buoy displacement to prevent the guiding system damaging the structure. The laser sensor Micro-Epsilon optoNCDT 1420 (Micro-Epsilon, Ortenburg, Germany) was installed on one WEC as a backup for the motor encoder to measure the displacement of the WEC buoy relative to the SWL. Moreover, the laser sensor could be used for displacement measurements for tests without a motor.

The torque request in the Simulink model is sent by the EtherCAT (Ethernet for Control Automation Technology) communication protocol to the Beckhoff motor drive as a master data telegram (MDT) process parameter. On the other hand, the Speedgoat target machine can receive by EtherCAT communication amplifier telegram (AT) process parameters from the Beckhoff motor drive. Normally, a Beckhoff motor drive is controlled by a Beckhoff's industrial PC, serving as a programmable logic controller (PLC). For the WECfarm project, it was decided to configure the Speedgoat target machine as a master and the Beckhoff drive as a slave. This allowed us to benefit from the control and postprocessing possibilities within the MATLAB-Simulink environment. To establish the communication between the Speedgoat target and the Beckhoff drive, the EtherCAT network needs to be initialised in the Simulink model with the EtherCAT initialisation block. This block requires to load an EtherCAT network information (ENI) file containing the motor topology and process parameters. To create this ENI file, a Beckhoff industrial PC was used. When this Beckhoff industrial PC is connected by an EtherCAT cable to the Beckhoff drive with motor, Beckhoff TwinCAT 3 software can scan the network to define the motor topology. The user can configure the operation modes and select the required MDT and AT process parameters. Consequently, the result of this process can be exported as an .xml ENI file to initialise the EtherCAT network in the Simulink model. On their turn, the Beckhoff PMSM input and output signals are processed by the Beckhoff motor drive by the "One Cable Technology", allowing to send commands and feedback using the same cable. The drive receives the 18-bit resolution position values from the single-turn absolute encoder. This encoder allows the real-time determination of the state (position and velocity) of the WEC buoy. The drive provides the motor with a certain current, corresponding with a torque multiplication with the torque constant of 1.43 Nm/A.

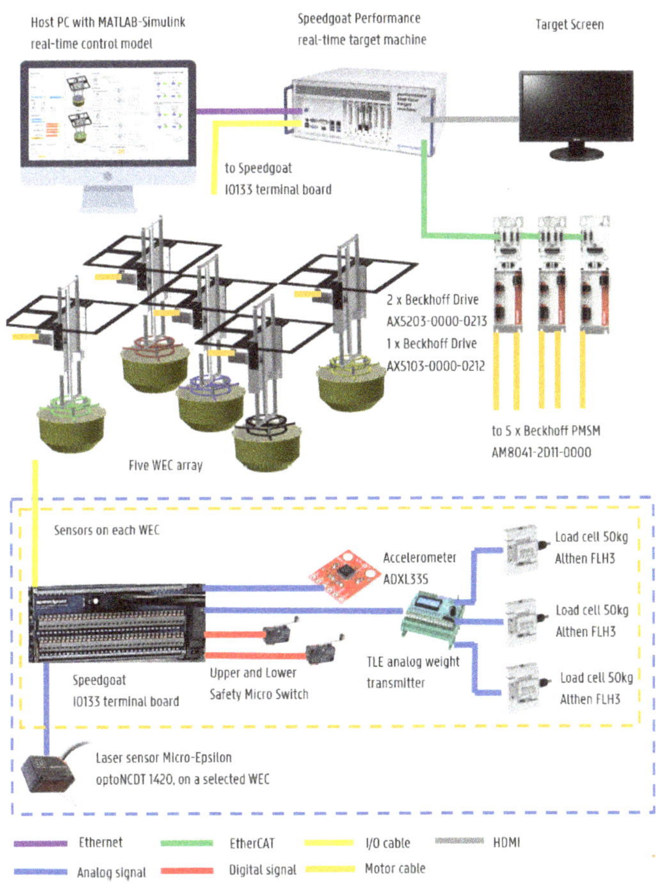

Figure 20. General data acquisition and control flow for the "WECfarm" five-WEC array.

4.5. The "WECfarm" WEC Array Approach

To obtain the experimental dataset on point-absorber WEC array tests, the "WECfarm" project consisted of the following consecutive steps, involving a gradual increase in complexity. In 2019, the "WECfarm" WEC was designed and numerically modelled. In 2020–2021, the first "WECfarm" WEC was constructed and tested in a small water basin (see Figure 21a) at the Coastal Engineering Research Group of Ghent University (Belgium) to evaluate all structural, mechanical, electronic, data acquisition and control aspects before deploying the WEC in a wave basin to evaluate the hydrodynamic performance.

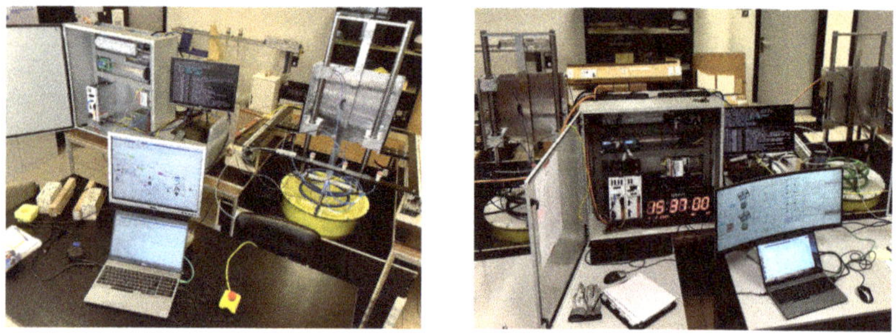

(**a**) (**b**)

Figure 21. "WECfarm" experimental setup at the Coastal Engineering Research Group of Ghent University (Belgium): (**a**) single WEC; (**b**) two WEC array.

In April 2021, this first "WECfarm" WEC was tested in the wave basin of Aalborg University (Denmark). Figure 22a shows a picture of the experimental setup. The hydrodynamical performance was assessed by the execution of system identification (SID) tests (radiation, excitation and free decay tests). The passive, reactive and latching controller were tested for a selection of regular and irregular long-crested waves.

In late 2021, an array of two "WECfarm" WECs was constructed and tested in two small water basins (see Figure 21b) at the Coastal Engineering Research Group of Ghent University (Belgium) to test the two-WEC array Simulink control model prior to wave basin testing.

In February 2022, an array of two "WECfarm" WECs was tested in the wave basin of Aalborg University (Denmark) to quantify wave–WEC and WEC–WEC interactions. Figure 22b shows a picture of the experimental setup. The WECs were placed in a row layout, with an axis-to-axis separation distance equal to 1.20 m, equivalent to 2 D. Based on SID radiation tests, causal impedance matching proportional (P) and proportional–integral (PI) controllers were designed [70,71]. These controllers, corresponding to resistive and reactive control, respectively, were tested for a selection of irregular long-crested waves. It should be mentioned that in the test campaign with a singe WEC, a Beckhoff PMSM "AM8542-2E11-0000" with an inertia of 6.17 kg cm^2 was used, updated to a Beckhoff PMSM "AM8041-2D11-0000" with an inertia of 1.73 kg cm^2 in the test campaign with the two-WEC array [69]. Theoretically, maximum accuracy with maximum dynamics can be reached when the ratio of system inertia to motor inertia is one. The motor for the first test campaign was designed to have this ratio lower than two, which was increased to ten based on the evaluation of the dynamics after the first test campaign. This lower motor inertia resulted in a higher natural frequency of the WEC.

Experimental testing of arrays of up to five "WECfarm" WECs is planned in 2023, in the new wave basin, the COB in Ostend (Belgium) [65]. This planned test campaign will deliver the main publicly available dataset for wave–WEC and WEC–WEC interactions. Moreover, the "WECfarm" WEC array experimental setup is meant to serve as a test-bed for future research, to deal with challenges defined within requirement 3 in Section 4.1.

(**a**) (**b**)

Figure 22. "WECfarm" experimental setup at the Aalborg University wave basin:(**a**) single WEC (April 2021); (**b**) two WEC array (February 2022).

5. Discussion and Conclusions

Experimental wave basin testing with arrays of two to five heaving point-absorber WECs within the "WECfarm" project aims to address the need for publicly available real-life data for the validation of WEC array numerical modelling and optimisation by establishing a dataset for the research community. This article provided a comprehensive literature review on experimental modelling of point-absorber WEC arrays, which allowed us to identify research gaps, to be addressed in the "WECfarm" project. These research gaps were reformulated as design requirements for the "WECfarm" experimental setup.

The designed truncated cylindrical "WECfarm" WEC buoy addressed the need for WEC array tests with a generic WEC concept on a suitable scale, resulting in data useful for a wide range of point-absorber concepts. The hydrodynamic design was optimised to enhance near-field interactions, based on literature findings and linear potential flow numerical simulations. The truncated cylindrical buoy had a draft of 0.16 m and a radius of 0.30 m. The mass of the hydrodynamic activated part was 36.83 kg. The RAO showed a resonance peak around 1.0 s. The high diameter-to-draft ratio of 3.75 resulted in a flat RAO response and high resonance bandwidth. A limited number of five closely spaced WECs was fundable and confirmed to be sufficient to study near-field interactions based on linear potential flow numerical simulations.

The WECs were equipped with a PMSM, addressing the need for WEC array tests with a highly accurate, real-time actively controllable PTO system with bidirectional power flow. The EtherCAT communication protocol allowed us to send motor commands and receive motor feedback at high resolution and sample rate, enabling advanced WEC control. This PTO system will overcome the shortcoming of uncertainties introduced by simplified PTO systems (e.g., friction brakes and hydraulic dampers) used within previously performed WEC array experimental campaigns. Moreover, the "WECfarm" WEC was designed with an air bushings linear guiding system to exclude guiding friction in the power absorption measurements.

The control with a Speedgoat Performance real-time target machine offers the possibility to implement GC and advanced WEC array control strategies in the MATLAB-Simulink model and test accordingly. This centralised control platform overcomes the limitation of former WEC array experimental setups where control could only be implemented at the level of individual WECs. Besides the motor feedback to determine the WEC position and velocity, each WEC was equipped with an accelerometer and a configuration of three load cells, providing accurate measurements.

Wave basin testing within the "WECfarm" project will cover a wide range of wave conditions, including regular waves, irregular long- and short-crested waves, focused waves and extreme wave conditions. This overcomes the shortcoming of previously performed WEC array experimental campaigns only considering small amplitude regular and irregular long-crested waves, and not meeting the current requirements to validate CNWTs with WEC arrays.

Within the "WECfarm" project, two test campaigns were performed at the Aalborg University wave basin: (a) a testing of the first WEC in April 2021 and (b) a testing of a two-WEC array in February 2022. An experimental campaign with arrays of up to five "WECfarm" WECs is planned in the COB in Ostend (Belgium) in 2023. The dataset of each test campaign will be made available to the research community for numerical model validation and WEC array optimisation purposes.

Author Contributions: Conceptualisation, methodology and analysis, T.V.; writing—original draft preparation, T.V.; electromechanical design of the WECfarm's WEC: T.V., B.D.B. and K.S.; writing—review and editing, V.S., M.V. and P.T.; funding acquisition for the WECfarm's experimental setup, V.S. and P.T. All authors have read and agreed to the published version of the manuscript.

Funding: This work is supported by the FWO (Fonds Wetenschappelijk Onderzoek - Research Foundation Flanders), Belgium, through the following funding: (1) Timothy Vervaet is a Ph.D. fellow (fellowship 11A6919N); (2) Vasiliki Stratigaki is a postdoctoral researcher (fellowship 1267321N) and has been granted the "FWO Research Grant" for constructing the WEC experimental setup (FWO-KAN-DPA376).

Institutional Review Board Statement: Not applicable.

Informed Consent Statement: Not applicable.

Data Availability Statement: The datasets resulting from the WECfarm project will be made available in due time at https://www.awww.ugent.be, accessed on 29 May 2022.

Acknowledgments: OAV (https://www.oavco.com/, accessed on 29 May 2022) is acknowledged for providing educational discount on the OAV 40 mm air bushings used in the experimental setup as well as for providing technical support [68]. Beckhoff, Wittenstein and Speedgoat are acknowledged for providing technical support.

Conflicts of Interest: The authors declare no conflict of interest. The funders had no role in the design of the study; in the collection, analyses, or interpretation of data; in the writing of the manuscript, or in the decision to publish the results.

Abbreviations

The following abbreviations are used in this manuscript:

ABS	Acrylonitrile butadiene styrene
AT	Amplifier telegram
BEM	Boundary element method
CFD	Computational fluid dynamics
CNWT	CFD-based numerical wave tank
COB	Coastal and Ocean Basin
DAQ	Data acquisition system
DOFs	Degrees of freedom
ENI	EtherCAT network information
EtherCAT	Ethernet for Control Automation Technology
FNT	Float-neck-tank
GC	Global control
IC	Independent control
I/O	Input and output
JONSWAP	Joint North Sea Wave Project
LTI	Linear time-invariant
MDT	Master data telegram
MPAWEC	Multipoint-absorber wave energy converter
MPC	Model predictive control
MWL	Mean water level
OWC	Oscillating water column
P	Proportional
PerAWaT	Performance Assessment of Wave and Tidal array systems

PI	Proportional–integral
PID	Proportional–integral–derivative
PLC	Programmable logic controller
PMDC	Permanent magnet direct current
PMSM	Permanent magnet synchronous motor
PTFE	Polytetrafluoroethylene
PTO	Power take-off
QUB	Queen's University Belfast
RAO	Response amplitude operator
RIMS	Research Institute of Small and Medium Shipbuilding
SID	System identification
SWL	Still water level
UCC	University College Cork
WEC	Wave energy converter

References

1. Falnes, J.; Lillebekken, P.M. Budal's latching-controlled-buoy type wave-power plant. In Proceedings of the 5th European Wave Energy Conference, Cork, Ireland, 17–20 September 2003; pp. 233–244.
2. Yang, S.H.; Ringsberg, J.W.; Johnson, E. Wave energy converters in array configurations—Influence of interaction effects on the power performance and fatigue of mooring lines. *Ocean. Eng.* **2020**, *211*, 107294. [CrossRef]
3. Budal, K. Theory for absorption of wave power by a system of interacting bodies. *J. Ship Res.* **1977**, *21*, 248–253. [CrossRef]
4. Xie, J.; Zuo, L. *Dynamics and Control of Ocean Wave Energy Converters*; Mechanical Engineering Department, State University of New York: Stony Brook, NY, USA, 2013.
5. Coe, R.G.; Bacelli, G.; Wilson, D.G.; Abdelkhalik, O.; Korde, U.A.; Robinett, R.D. A comparison of control strategies for wave energy converterss. *Int. J. Mar. Energy* **2017**, *20*, 45–63. [CrossRef]
6. Garcia-Rosa, P.B.; Bacelli, G.; Ringwood, J.V. Control-Informed Optimal Array Layout for Wave Farms. *IEEE Trans. Sustain. Energy* **2011**, *6*, 575–582. [CrossRef]
7. Göteman, M.; Giassi, M.; Engström, J.; Isberg, J. Advances and Challenges in Wave Energy Park Optimization—A Review. *Front. Energy Res.* **2020**, *8*, 26. [CrossRef]
8. Giassi, M.; Thomas, S.; Shahroozi, Z.; Engström, J.; Isberg, J.; Tosdevin, T.; Hann, M.; Göteman, M. Preliminary results from a scaled test of arrays of point-absorbers with 6 DOF. In Proceedings of the 13th EuropeanWave and Tidal Conference (EWTEC), Napoli, Italy, 1–6 September 2019.
9. Nader, J.R.; Fleming, A.; Macfarlane, G.; Penesis, I.; Manasseh, R. Novel experimental modelling of the hydrodynamic interactions of arrays of wave energy converters. *Int. J. Mar. Energy* **2017**, *20*, 109–124. [CrossRef]
10. Stratigaki, V.; Troch, P.; Stallard, T.; Forehand, D.; Kofoed, J.P.; Folley, M.; Benoit, M.; Babarit, A.; Kirkegaard, J. Wave basin experiments with large wave energy converter arrays to study interactions between the converters and effects on other users in the sea and the coastal area. *Energies* **2014**, *7*, 701–734. [CrossRef]
11. Stratigaki, V.; Troch, P.; Stallard, T.; Forehand, D.; Folley, M.; Kofoed, J.P.; Benoit, M.; Babarit, A.; Vantorre, M.; Kirkegaard, J. Sea-state modification and heaving float interaction factors from physical modelling of arrays of wave energy converters. *J. Renew. Sustain. Energy* **2015**, *7*, 061705. [CrossRef]
12. Stratigaki, V. Experimental Study and Numerical Modelling of Intra-Array Interactions and Extra-Array effects of Wave Energy Converter Arrays. Ph.D. Thesis, Ghent University, Faculty of Engineering and Architecture, Ghent, Belgium, 2014.
13. WECwakes. Large Scale Experiments on Wave Energy Converter Farms to Study the Near-Field Effects between the Converters and the Far-Field Effects on Other Users in the Coastal Area. 2014. Available online: https://hydralab.eu/research--results/ta-projects/project/16/ (accessed on 20 May 2022).
14. Balitsky, P.; Verao Fernandez, G.; Stratigaki, V.; Troch, P. Assessment of the power output of a two array clustered WEC farm using a BEM solver coupling and a wave propagation model. *Energies* **2018**, *11*, 2907. [CrossRef]
15. Devolder, B.; Stratigaki, V.; Troch, P.; Rauwoens, P. CFD simulations of floating point absorber wave energy converter arrays subjected to regular waves. *Energies* **2018**, *11*, 641. [CrossRef]
16. Verao Fernandez, G.; Stratigaki, V.; Troch, P. Irregular wave validation of a coupling methodology for numerical modelling of near and far field effects of wave energy converter arrays. *Energies* **2019**, *12*, 538. [CrossRef]
17. Verbrugghe, T.; Stratigaki, V.; Altomare, C.; Domínguez, J.; Troch, P.; Kortenhaus, A. Implementation of open boundaries within a two way coupled SPH model to simulate nonlinear wave structure interactions. *Energies* **2019**, *12*, 697. [CrossRef]
18. Sheng, W.; Tapoglou, E.; Ma, X.; Taylor, C.J.; Dorrell, R.M.; Parsons, D.R.; Aggidis, G. Hydrodynamic studies of floating structures: Comparison of wave-structure interaction modelling. *Ocean. Eng.* **2022**, *249*, 110878. [CrossRef]
19. Sheng, W.; Tapoglou, E.; Ma, X.; Taylor, C.J.; Dorrell, R.M.; Parsons, D.R.; Aggidis, G. Time-Domain Implementation and Analyses of Multi-Motion Modes of Floating Structures. *J. Mar. Sci. Eng.* **2022**, *10*, 662. [CrossRef]
20. Falnes, J. *Ocean Waves and Oscillating systems: Linear Interactions Including Wave-Energy Extraction*; Cambridge University Press: Cambridge, UK, 2004.

21. Falnes, J.; Hals, J. Heaving buoys, point absorbers and arrays. *Philos. Trans. Math. Phys. Eng. Sci.* **2012**, *370*, 246–277. [CrossRef]
22. Sheng, W.; Alcorn, R.; Lewis, T. Physical modelling of wave energy converters. *Ocean. Eng.* **2014**, *84*, 29–36. [CrossRef]
23. Gaebele, D.T.; Magana, M.E.; Brekken, T.K.A.; Sawodny, O. State space model of an array of oscillating water column wave energy converters with inter-body hydrodynamic coupling. *Ocean. Eng.* **2020**, *195*, 106668. [CrossRef]
24. Kelly, T.; Dooley, T.; Campbell, J.; Ringwood, J. Modelling and Results for an Array of 32 Oscillating Water Columns. *Ewtec Proc.* **2013**. Available online: http://eprints.maynoothuniversity.ie/6788/ (accessed on 29 May 2022).
25. Lamont-Kane, P. Physical and Numerical Modelling of Wave Energy Converter Arrays. Ph.D. Thesis, Queen's University Belfast, Faculty of Engineering and Physical Sciences, Belfast, UK, 2015.
26. Doyle, S.; Aggidis, G.A. Experimental investigation and performance comparison of a 1 single OWC, array and M-OWC. *Renew. Energy* **2020**, *168*, 365–374. [CrossRef]
27. Doyle, S. Experimental and Numerical Investigation into the Energy Conversion Process in Single and Multi Chamber Oscillating Water Column Wave Energy Converters. Ph.D. Thesis, Lancaster University, Lancaster, UK, 2021.
28. Budal, K.; Falnes, J.; Kyllingstad, A.; Oltedal, G. Experiments with point absorbers. In Proceedings of the First Symposium on Wave Energy Utilization, Gothenburg, Sweden, 30 October–1 November 1979; pp. 253–282.
29. Count, B.M.; Jefferys, E.R. Wave power, the primary interface. In Proceedings of the 13th Symp. Naval Hydrodynamics, Tokyo, Japan, 6–10 October 1980; pp. 1–10.
30. Stallard, T.; Stansby, P.K.; Williamson, A.J. An experimental study of closely spaced point absorber arrays. In Proceedings of the International Offshore and Polar Engineering Conference, Vancouver, BC, Canada, 6–11 July 2008.
31. Thomas, S.; Weller, S.; Stallard, T. Float response within an array: Numerical and experimental comparison. In Proceedings of the 2nd International Conference on Ocean Energy (ICOE), Brest, France, 15–17 October 2008; Volume 1517.
32. Alexandre, A.; Stallard, T.; Stansby, P.K. Transformation of wave spectra across a line of wave devices. In Proceedings of the 8th European Wave and Tidal Energy Conference (EWTEC), Uppsala, Sweden, 7–10 September 2009.
33. Weller, S.D.; Stallard, T.; Stansby, P. Experimental measurements of irregular wave interaction factors in closely spaced arrays. *IET Renew. Power Gener.* **2010**, *4*, 628–637. [CrossRef]
34. Weller, S.D. Wave Energy Extraction from Device Arrays: Experimental Investigation in a Large Wave Facility. Ph.D. Thesis, University of Manchester, Faculty of Engineering and Physical Sciences, Manchester, UK, 2010.
35. Manchester University. Manchester Bobber Array. 2008. Available online: https://personalpages.manchester.ac.uk/staff/robert.j.brown/Default_files/Manchester_Bobber.htm (accessed on 12 August 2021).
36. Child, B.; Laporte Weywada, P. Verification and validation of a wave farm planning tool. In Proceedings of the 10th European Wave and Tidal Energy Conference Series-EWTEC, Aalborg, Denmark, 2–5 September 2013.
37. Folley, M.; Whittaker, T. Preliminary Cross-Validation of Wave Energy Converter Array Interactions. In Proceedings of the International Conference on Offshore Mechanics and Arctic Engineering–OMAE, Nantes, France, 9–14 June 2013; Volume 10837.
38. Lamont-Kane, P.; Folley, M.; Whittaker, T. Investigating Uncertainties in Physical Testing of Wave Energy Converter Arrays. In Proceedings of the 10th European Wave and Tidal Energy Conference Series-EWTEC, Aalborg, Denmark, 2–5 September 2013.
39. Mackay, E.; Cruz, J.; Livingstone, M.; Arnold, P. Validation of a Time-Domain Modelling Tool for Wave Energy Converter Arrays. In Proceedings of the 10th European Wave and Tidal Energy Conference, Aalborg, Denmark, 2–5 September 2013.
40. Ruiz, P.M.; Ferri, F.; Kofoed, J.P. Experimental Validation of a Wave Energy Converter Array Hydrodynamics Tool. *Sustainability* **2017**, *9*, 115 . [CrossRef]
41. Wave Star A/S. Wavestar. 2013. Available online: http://wavestarenergy.com/ (accessed on 15 June 2022).
42. Boere, R.; Goudswaard, R.; Schneider, T.; van Vlijmen, B. Interaction of Ocean Wave Energy Converters. Bachelor Thesis, TU Delft, Delft, The Netherlands, 2018.
43. Do, H.T.; Dang, T.D.; Ahn, K.K. A multi-point-absorber wave-energy converter for the stabilization of output power. *Ocean. Eng.* **2018**, *161*, 337–349. [CrossRef]
44. Binh, P.C.; Tri, N.M.; Dung, D.T.; Ahn, K.K.; Kim, S.J.; Koo, W. Analysis, design and experiment investigation of a novel wave energy converter. *IET Gener. Transm. Distrib.* **2015**, *10*, 460–469. [CrossRef]
45. Giassi, M.; Engström, J.; Isberg, J.; Göteman, M. Comparison of Wave Energy Park Layouts by Experimental and Numerical Methods. *J. Mar. Sci. Eng.* **2020**, *8*, 750. [CrossRef]
46. Thomas, S.; Giassi, M.; Göteman, M.; Hann, M.; Ransley, E.; Isberg, J.; Engström, J. Performance of a Direct-Driven Wave Energy Point Absorber with High Inertia Rotatory Power Take-off. *Energies* **2018**, *11*, 2332. [CrossRef]
47. Moreno, E.C.; Stansby, P. The 6-float wave energy converter M4: Ocean basin tests giving capture width, response and energy yield for several sites. *Renew. Sustain. Energy Rev.* **2019**, *104*, 307–318. [CrossRef]
48. Stansby, P.; Moreno, E.C.; Stallard, T.; Maggi, A. Three-float broad-band resonant line absorber with surge for wave energy conversion. *Renew. Energy* **2015**, *78*, 132–140. [CrossRef]
49. Stansby, P.; Moreno, E.C.; Stallard, T. Capture width of the three-float multi-mode multi-resonance broadband wave energy line absorber M4 from laboratory studies with irregular waves of different spectral shape and directional spread. *J. Ocean Eng. Mar. Energy* **2015**, *1*, 287–298. [CrossRef]
50. Santo, H.; Taylor, P.H.; Stansby, P.K. The performance of the three-float M4 wave energy converter off Albany, on the south coast of western Australia, compared to Orkney (EMEC) in the U.K. *Renew. Energy* **2020**, *146*, 444–459. [CrossRef]

51. Liao, Z.; Stansby, P.; Li, G.; Moreno, E.C. High-Capacity Wave Energy Conversion by Multi-Float, Multi-PTO, Control and Prediction: Generalized State-Space Modelling With Linear Optimal Control and Arbitrary Headings. *IEEE Trans. Sustain. Energy* **2021**, *12*, 2123–2131. [CrossRef]
52. Stansby, P.; Moreno, E.C.; Stallard, T. Large capacity multi-float configurations for the wave energy converter M4 using a time-domain linear diffraction model. *Appl. Ocean. Res.* **2017**, *68*, 53–64. [CrossRef]
53. Kamarlouei, M.; Gaspar, J.F.; Calvario, M.; Hallak, T.S.; Mendes, M.J.G.C.; Thiebaut, F.; Guedes Soares, C. Experimental analysis of wave energy converters concentrically attached on a floating offshore platform. *Renew. Energy* **2020**, *152*, 1171–1185. [CrossRef]
54. Sun, K.; Yang, Y.; Zheng, X.; Cui, L.; Zhao, C.; Liu, M.; Rao, X. Experimental investigation of semi-submersible platform combined with point-absorber array. *Energy Convers. Manag.* **2021**, *245*, 114623. [CrossRef]
55. Principle Power. WindFloat. 2022. Available online: https://www.principlepower.com/windfloat (accessed on 15 June 2022).
56. Beatty, S.; Ferri, F.; Bocking, B.; Kofoed, J.P.; Buckham, B. Power Take-Off Simulation for Scale Model Testing of Wave Energy Converters. *Energies* **2017**, *10*, 973. [CrossRef]
57. Windt, C.; Davidson, J.; Ringwood, J. High-fidelity numerical modelling of ocean wave energy systems: A review of computational fluid dynamics-based numerical wave tanks. *Renew. Sustain. Energy Rev.* **2018**, *93*, 610–630. [CrossRef]
58. García-Violini, D.; Peña Sanchez, Y.; Faedo, N.; Windt, C.; Ferri, F.; Ringwood, J.V. Experimental implementation and validation of a broadband LTI energy-maximising control strategy for the Wavestar device. *IEEE Trans. Control. Syst. Technol.* **2020**, *29*, 2609–2621. [CrossRef]
59. Bacelli, G.; Ringwood, J. Constrained control of arrays of wave energy devices. *Int. J. Mar. Energy* **2013**, *3–4*, 53–69. [CrossRef]
60. De Backer, G.; Vantorre, M.; Frigaard, P.; Beels, C.; De Rouck, J. Bottom slamming on heaving point absorber wave energy devices. *J. Mar. Sci. Technol.* **2010**, *15*, 119–130. [CrossRef]
61. Giorgi, G.; Ringwood, J.V. Nonlinear Froude-Krylov and viscous drag representations for wave energy converters in the computation/fidelity continuum. *Ocean. Eng.* **2017**, *141*, 164–175. [CrossRef]
62. Child, B.; Cruz, J.; Livingstone, M. The development of a tool for optimising of arrays of wave energy converters. In Proceedings of the 9th European Wave and Tidal Energy Conference, Southampton, UK, 5–9 September 2011.
63. Shadman, M.; Estefen, S.F.; Rodriguez, C.A.; Nogueira, I.C.M. A geometrical optimization method applied to a heaving point absorber wave energy converter. *Renew. Energy* **2017**, *115*, 533–546. [CrossRef]
64. Bacelli, G.; Spencer, S.J.; Patterson, D.C.; Coe, R.G. Wave tank and bench-top control testing of a wave energy converter. *Appl. Ocean. Res.* **2019**, *86*, 351–366. [CrossRef]
65. Troch, P.; Stratigaki, V.; Devriese, P.; Kortenhaus, A.; De Maeyer, J.; Monballiu, J.; Toorman, E.; Rawoens, P.; Vanneste, D.; Suzuki, T.; et al. Design Features of the Upcoming Coastal and Ocean Basin in Ostend, Belgium. 2018. Available online: https://icce-ojs-tamu.tdl.org/icce/index.php/icce/article/view/8721 (accessed on 29 May 2022).
66. Cargo, C.J.; Plummer, A.R.; Hillis, A.J.; Schlotter, M. Determination of optimal parameters for a hydraulic power take-off unit of a wave energy converter in regular waves. *Proc. IMechE Part A J. Power Energy* **2011**, *226*, 98–111. [CrossRef]
67. Verbrugghe, T. openWEC: Open Source Wave Energy Converter (WEC) Simulation Tool. Ghent University. 2018. Available online: https://users.ugent.be/~tverbrug/ (accessed on 2 June 2022).
68. OAV. OAV Air Bearing Product Book & Design Guide: Where Aerospace Techonology Meets Air Bearing Systems. 2021. Available online: https://www.oavco.com/brochures (accessed on 7 July 2021).
69. Beckhoff Automation. Beckhoff New Automation Technology, Operation Instructions, AM8000 and AM8500: Synchronous Servomotors, Version 4.9. 2022. Available online: https://download.beckhoff.com/download/document/motion/am8000-am8500-ba-en.pdf (accessed on 24 June 2022).
70. Coe, R.G.; Bacelli, G.; Forbush, D. A practical approach to wave energy modeling and control. *Renew. Sustain. Energy Rev.* **2021**, *142*, 110791. [CrossRef]
71. Faedo, N.; Carapellese, F.; Pasta, E.; Mattiazzo, G. On the principle of impedance-matching for underactuated wave energy harvesting systems. *Appl. Ocean. Res.* **2022**, *118*, 102958. [CrossRef]

Journal of
Marine Science and Engineering

MDPI

Article

Predictive Control for a Wave-Energy Converter Array Based on an Interconnected Model

Bo Zhang [1,2], Haixu Zhang [1,2], Sheng Yang [1,2], Shiyu Chen [1,2], Xiaoshan Bai [1,2,*] and Awais Khan [1,2]

[1] College of Mechatronics and Control Engineering, Shenzhen University, Shenzhen 518060, China; zhangbo@szu.edu.cn (B.Z.); 2110296047@email.szu.edu.cn (H.Z.); yangsheng.yang@lazada.com (S.Y.); rebeccacsy23@hotmail.com (S.C.); awaiskhan@szu.edu.cn (A.K.)

[2] Shenzhen City Joint Laboratory of Autonomous Unmanned Systems and Intelligent Manipulation, Shenzhen University, Shenzhen 518060, China

* Correspondence: baixiaoshan@szu.edu.cn

Abstract: This paper proposes a model predictive control (MPC) method based on an interconnected model to maximize the ocean wave energy extracted by a wave-energy converter (WEC) array. In the proposed method, a formally uniform interconnected model is applied to represent the dynamics of an array consisting of an arbitrary quantity of WECs, simultaneously considering the hydrodynamic interaction among all the WEC devices. First, the WEC devices and their hydrodynamic interaction are represented in an interconnected model that describes the networked dynamics of a variety of WEC arrays with distinct spatial geometry layout of the WEC devices deployed in the sea field. Second, based on the presented model, an MPC method is applied to achieve the coordinated control of the WEC array to improve its energy conversion efficiency under the constraints of buoy position and control force. Third, a hardware-in-the-loop (HIL) platform is developed to simulate the WEC array's physical operating conditions, and the proposed method is implemented on the platform to test its performance. The test results show that the proposed MPC method using the interconnected model has a higher energy harvesting efficiency than the classic MPC method.

Keywords: wave-energy converter array; interconnected model; model predictive control; hardware-in-the-loop

Citation: Zhang, B.; Zhang, H.; Yang, S.; Chen, S.; Bai, X.; Khan, A. Predictive Control for a Wave-Energy Converter Array Based on an Interconnected Model. *J. Mar. Sci. Eng.* **2022**, *10*, 1033. https://doi.org/10.3390/jmse10081033

Academic Editors: Liliana Rusu and Vicky Stratigaki

Received: 9 June 2022
Accepted: 22 July 2022
Published: 27 July 2022

1. Introduction

Using ocean wave energy is a novel approach to solving the problem of resource depletion and environmental damage associated with traditional energy sources. However, the exploitation of ocean wave energy is not as well commercialized as other sustainable energy sources, such as solar and wind energy. One of the key reasons is that ocean wave energy is distributed homogeneously on the ocean surface, which makes it difficult for any wave-energy converter (WEC) to harvest a great deal of wave energy by operating a single machine over a small range. Using an array system composed of multiple point-absorber-type WECs can expand the range of wave energy harvesting options. Another advantage of WEC arrays is that they effectively reduce and eliminate power fluctuations, so that a smoother and more stable power supply for power grids and power consumers can be provided [1]. In recent years, WEC arrays have been intensively investigated due to their advantages [2].

The existence of hydrodynamic interactions among WEC devices has a significant impact on the performance of a WEC array. In [3], a wave interaction theory, including the diffraction interaction of evanescent waves, was presented to enable the calculation of the wave hydrodynamics of a multimember structure with arbitrary configurations and spacings. The effect of interactions within basic two-unit WEC arrays with different separating distances was studied in [4]. Subsequently, the influence of interactions among bodies on the overall yearly energy production of an array was assessed [5]. A CFD motion solver based on a coupled model was developed in [6] to analyze the interactions ocurring in a WEC array. Due to diffracted and radiated waves, interactions among the WECs can occur which may be constructive or destructive [7]. If such hydrodynamic interactions are effectively utilized, they have the potential to increase the energy production efficiency dramatically.

Optimizing the geometry layout allows for the effective use of constructive interaction effects. A topological method was presented in [7] to enable the design of an optimized single-row wave power plant established by a triangle array unit. An improved differential evolution algorithm was proposed in [8] to optimize the layout of a WEC array for faster speed of convergence and improved accuracy. A genetic algorithm that included a cost model was proposed in [9] to enable consideration of economic factors. For the example introduced in the paper, the distance between the devices is reduced as much as possible to diminish the maintenance cost. However, to be more efficient for wave energy harvesting in particular sea environments, it is necessary to specify an optimized spatial geometry layout of a WEC array corresponding to the ocean wave features and conditions in the area in which it is located. Therefore, a uniform modeling approach that can represent arbitrary layouts of WEC arrays is crucial for the layout optimization of WEC arrays.

To maximize the wave energy harvesting of wave energy by a WEC, efficient control of the array is necessary. A comparison between MPC and causal control was made from the perspective of performance and constraint handling in [10], with MPC being found to have significant advantages over optimal causal control. Since MPC has shown remarkable control performance when dealing with the constraints in realistic control processes, it has received increasing attention in recent years [11]. In the early literature, researchers focused mainly on the application of MPC in the control of a single WEC. A real-time non-linear model predictive controller (NMPC) was utilized to increase the WEC efficiency of a two-degrees-of-freedom WEC with highly non-linear power take-off (PTO) characteristics in [12]. An economic model predictive control (EMPC) strategy with a general economic cost function that directly reflects the economic objective of the system was proposed in [13] to maximize the energy extracted from ocean waves and to minimize the operational cost. Two MPC control strategies for a two-body point absorber were designed to extract the maximum energy from waves through unidirectional power flow in [14]. A non-quadratic piecewise discontinuous functional was applied as a cost index of NMPC to maximize the energy produced by a WEC in [15]. Various MPC strategies based on a single-unit model were successfully applied to maximize the wave energy harvested by WECs [16]. MPC methods based on WEC arrays have also attracted the attention of researchers. A model predictive control method was developed in [17], where each MPC effectively optimizes the energy absorption of its own WEC under state constraints. Nevertheless, because of the non-negligible interference generated by the movement of unit buoys, if the WEC units in the WEC array are closely spaced, implementing an MPC with an isolated unit model can result in degraded MPC control performance caused by model mismatch. Therefore, the proposed model not only takes into account the individual devices, but also the complex coupling relationships between the WEC devices, represented as hydrodynamic interactions among each pair of WEC devices. Centralized MPCs were described in [18] which led to a potential 10% improvement in converted energy. These studies investigated the impact of interference between WECs operating close to each other, which can increase wave energy harvesting efficiency. The majority of the studies reported derive sets of equations representing interference relationships directly from a particular geometry (e.g., equilateral triangles) [19]. It is challenging to directly extend the methods discussed to the modeling of WEC arrays with diverse spatial geometry layouts.

Since hydrodynamic interactions vary with local wave conditions in the ocean, it is unrealistic for a fixed model to maintain accuracy all the time. Therefore, in future commercial systems, not only does the system's layout need to be regularly adjusted, but the model also needs to be regularly modified and adjusted to ensure the performance of MPC. This paper proposes a modeling method to rapidly adapt to arbitrary geometric layouts of a WEC array by considering a WEC array with coupled interconnection relationships. The model can be broken down into two parts: the dynamics of the WEC unit and the dynamics of the interfering effects of the coupled interconnection relationships. Generalization of the model for the WEC array can thus be addressed through the modeling method. This method reduces modeling difficulty and improves MPC survival adaptability for practical applications. We design an MPC strategy to control a WEC array to yield a globally optimal wave energy capture efficiency despite interference effects occurring between the WEC units.

There are two key contributions of this paper: First, the WEC array is depicted as an interconnected system model, i.e., a network model with dynamic edges, in which the node dynamics are the dynamics of the WEC units and the edge dynamics are the dynamics of the interfering waves formed between neighboring units. Second, underpinned by an interconnected system model, an MPC method is developed to control each WEC unit in a WEC array to maximize wave energy harvesting.

After reviewing related research in Section 1, an interconnected system model for WEC arrays is developed in Section 2. An MPC strategy for a generalized network model is presented in Section 3. In Section 4, simulations and experiments conducted are described for validating the proposed modeling and MPC methods, respectively. The results are evaluated and discussed in Section 4.2. The conclusions are presented in Section 5.

2. Model of WEC Array

2.1. Model of WEC Unit

A point-absorber-type WEC is primarily composed of a buoy and a PTO unit, as shown in Figure 1a. The WEC array is represented as an interconnected system model in which the topology structure describes the interfering relationship between all WEC units, as shown in Figure 1b. The symbol ∂ in Figure 1b is the angle of the wave propagation direction, and $\ell_1, \ldots, \ell_6$ denote the directed interfering relationship between each pair of the three units, Unit 1, Unit 2 and Unit 3. Buoys are used to harvest the kinetic energy of ocean waves, and PTO converts the kinetic energy of the buoy into electric energy. Meanwhile, PTO produces electric power and a damping force, which affects the motion of the buoy simultaneously excited by ocean waves.

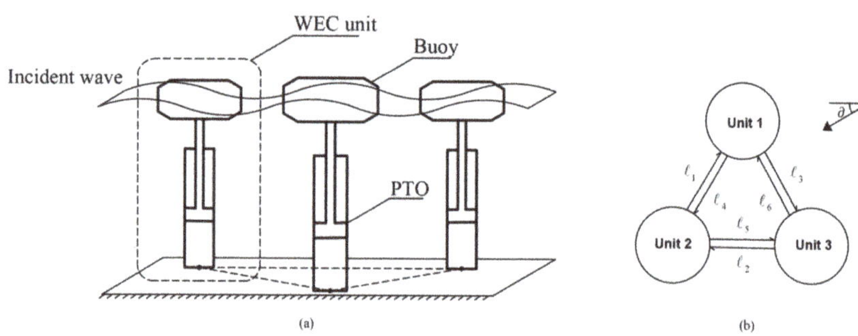

Figure 1. Point absorption WEC array: (**a**) the structure of the point-absorber-type WEC; (**b**) the interfering relationship between three units.

The dynamics of a point-absorber WEC are expressed as

$$m_f \ddot{z}(t) = F_w(t) + F_m(t), \tag{1}$$

where m_f is the mass of the buoy, $z(t)$ is the vertical displacement of the buoy, $F_w(t)$ is the hydrodynamic force, and $F_m(t)$ is the mechanical force.

The variables $F_w(t)$ and $F_m(t)$ satisfy

$$\begin{cases} F_w(t) = F_h(t) + F_r(t) + F_e(t), \\ F_m(t) = F_o(t) + F_u(t), \end{cases} \tag{2}$$

where $F_h(t)$, $F_r(t)$, $F_e(t)$, $F_o(t)$ and $F_u(t)$ are the buoyancy, radiation force, excitation force, friction force of the mechanical device, and reaction force, respectively. The buoyancy $F_h(t)$ is

$$F_h(t) = S_h[\mathrm{w}(t) - z(t)], \tag{3}$$

where $\mathrm{w}(t)$ and S_h are the vertical height of the incident wave and constant buoyancy coefficient, respectively. The radiation force $F_r(t)$ is

$$F_r(t) = F_D(t) - m_\infty \ddot{z}(t), \tag{4}$$

where m_∞ denotes the additional mass related to the quantity of the water being moved and is used as the coefficient of the linear term of the radiation force in the case of infinite frequency. Inspired by [20], the convolution term $F_D(t)$ is approximated as $D \cdot (\dot{\mathrm{w}}(t) - \dot{z}(t))$, where D is a constant damping coefficient to roughly approximate the Fourier transform $\widehat{D}(jw)$ of $F_D(t)$. $\widehat{D}(jw)$ can be regarded as a frequency-dependent damping coefficient.

The excitation force $F_e(t)$ is a non-causal force that is approximated by a fifth-order state-space representation of the relationship between the wave height and excitation force [21]. Since the influence of $F_o(t)$ is relatively slight, we assume $F_o(t) = 0$ [22]. The proposed control algorithm calculates the reaction force $F_u(t)$.

Let $m = m_f + m_\infty$ and $\xi(t) = [F_h(t), \dot{z}(t)]^T$. Then, the WEC unit model is achieved as

$$\begin{aligned} \dot{\xi}(t) &= A\xi(t) + B_w\dot{\mathrm{w}}(t) + B[F_u(t) + F_e(t)], \\ y(t) &= C\xi(t), \\ y_z(t) &= C_z\xi(t), \end{aligned} \tag{5}$$

where

$$A = \begin{bmatrix} 0 & -S_h \\ -\frac{1}{m} & -\frac{D}{m} \end{bmatrix},\ B_w = \begin{bmatrix} S_h \\ \frac{D}{m} \end{bmatrix},\ B = \begin{bmatrix} 0 \\ \frac{1}{m} \end{bmatrix},\ C = \begin{bmatrix} 0 & 1 \end{bmatrix}, \text{ and } C_z = \begin{bmatrix} 1 & 0 \end{bmatrix}.$$

Remark 1. *According to (5), it holds that $y(t) = \dot{z}(t)$, and $y_z(t) = F_h(t)$.*

2.2. WEC Array

2.2.1. WEC Units in Array

When the WEC units are close together, their interactions are enhanced, thus affecting the movement of the buoys more significantly. Radiation waves are the main source of interactions and therefore cannot be ignored. Considering the interaction effect, the i-th ($i = 1, \ldots, N$) WEC unit receives the cumulative effect of the radiation wave velocity $\dot{\mathrm{w}}_{ij}$ generated by the other j-th ($j = 1, \ldots, N-1, j \neq i$) WEC units, in addition to the effect of the incident wave velocity $\dot{\mathrm{w}}_i(t)$ [23]. The model of the i-th WEC unit is formulated as follows,

$$\begin{aligned} \dot{\xi}_i(t) &= A_i\xi_i(t) + B_{wi}\left[\dot{\mathrm{w}}_i(t) + \sum_{j\neq i}^{N-1} \dot{\mathrm{w}}_{ij}(t)\right] + B_i(F_{ui}(t) + F_{ei}(t)), \\ y_i(t) &= C_i\xi_i(t), \\ y_{zi}(t) &= C_{zi}\xi_i(t), \end{aligned} \tag{6}$$

where $\xi_i(t) = [F_{hi}(t), \dot{z}_i(t)]^T$ is the state of the i-th WEC unit, $F_{hi}(t)$ is the buoyancy of the i-th unit, $\dot{z}_i(t)$ is the velocity of the i-th buoy. In this model, the input to the WEC unit is the total of the velocity of the incident wave and the radiated waves generated by the other WEC units.

Moreover, the array composed of N WEC units is considered as a continuous interconnected system modeled as follows:

$$\begin{aligned}\dot{\tilde{\xi}}(t) &= \mathbb{A}_c\tilde{\xi}(t) + \mathbb{B}_{wc}\dot{\tilde{\mathrm{w}}}(t) + \mathbb{B}_{wc}\dot{\tilde{\mathrm{w}}}_{\mathrm{ij}}(t) + \mathbb{B}_c\left[\tilde{F}_u(t) + \tilde{F}_e(t)\right], \\ \tilde{y}(t) &= \mathbb{C}_c\tilde{\xi}(t), \\ \tilde{y}_z(t) &= \mathbb{C}_{zc}\tilde{\xi}(t),\end{aligned} \tag{7}$$

where

$$\tilde{\xi}(t) = \begin{bmatrix}\xi_1\\ \xi_2\\ \vdots\\ \xi_N\end{bmatrix},\ \dot{\tilde{\mathrm{w}}}(t) = \begin{bmatrix}\dot{\mathrm{w}}_1\\ \dot{\mathrm{w}}_2\\ \vdots\\ \dot{\mathrm{w}}_N\end{bmatrix},\ \dot{\tilde{\mathrm{w}}}_{\mathrm{ij}} = \begin{bmatrix}\sum_{j\neq 1}^{N-1}\dot{\mathrm{w}}_{1j}\\ \sum_{j\neq 2}^{N-1}\dot{\mathrm{w}}_{2j}\\ \vdots\\ \sum_{j\neq N}^{N-1}\dot{\mathrm{w}}_{Nj}\end{bmatrix},\ \tilde{F}_u(t) = \begin{bmatrix}F_{u1}\\ F_{u2}\\ \vdots\\ F_{uN}\end{bmatrix},$$

$$\tilde{F}_e(t) = \begin{bmatrix}F_{e1}\\ F_{e2}\\ \vdots\\ F_{eN}\end{bmatrix},\ \tilde{y}(t) = \begin{bmatrix}y_1\\ y_2\\ \vdots\\ y_N\end{bmatrix},\ \tilde{y}_z(t) = \begin{bmatrix}y_{z1}\\ y_{z2}\\ \vdots\\ y_{zN}\end{bmatrix},$$

$$\mathbb{A}_c = \mathbb{I}_N \otimes A,\ \mathbb{B}_{wc} = \mathbb{I}_N \otimes B_w,\ \mathbb{B}_c = \mathbb{I}_N \otimes B,\ \mathbb{C}_c = \mathbb{I}_N \otimes C,\ \mathbb{C}_{zc} = \mathbb{I}_N \otimes C_z,$$

and the subscript c indicates the term of continuous system. $\mathbb{I}$ is an identity matrix.

Remark 2. *In this paper, a system connecting several independent WEC units with a particular network structure is called an interconnected system, where the WEC units are coupled with each other after being connected by the network. The primary distinction between the network of an interconnected system and a typical network is that the former processes dynamical edges, which can obtain system inputs based on node outputs and transmit outputs to the inputs of other nodes [24].*

2.2.2. Interactions among WEC Units

According to Equation (6), $\dot{\mathrm{w}}_{ij}(t)$ indicates a variation in the vertical displacement of WEC unit i excited by the interference wave generated by WEC unit j. $\dot{\mathrm{w}}_{ij}(t)$ is associated to the locations of WEC unit i and j, and the heave motion $\dot{z}_j(t)$ of WEC unit j. Using experimental data, and ignoring the non-linearity relationship between $\dot{\mathrm{w}}_{ij}(t)$ and $\dot{z}_j(t)$, this transfer function $\mathrm{H}_{ij}(s)$ of a linear dynamic system with $\dot{z}_j(t)$ and $\dot{\mathrm{w}}_{ij}(t)$ as input and output, respectively, can be derived by standard linear system identification techniques [23]. In addition to being an edge with dynamic properties in the network topology of the interconnected system, this linear dynamical model can be expressed simultaneously as a frequency characteristics model $\mathrm{H}_{ij}(\mathrm{j}\omega)$ (simplified as H_{ij}) as follows:

$$\dot{w}_{ij}(\omega) = \mathrm{H}_{ij}\dot{z}_j(\omega), \tag{8}$$

$$\mathrm{H}_{ij} = \mathrm{j}\omega\sqrt{\frac{\pi|\omega|^4}{8g^3}}a^2\frac{1}{\sqrt{d_{ij}}}e^{-\mathrm{j}\left\{\frac{\omega^2}{g}d_{ij}+\frac{3\pi}{4}\right\}}, \tag{9}$$

where $\dot{z}_j(\omega)$ and $\dot{w}_{ij}(\omega)$ are defined as the Fourier transform of $\dot{z}_j(t)$ and $\dot{\mathrm{w}}_{ij}(t)$; g, a and d_{ij} are the gravity constant, the diameter of the buoy, and the distance between the i-th WEC unit and the j-th WEC unit, respectively; ω is the frequency and j is the imaginary number. For $\omega > 0$, the frequency characteristics model Equation (8) is equivalently formulated as [25]:

$$\mathrm{H}_{ij} = \mathrm{A}(\omega)e^{-\mathrm{j}\phi(\omega)} \tag{10}$$

where $A(\omega) = \omega^3 a^2 \sqrt{\frac{\pi}{8g^3 d_{ij}}}$ and $\phi(\omega) = \frac{d_{ij}\omega^2}{g} + \frac{\pi}{4}$. The frequency characteristics model Equation (10) as a linear system is rewritten as a state-space formula as follows:

$$\begin{aligned} \dot{\zeta}_\ell(t) &= \mathcal{E}_\ell \zeta_\ell(t) + \mathcal{F}_\ell u_\ell(t), \\ v_\ell(t) &= \mathcal{H}_\ell \zeta_\ell(t), \end{aligned} \tag{11}$$

where $\mathcal{E}_\ell \in \mathbb{R}^{(5\times5)}$, $\mathcal{F}_\ell \in \mathbb{R}^{(5\times1)}$, $\mathcal{H}_\ell \in \mathbb{R}^{(1\times5)}$, $\zeta_\ell(t) \in \mathbb{R}^{(5\times1)}$ are the system matrix, input matrix, output matrix and state, respectively; $u_\ell(t)$ and $v_\ell(t)$ are the input and output of the ℓ-th interaction state-space model, respectively. Equation (11) denotes the ℓ-th ($\ell = 1,\ldots,2M, M = \frac{N(N-1)}{2}$) directed interaction effect between any two WEC units. This interaction effect is the weight of the corresponding edge of the interconnected system, as shown in Figure 1b.

In the WEC array, the integrated interaction model is expressed as

$$\begin{aligned} \dot{\tilde{\zeta}}_\ell(t) &= \mathbb{E}_c \tilde{\zeta}_\ell(t) + \mathbb{F}_c \tilde{u}_\ell(t), \\ \tilde{v}_\ell(t) &= \mathbb{H}_c \tilde{\zeta}_\ell(t), \end{aligned} \tag{12}$$

where $\mathbb{E}_c = \mathbb{I}_{2M} \otimes \mathcal{E}_\ell$, $\mathbb{F}_c = \mathbb{I}_{2M} \otimes \mathcal{F}_\ell$, $\mathbb{H}_c = \mathbb{I}_{2M} \otimes \mathcal{H}_\ell$, and

$$\tilde{\zeta}_\ell(t) = \begin{bmatrix} \zeta_1 \\ \zeta_2 \\ \vdots \\ \zeta_{2M} \end{bmatrix}, \ \tilde{u}_\ell(t) = \begin{bmatrix} u_1 \\ u_2 \\ \vdots \\ u_{2M} \end{bmatrix}, \ \tilde{v}_\ell(t) = \begin{bmatrix} v_1 \\ v_2 \\ \vdots \\ v_{2M} \end{bmatrix}.$$

2.2.3. WEC Array System

Since a WEC array is an interconnected system, it can be represented as a graph $\mathcal{G}$. The incidence matrix $\mathcal{D}$ of the graph $\mathcal{G}$ is applied to denote the interconnected system with N nodes and $2M$ directed edges. The element located at the i-th row and ℓ-th column of $\mathcal{D}$ is defined as

$$\mathcal{D}_{i\ell} = \begin{cases} +1, & \text{node } i \text{ is the head of edge } \ell, \\ -1, & \text{node } i \text{ is the tail of edge } \ell, \\ 0, & \text{otherwise.} \end{cases} \tag{13}$$

The incidence matrix $\mathcal{D}$ describes the connection relationship between the edges and nodes in graph $\mathcal{G}$, and is used to determine the edge on which the output of the node is located and the node at which the input of the edge arrives. From the perspective of the node inputs and outputs, the incidence matrix satisfies

$$\mathcal{D} = \mathbb{D}_{\mathrm{O}} - \mathbb{D}_{\mathrm{I}}^{\mathrm{T}}, \tag{14}$$

where $\mathbb{D}_{\mathrm{O}}$ and $\mathbb{D}_{\mathrm{I}}$ are the input and output matrices of the nodes, respectively, as

$$\mathbb{D}_{\mathrm{O}i\ell} = \begin{cases} 1, & \text{edge } \ell \text{ is the input of node } i, \\ 0, & \text{otherwise.} \end{cases}, \ \mathbb{D}_{\mathrm{I}\ell i} = \begin{cases} 1, & \text{edge } \ell \text{ is the output of node } i, \\ 0, & \text{otherwise.} \end{cases}$$

Furthermore, considering Equations (6), (11) and (13), $\sum_{j\neq i}^{N-1} \dot{\mathrm{w}}_{ij}(t)$ in Equation (6) and $u_\ell(t)$ in Equation (11) are denoted as

$$\begin{cases} \sum\limits_{j\neq i}^{N-1} \dot{\mathrm{w}}_{ij}(t) = \sum\limits_{\ell=1}^{M} \mathbb{D}_{\mathrm{O}i\ell} v_\ell(t), & i = 1,2,\ldots,N, \\ u_\ell(t) = \sum\limits_{i=1}^{N} \mathbb{D}_{\mathrm{I}\ell i} y_i(t), & \ell = 1,2,\ldots,2M. \end{cases} \tag{15}$$

Moreover, Equation (15) can be written in matrix form as follows

$$\begin{cases} \dot{\tilde{\mathrm{w}}}_{\mathrm{ij}}(t) = \mathbb{D}_{\mathrm{O}}\tilde{v}_{\ell}(t), \\ \tilde{u}_{\ell}(t) = \mathbb{D}_{\mathrm{I}}\tilde{y}(t). \end{cases} \tag{16}$$

For example, for the three-unit WEC array interconnected system shown in Figure 1b, the incidence matrix $\mathcal{D}$ is

$$\mathcal{D} = \begin{bmatrix} 1 & 0 & -1 & -1 & 0 & 1 \\ -1 & 1 & 0 & 1 & -1 & 0 \\ 0 & -1 & 1 & 0 & 1 & -1 \end{bmatrix}. \tag{17}$$

Considering Equation (14), the incidence matrix $\mathcal{D}$ is decomposed into $\mathbb{D}_{\mathrm{O}}$ and $\mathbb{D}_{\mathrm{I}}$. According to Equations (15) and (16), we can obtain

$$\dot{\tilde{\mathrm{w}}}_{\mathrm{ij}}(t) = \begin{bmatrix} \sum_{j\neq 1}^{N-1} \dot{\mathrm{w}}_{1j}(t) \\ \sum_{j\neq 2}^{N-1} \dot{\mathrm{w}}_{2j}(t) \\ \sum_{j\neq 3}^{N-1} \dot{\mathrm{w}}_{3j}(t) \end{bmatrix} = \begin{bmatrix} v_1 + v_6 \\ v_2 + v_4 \\ v_3 + v_5 \end{bmatrix}, \ \tilde{u}_{\ell}(t) = \begin{bmatrix} u_1 \\ u_2 \\ u_3 \\ u_4 \\ u_5 \\ u_6 \end{bmatrix} = \begin{bmatrix} y_2 \\ y_3 \\ y_1 \\ y_1 \\ y_2 \\ y_3 \end{bmatrix}. \tag{18}$$

By substituting Equation (16) into Equations (7) and (12), the model of a three-unit WEC array as an interconnected system is obtained. More generally, a formally uniform state-space model is achieved to express a WEC array system with N units as

$$\begin{aligned} \begin{bmatrix} \dot{\tilde{\xi}}(t) \\ \dot{\tilde{\zeta}}_{\ell}(t) \end{bmatrix} &= \begin{bmatrix} \mathbb{A}_c & \mathbb{B}_{wc}\mathbb{D}_{\mathrm{O}}\mathbb{H}_c \\ \mathbb{F}_c\mathbb{D}_{\mathrm{I}}\mathbb{C}_c & \mathbb{E}_c \end{bmatrix} \begin{bmatrix} \tilde{\xi}(t) \\ \tilde{\zeta}_{\ell}(t) \end{bmatrix} + \begin{bmatrix} \mathbb{B}_{wc} \\ \mathbf{0} \end{bmatrix} \dot{\tilde{\mathrm{w}}}(t) + \begin{bmatrix} \mathbb{B}_c \\ \mathbf{0} \end{bmatrix} (\tilde{F}_u(t) + \tilde{F}_e(t)), \\ \tilde{y}(t) &= [\mathbb{C}_c, \mathbf{0}] \begin{bmatrix} \tilde{\xi}(t) \\ \tilde{\zeta}_{\ell}(t) \end{bmatrix}, \\ \tilde{y}_z(t) &= [\mathbb{C}_{zc}, \mathbf{0}] \begin{bmatrix} \tilde{\xi}(t) \\ \tilde{\zeta}_{\ell}(t) \end{bmatrix}, \end{aligned} \tag{19}$$

where $\dot{\tilde{w}}(t) = \dot{\mathrm{w}}(t) + \dot{\mathrm{w}}_{\mathrm{ij}}(t)$.

3. MPC Algorithm Design

3.1. Control Problem Statement

Equation (19) is discretized by a zero-order hold method as

$$\begin{aligned} \begin{bmatrix} \tilde{\xi}(k+1) \\ \tilde{\zeta}_{\ell}(k+1) \end{bmatrix} &= \begin{bmatrix} \mathbb{A} & \mathbb{B}_w\mathbb{D}_{\mathrm{O}}\mathbb{H} \\ \mathbb{F}\mathbb{D}_{\mathrm{I}}\mathbb{C} & \mathbb{E} \end{bmatrix} \begin{bmatrix} \tilde{\xi}(k) \\ \tilde{\zeta}_{\ell}(k) \end{bmatrix} + \begin{bmatrix} \mathbb{B}_w \\ \mathbf{0} \end{bmatrix} \dot{\mathrm{w}}(k) + \begin{bmatrix} \mathbb{B} \\ \mathbf{0} \end{bmatrix} (\tilde{F}_u(k) + \tilde{F}_e(k)), \\ \tilde{y}(k) &= [\mathbb{C}, \mathbf{0}] \begin{bmatrix} \tilde{\xi}(k) \\ \tilde{\zeta}_{\ell}(k) \end{bmatrix}, \\ \tilde{y}_z(k) &= [\mathbb{C}_z, \mathbf{0}] \begin{bmatrix} \tilde{\xi}(k) \\ \tilde{\zeta}_{\ell}(k) \end{bmatrix}, \end{aligned} \tag{20}$$

where

$$\mathbb{A} = e^{\mathbb{A}_c T}, \ \mathbb{B} = \int_0^T e^{\mathbb{A}_c \tau} d\tau \mathbb{B}_c, \ \mathbb{B}_w = \int_0^T e^{\mathbb{A}_c \tau} d\tau \mathbb{B}_{wc}, \ \mathbb{C} = \mathbb{C}_c, \ \mathbb{C}_z = \mathbb{C}_{zc}, \ \mathbb{E} = e^{\mathbb{E}_c T},$$

$$\mathbb{F} = \int_0^T e^{\mathbb{E}_c \tau} d\tau \mathbb{F}_c, \ \mathbb{H} = \mathbb{H}_c,$$

$$\dot{\tilde{w}}(k) = \begin{bmatrix} \dot{w}_1(k) \\ \dot{w}_2(k) \\ \vdots \\ \dot{w}_N(k) \end{bmatrix}, \tilde{F}_u(k) = \begin{bmatrix} F_{u1}(k) \\ F_{u2}(k) \\ \vdots \\ F_{uN}(k) \end{bmatrix}, \tilde{F}_e(k) = \begin{bmatrix} F_{e1}(k) \\ F_{e2}(k) \\ \vdots \\ F_{eN}(k) \end{bmatrix}, \tilde{y}(k) = \begin{bmatrix} y_1(k) \\ y_2(k) \\ \vdots \\ y_N(k) \end{bmatrix}, \tilde{y}_z(k) = \begin{bmatrix} y_{z1}(k) \\ y_{z2}(k) \\ \vdots \\ y_{zN}(k) \end{bmatrix}.$$

In Equation (20), $\tilde{\xi}(k) \in \mathbb{R}^{(\alpha \times N) \times 1}$ is the states set of all the WEC units at instant k, where α is the model order of each WEC unit. In addition, $\tilde{\zeta}_\ell(k) \in \mathbb{R}^{(\beta \times 2M) \times 1}$ is the states set of all interactions at time instant k, where β is the model order of each interaction. The wave excitation force and wave velocity are assumed to be predictable in this paper.

The MPC method controls the WEC array to maximize the harvesting of wave energy and convert it into electrical energy while satisfying the constraints. Since the buoy of the point-absorbing WEC is directly connected to PTO, the inner product of the buoy velocity and the corresponding reaction force is equal to the total power of the wave energy captured by the WEC array. As the velocity of the buoy is opposite to the positive direction of the reaction force, it holds that

$$P(k) = -\tilde{F}_u^{\mathrm{T}}(k)\tilde{y}(k), \tag{21}$$

and

$$\max_{\tilde{F}_{uk}^{k+N_p}} \sum_{T=0}^{N_p} P(k+T) = \min_{\tilde{F}_{uk}^{k+N_p}} \sum_{T=0}^{N_p} \tilde{F_u}^{\mathrm{T}}(k+T)\tilde{y}(k+T), \tag{22}$$

where N_p is the predicted time domain, and T is the time step. To ensure the convexity of the optimization problem, Equation (22) is modified as the cost function

$$J = \sum_{T=0}^{N_p} \left[\tilde{F}_u^{\mathrm{T}}(k+T)\tilde{y}(k+T) + r\tilde{F}_u^{\mathrm{T}}(k+T)\tilde{F}_u(k+T) + q\tilde{y}_z^{\mathrm{T}}(k+T)\tilde{y}_z(k+T) \right], \tag{23}$$

where r and q are two artificially determined weighting factors. The first part of the cost function Equation (23) is the wave energy harvesting power in Equation (22). The second part of Equation (23) represents a soft constraint on the control force, primarily to guarantee that the solution of the MPC is to solve a convex optimization problem and the control force is within the PTO's operating range. The third part represents the penalty term of the relative displacement of the buoy and the wave surface, ensuring the PTO operates within an effective operating range. To guarantee the safe operation of the WEC array and reduce the loss and maintenance cost, it is necessary to restrict the amplitude of the reaction force. Moreover, the buoy needs to be constrained to keep the proportional displacement between the midpoint of the buoy and the wave surface within a safe range. These constraints are formulated as follows,

$$\begin{aligned} |F_u(k+T)| &\leq F_{u,max}, \text{ for } T = 0, 1, \ldots, N_p, \\ |y_z(k+T)| &\leq y_{z,max}, \text{ for } T = 0, 1, \ldots, N_p, \end{aligned} \tag{24}$$

where $F_{u,max}$ and $y_{z,max}$ are the maximum amplitude of the control force and buoyancy, respectively.

The MPC aims to solve the optimal control input sequence that minimizes the cost function Equation (23) in the prediction time domain N_p, under the premise that the constraint Equation (24) is satisfied.

Remark 3. *The weighting factors r and q in Equation (23) are specified as two specific values independently to ensure that the MPC is a convex optimization problem.*

3.2. Prediction Model

The cost function Equation (23) can be expressed in matrix form as

$$J = \left(\tilde{U}_k^{k+N_p}\right)^{\mathrm{T}} \tilde{Y}_k^{k+N_p} + \left(\tilde{U}_k^{k+N_p}\right)^{\mathrm{T}} R \tilde{U}_k^{k+N_p} + \left(\tilde{Y}_{zk}^{k+N_p}\right)^{\mathrm{T}} Q \tilde{Y}_{zk}^{k+N_p}, \tag{25}$$

where $R = r \times I_{N \times (N_p+1)}$, $Q = q \times I_{N \times (N_p+1)}$,

$$\tilde{U}_k^{k+N_p} = \begin{bmatrix} \tilde{F}_u(k) \\ \tilde{F}_u(k+1) \\ \vdots \\ \tilde{F}_u(k+N_p) \end{bmatrix}, \tilde{Y}_k^{k+N_p} = \begin{bmatrix} \tilde{y}(k) \\ \tilde{y}(k+1) \\ \vdots \\ \tilde{y}(k+N_p) \end{bmatrix}, \tilde{Y}_{zk}^{k+N_p} = \begin{bmatrix} \tilde{y}_z(k) \\ \tilde{y}_z(k+1) \\ \vdots \\ \tilde{y}_z(k+N_p) \end{bmatrix},$$

and I is a unit diagonal matrix.

In the cost function, it is necessary to know the buoy velocity and buoyancy of each unit in the predicted time domain. By the discrete model of the interconnected system, we derive a prediction model from k to $k+N_p$ as follows:

$$\begin{aligned} \tilde{Y}_k^{k+N_p} &= \Lambda X(k) + \Phi_u\left(\tilde{U}_k^{k+N_p} + \tilde{F}_{ek}^{k+N_p}\right) + \Phi_w \dot{\tilde{W}}_k^{k+N_p}, \\ \tilde{Y}_{zk}^{k+N_p} &= \Lambda_z X(k) + \Phi_{uz}\left(\tilde{U}_k^{k+N_p} + \tilde{F}_{ek}^{k+N_p}\right) + \Phi_{wz} \dot{\tilde{W}}_k^{k+N_p}, \end{aligned} \tag{26}$$

where

$$X(k) = \begin{bmatrix} \tilde{\xi}(k) \\ \tilde{\zeta}_\ell(k) \end{bmatrix}, \tilde{F}_{ek}^{k+N_p} = \begin{bmatrix} \tilde{F}_e(k) \\ \tilde{F}_e(k+1) \\ \vdots \\ \tilde{F}_e(k+N_p) \end{bmatrix}, \dot{\tilde{W}}_k^{k+N_p} = \begin{bmatrix} \dot{\tilde{w}}(k) \\ \dot{\tilde{w}}(k+1) \\ \vdots \\ \dot{\tilde{w}}(k+N_p) \end{bmatrix}, \Lambda = \begin{bmatrix} \mathcal{C} \\ \mathcal{C}\mathcal{A} \\ \mathcal{C}\mathcal{A}^2 \\ \vdots \\ \mathcal{C}\mathcal{A}^{N_p} \end{bmatrix},$$

$$\Phi_u = \begin{bmatrix} \mathbf{0} & & & & \\ \mathcal{C}\mathcal{B}_u & \mathbf{0} & & & \\ \mathcal{C}\mathcal{A}\mathcal{B}_u & \mathcal{C}\mathcal{B}_u & \mathbf{0} & & \\ \vdots & \vdots & \ddots & \ddots & \\ \mathcal{C}\mathcal{A}^{N_p-1}\mathcal{B}_u & \mathcal{C}\mathcal{A}^{N_p-2}\mathcal{B}_u & \cdots & \mathcal{C}\mathcal{B}_u & \mathbf{0} \end{bmatrix},$$

$$\Phi_w = \begin{bmatrix} \mathbf{0} & & & & \\ \mathcal{C}\mathcal{B}_w & \mathbf{0} & & & \\ \mathcal{C}\mathcal{A}\mathcal{B}_w & \mathcal{C}\mathcal{B}_w & \mathbf{0} & & \\ \vdots & \vdots & \ddots & \ddots & \\ \mathcal{C}\mathcal{A}^{N_p-1}\mathcal{B}_w & \mathcal{C}\mathcal{A}^{N_p-2}\mathcal{B}_w & \cdots & \mathcal{C}\mathcal{B}_w & \mathbf{0} \end{bmatrix},$$

$$\Lambda \in \mathbb{R}^{(N\times(N_p+1))\times(\alpha N+2\beta M)}, \Phi_u \in \mathbb{R}^{(N\times(N_p+1))\times(N\times(N_p+1))},$$
$$\Phi_w \in \mathbb{R}^{(N\times(N_p+1))\times(N\times(N_p+1))},$$

with

$$\mathcal{A} = \begin{bmatrix} \mathbb{A} & \mathbb{B}_w\mathbb{D}_{i\ell}\mathbb{H} \\ \mathbb{F}\mathbb{D}_{\ell i}\mathbb{C} & \mathbb{E} \end{bmatrix}, \mathcal{B}_u = \begin{bmatrix} \mathbb{B} \\ \mathbf{0} \end{bmatrix}, \mathcal{B}_w = \begin{bmatrix} \mathbb{B}_w \\ \mathbf{0} \end{bmatrix}, \mathcal{C} = [\mathbb{C}, \mathbf{0}], \mathcal{C}_z = [\mathbb{C}_z, \mathbf{0}].$$

Similarly, Λ_z, Φ_{uz} and Φ_{wz} need to replace $\mathcal{C}$ in the corresponding matrix element by $\mathcal{C}_z$. Using prediction models Equation (24) to represent their constraints in the predicted time domain achieves

$$\begin{aligned} &\left|\tilde{U}_k^{k+N_p}\right| \le U_{\max}, \\ &\left|\Lambda_z x(k) + \Phi_{uz}\left(\tilde{U}_k^{k+N_p} + \tilde{F}_{ek}^{k+N_p}\right) + \Phi_{wz}\dot{\tilde{W}}_k^{k+N_p}\right| \le Y_{z,\max}, \end{aligned} \tag{27}$$

with

$$U_{\max} = \begin{bmatrix} F_{u,\max} \\ F_{u,\max} \\ \vdots \\ F_{u,\max} \end{bmatrix} \in \mathbb{R}(N \times N_p) \times 1, Y_{z,\max} = \begin{bmatrix} y_{z,\max} \\ y_{z,\max} \\ \vdots \\ y_{z,\max} \end{bmatrix} \in \mathbb{R}(N \times N_p) \times 1.$$

3.3. Quadratic Programming Problem

We need to solve the optimal reaction force sequence so that the cost function J can obtain the minimum value as

$$\tilde{U}_k^{*k+N_p} = \arg \min_{\tilde{U}_k^{k+N_p}} J. \tag{28}$$

Equation (25) is reorganized into a general form to utilize the standard QP solver.

Substituting the prediction model Equation (26) into the cost function Equation (25), and ignoring terms unrelated to $\tilde{U}_k^{k+N_p}$, leads to

$$J = \frac{1}{2}\left(\tilde{U}_k^{k+N_p}\right)^{\mathrm{T}} \mathcal{H} \tilde{U}_k^{k+N_p} + \mathcal{F}^{\mathrm{T}} \tilde{U}_k^{k+N_p}, \tag{29}$$

where

$$\begin{aligned} \mathcal{H} &= \Phi_u + \Phi_u^{\mathrm{T}} + 2R + 2\Phi_{uz}^{\mathrm{T}} Q \Phi_{uz}, \\ \mathcal{F} &= \left(\Lambda + 2\Phi_{uz}^{\mathrm{T}} Q \Lambda_z\right) X(k) + \left(\Phi_w + 2\Phi_{uz}^{\mathrm{T}} Q \Phi_{wz}\right) \dot{\tilde{W}}_k^{k+N_p} + \left(\Phi_u + 2\Phi_{uz}^{\mathrm{T}} Q \Phi_{uz}\right) \tilde{F}_{ek}^{k+N_p}. \end{aligned} \tag{30}$$

The constraints Equation (27) are reorganized into

$$\mathcal{S} \tilde{U}_k^{k+N_p} \le b_u, \tag{31}$$

where $\mathcal{S} = \begin{bmatrix} I \\ -I \\ \Phi_{uz} \\ -\Phi_{uz} \end{bmatrix}$, and $b_u = \begin{bmatrix} U_{\max} \\ U_{\max} \\ Y_{z,\max} - \Lambda_{xz} x(k) - \Phi_{wz} \dot{\tilde{W}}_k^{k+N_p} - \Phi_{uz} \tilde{F}_{ek}^{k+N_p} \\ Y_{z,\max} + \Lambda_{xz} x(k) + \Phi_{wz} \dot{\tilde{W}}_k^{k+N_p} + \Phi_{uz} \tilde{F}_{ek}^{k+N_p} \end{bmatrix}$.

Finally, the general form of the QP problem is summarized as

$$\begin{aligned} \tilde{U}^* &= \arg \min_{U_0^k} \frac{1}{2}\left(\tilde{U}_k^{k+N_p}\right)^{\mathrm{T}} \mathcal{H} \tilde{U}_k^{k+N_p} + \mathcal{F}^{\mathrm{T}} \tilde{U}_k^{k+N_p}, \\ &\text{S.t. } \mathcal{S} \tilde{U}_k^{k+N_p} \le b_u. \end{aligned} \tag{32}$$

In each control horizon, the first term $\tilde{U}_k^{k+1}$ in the sequence of optimal control forces $\tilde{U}_k^{*k+N_p}$ is adopted as the optimal control force at time k.

4. Experimental Results and Discussion

4.1. Simulation of Wave Energy Harvesting

We designed a semi-physical hardware-in-the-loop (HIL) experiment to simulate the wave energy capture more realistically than the digital simulation, and to reduce the high cost and complexity of the real ocean or large pool environment. The semi-physical experimental platform consisted of a front-end part and a back-end part as shown in Figure 2a. The front-end part comprised a desktop computer running MATLAB® and an RT-LAB simulator (a rapid control prototyping (RCP) platform from OPAL-RT Technologies Canada), as shown in the upper part of Figure 2a. The front-end part was mainly dedicated to simulating realistic sea conditions to implement a fully digital simulation of the WEC array to validate the established interconnected system model and the MPC control strategy mapped to the back-end part for the physical experiment bench operation.

The back-end part was a physical experiment bench as shown in the lower part of Figure 2a,b. The rotating motor was controlled by a commercial drive (Copley Controls ADP-090-36) and was used to imitate the rippling motion of the WEC buoys in the array under the effect of wave force. As the PTO device of the WEC, the linear generator (see Table 1 for the specific parameters) was dragged by a belt to convert the motion of the simulated buoy into an AC power output. The AC power generated by three PTO devices was converted to DC power using an AC-DC converter and rectifier (modules 6 and 7 in Figure 2b) developed in our laboratory and fed to the DC bus and electronic loads.

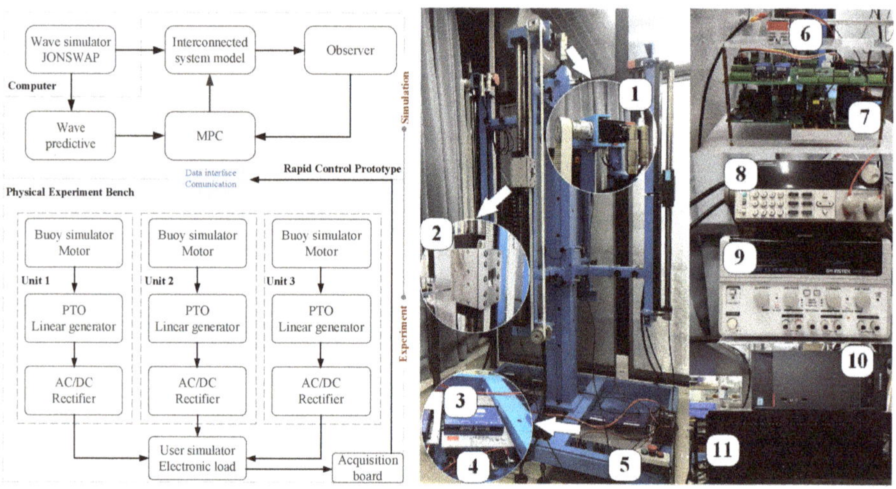

(a) (b)

Figure 2. Experimental bench: (**a**) Framework; (**b**) Hardware: 1. Rotating motor; 2. Linear motor; 3. Driver; 4. DC power; 5. Switch; 6. Rectifier; 7. Acquisition board; 8. Electronic load; 9. DC power; 10. Computer; 11. RT-LAB.

Table 1. Specifications for wave energy linear generators.

Electrical Parameters	Value	Mechanical Parameters	Value
Rated power (W)	50	Mover stroke (cm)	70
Rated current (A)	20	Mover speed (m/s)	3.5 m
Rated voltage (V)	20	Mover length (cm)	18
Conversion efficiency	85%	Mover weight (kg)	1.65

The purpose of constructing the above experimental system was to perform the following procedures:

(i) Verifying the correctness of the interconnected model and the control performance of the proposed MPC algorithm;
(ii) Comparing the difference in the motion characteristics affected by the interaction and the wave energy harvesting efficiency between the interconnected model and the isolated model of the WEC array, and exploring how to affect the wave energy harvesting efficiency by change in the incident wave propagation direction in the WEC interconnected system.

4.2. Simulation Parameter Setting

We built a three-unit WEC array in the front end with reference to the PB3 device described in [26,27] developed by Ocean Power Technologies Inc. All the parameters are shown in Table 2. The three WEC units were arranged in an equilateral triangle with a distance of 25 m from each other.

Table 2. Simulation parameters.

Description	Notation	Value
Diameter	a	2.65 m
Height	h	3.5 m
Damping	D	2×10^5 Nm/s
Buoy mass	m	8300 kg
Added mass	m_∞	6100 kg
Stiffness	S_h	2.21×10^5 N/m
Density of sea water	ρ	1020 kg/m^3
Gravity	g	9.81 m/s^2
Maximum heave motion	$y_{z,\max}$	1.75 m
Maximum input magnitude	$F_{u,\max}$	1×10^5 N

The JONSWAP function generated the incident wave [28]; the wave significant height was 2 m, and the peak period was 3 s. The wave velocity and the corresponding wave excitation force F_e are shown in Figure 3. To investigate the effect that the angle ∂ of wave propagation direction waves had on the WEC array, three sets of experiments with different wave propagation directions $\partial = 30°, 45°, 60°$ were set up.

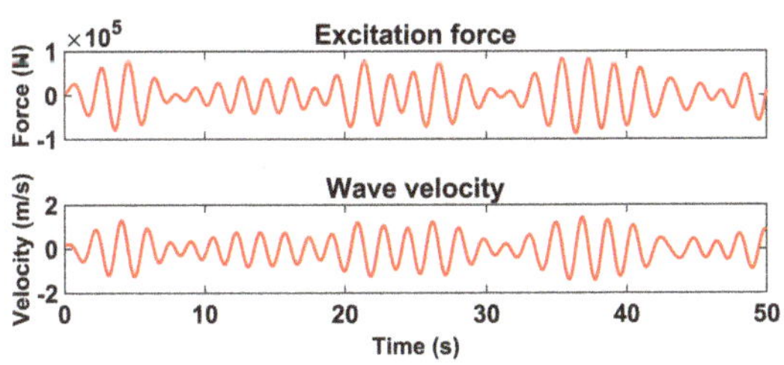

Figure 3. Excitation force and wave velocity.

4.3. Results and Discussion

The prediction horizon and the control horizon were both 10 s, the sampling time was 0.02 s, and the simulation time was 50 s. The weight value r was set to 2.5×10^{-6} to ensure that the Hermitian matrix was a positive definite matrix, and Q was also set to 2.5×10^{-6}. When the radius of the buoy was $5 \times a$ m and the distance d_{ij} was 25 m, the matrices of the state-space representation of the edge model were

$$\mathcal{E}_\ell = \begin{bmatrix} -1.3192 & -2.6404 & -1.8147 & -1.3222 & -0.2090 \\ 1 & & & & \\ & 1 & & & \\ & & 1 & & \\ & & & 1 & \end{bmatrix},$$

$$\mathcal{F}_\ell = \begin{bmatrix} 1 & 0 & 0 & 0 & 0 \end{bmatrix}^{\mathrm{T}}, \text{and } \mathcal{H}_\ell = \begin{bmatrix} 0.1470 & -0.1931 & -0.1043 & -0.0100 & -0.0053 \end{bmatrix}.$$

Figure 4 shows the velocity of each unit's radiation wave in the three layout cases. In particular, when $\partial = 30°$, Units 2 and 3 were affected by the same interference. This was because of the symmetrical structure of the equilateral triangle, and the propagation direction of the incident wave overlapping the symmetric axis, so that the waveform of the incident wave received by Units 2 and 3 was synchronized, and the identical edge model caused the same interference. Figure 5 shows that, under the control of the MPC algorithm, the control force of each WEC unit was limited to a range of $\pm 1 \times 10^5$ N, which satisfies the input constraint in Table 1.

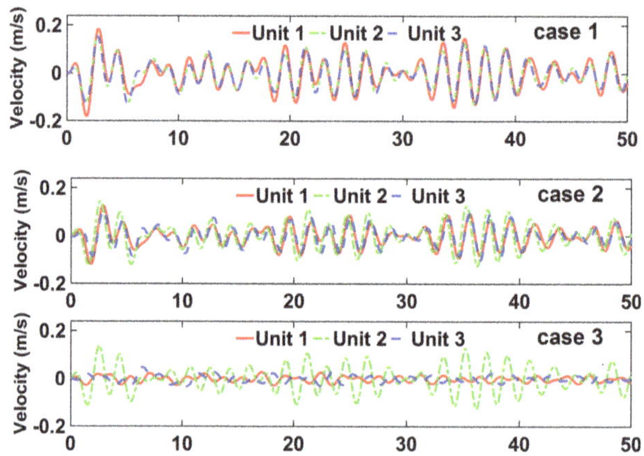

Figure 4. Interaction of radiation wave velocities received by each unit.

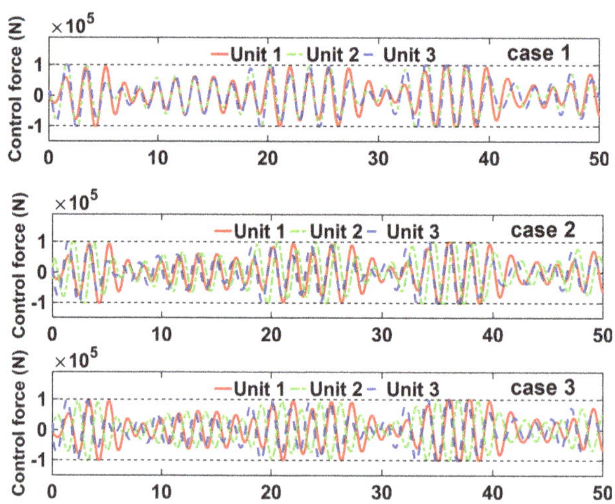

Figure 5. Reaction force controlled by the MPC algorithm.

In Figure 6, the waveform of each buoy velocity is almost the same as that of the corresponding incident wave velocity, and resonance is almost realized between the buoy motion and the wave motion. Due to the existence of other terms in the cost function, the execution delay, and the limitation of the control force and displacement, the resonance between the buoy and the incident wave is not fully realized. However, the wave energy capture efficiency is optimized by the control of the MPC algorithm.

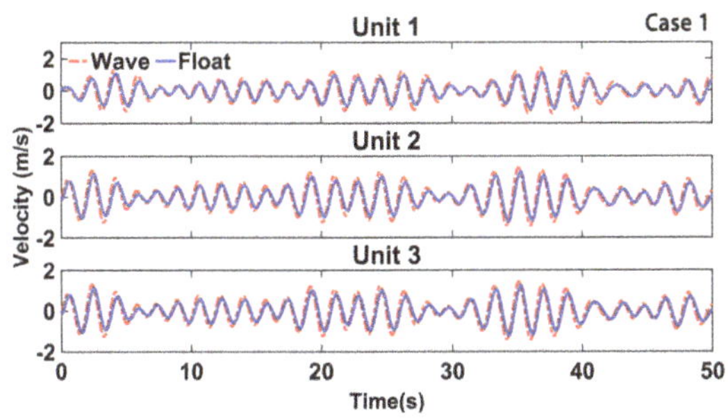

Figure 6. Buoy velocity and incident wave velocity.

To show the impact of interactions in the interconnected system, three units of isolated systems were set up as baseline cases. The isolated system had no interaction among the three WEC units. Thus, the radiation waves caused by the oscillating motion of units did not affect the neighborhood units. In the mathematical model Equation (20), the induction matrix $\mathbb{D}$ is an all-0 matrix. Because the interaction exists in the WEC array, the buoy's motion velocity is different between the interconnected and isolated systems without interaction. Furthermore, to adapt to the existence of the interaction, the MPC algorithm solves the optimal reaction force to make the units coordinate with each other to achieve the global optimal efficiency.

Figure 7 shows the variation in the total power generated by the corresponding three WEC units between the interconnected and the isolated system when the wave incidence angle $\partial = 45^{\circ}$. Each curve in Figure 7 illustrates the difference in wave energy captured by a unit, taking into account the impact of interfering waves between units compared with those not taking into account interfering waves. The energy error curves for the three units in Figure 7 highlight the various effects of interactions within the array on the energy harvesting of each unit. Figure 8 shows the waveform of the radiation wave and the incident wave when the incident wave angle $\partial = 45^{\circ}$. Note that in Figure 7, Unit 2 of the interconnected system harvests more wave energy than Unit 2 of the isolated system. In Figure 8, the waveforms of the incident wave received by Unit 2 are nearly the same as those of the radiation wave generated by the other units. The wave energy harvested by Units 1 and 3 of the interconnected system is not as high as the corresponding unit in the isolated system, and the waveforms of the incident wave and the radiation wave are significantly asynchronous.

When the waveform of the radiation wave received by the WEC unit is close to that of the incident wave, the unit can generate more energy. When their phase error is close to zero, so that the radiation wave synchronizes with the incident wave, just as when the buoy motion synchronizes with the wave motion, the efficiency of the wave energy captured by the WEC is optimized. At this time, the radiation wave also enables maximum optimization of the wave energy captured by the corresponding unit of the interconnected system.

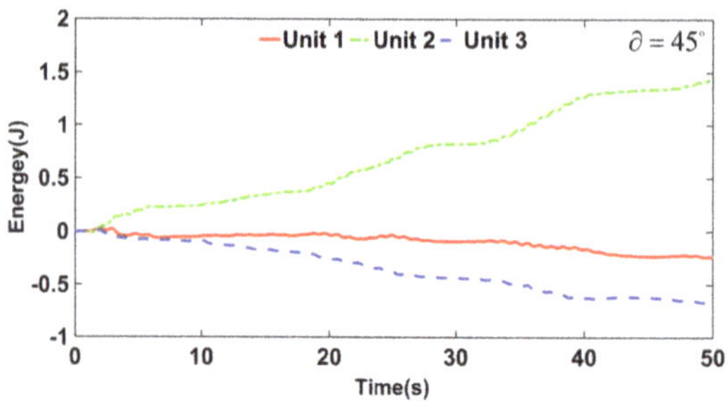

Figure 7. Variation in the wave energy harvested by each WEC unit between the interconnected and isolated systems.

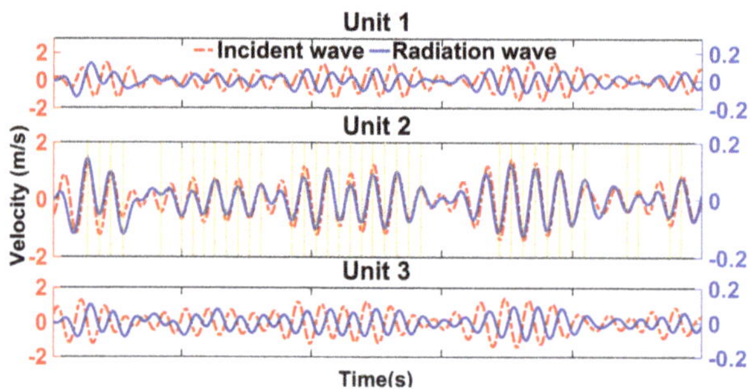

Figure 8. Wave velocity of the interacting radiation wave and the incident wave.

To show the influence of the interaction in different cases, we define the index ϱ as

$$\varrho = \frac{P_{\text{array},i}}{\sum\limits_{j=1}^{N} P^{j}_{\text{isolate},i}}, \tag{33}$$

where $P_{\text{array},i}$ and $P^{j}_{\text{isolate},i}$ are the power of the WEC array as an interconnected system and the power of each isolated unit at $t = i$, respectively.

In the three cases, the interaction of the interconnected system had a positive or negative effect at different instances within 50 s. The interaction in the first and second cases had more positive than negative effects, and $\varrho > 1$ was satisfied, respectively, 62.72% and 52.08% of the total time. In the third case, the negative effect was more than the positive effect, and $\varrho > 1$ was satisfied only 29.08% of the total time.

The results in Figure 9 show that in the first and second cases, the interconnected system controlled by the MPC algorithm had higher wave energy harvesting efficiency than the isolated system. In the third case, the interconnected system harvested less wave energy than the isolated system. In addition, when the incident waves had different propagation directions, the increase in the wave energy harvesting efficiency of the WEC interconnected system changed accordingly.

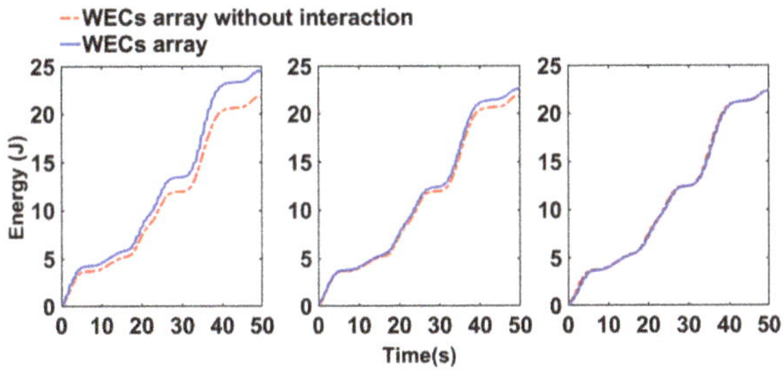

Figure 9. Electrical energy generated by the WEC array and WEC array without interaction.

5. Conclusions

In this paper, we proposed an interconnected system model of a WEC array which graphically integrates a WEC unit model and an interference model. The proposed model with wave interference dynamics describes the WEC array system more accurately. Each part of the interconnected system model has a well-defined structure that can better reflect uniformity when the array configuration changes. The interconnected model accurately describes the wave energy harvesting process under the influence of wave interference. The MPC algorithm based on the interconnected system model is applied to enable the WEC array to achieve optimal global performance of wave energy capture control, subject to several constraints. Through experimental tests, it was found that the phase error between the radiation and incident waves significantly affected the wave energy capture efficiency, and the wave energy capture efficiency was effectively improved as the error decreased. The experimental results show that the MPC with the interconnected system model had a higher wave energy capture efficiency compared with the isolated system model. However, the performance of the MPC relies on accurate and sufficient information on wave excitation forecasts. Future research will concentrate on more realistic hydrodynamics of the buoys and their coupling in the ocean environment, as well as inaccurate or insufficient wave excitation forecasts. Strengthening the robustness and adaptability of the MPC controller by combining intelligent control techniques is another promising direction.

Author Contributions: All authors contributed to the study conception and design. Administrative support and provision of study materials were provided by B.Z., X.B. and A.K.; collection and assembly of data were undertaken by H.Z., S.Y.; data analysis and interpretation were performed by B.Z., S.C., X.B. and A.K. All authors have read and agreed to the published version of the manuscript.

Funding: This work is supported in part by the National Natural Science Foundation of China under Grants 62173234 and 62003217.

Institutional Review Board Statement: Not applicable.

Informed Consent Statement: Not applicable.

Data Availability Statement: All data generated or analysed during the study of this manuscript are included in the article.

Conflicts of Interest: With respect to the content of this paper, all the authors have no conflict of interest to declare.

References

1. O'Sullivan, A.; Sheng, W.; Lightbody, G. An analysis of the potential benefits of centralised predictive control for optimal electrical power generation from wave energy arrays. *IEEE Trans. Sustain. Energy* **2018**, *9*, 1761–1771. [CrossRef]
2. Babarit, A. On the park effect in arrays of oscillating wave energy converters. *Renew. Energy* **2013**, *58*, 68–78. [CrossRef]
3. Kagemoto, H.; Yue, D.K. Interactions among multiple three-dimensional bodies in water waves: An exact algebraic method. *J. Fluid Mech.* **1986**, *166*, 189–209. [CrossRef]

4. Babarit, A.; Borgarino, B.; Ferrant, P.; Clément, A.H. Assessment of the influence of the distance between two wave energy converters on energy production. *IET Renew. Power Gener.* **2009**, *4*, 592–601. [CrossRef]
5. Borgarino, B.; Babarit, A.; Ferrant, P. Impact of wave interactions effects on energy absorption in large arrays of wave energy converters. *Ocean Eng.* **2012**, *41*, 79–88. [CrossRef]
6. Devolder, B.; Stratigaki, V.; Troch, P.A.; Rauwoens, P. CFD simulations of floating point absorber wave energy converter arrays subjected to regular waves. *Energies* **2018**, *11*, 641. [CrossRef]
7. Chen, W.; Gao, F.; Meng, X.; Fu, J. Design of the wave energy converter array to achieve constructive effects. *Ocean Eng.* **2016**, *124*, 13–20. [CrossRef]
8. Fang, H.; Feng, Y.; Li, G. Optimization of wave energy converter arrays by an improved differential evolution algorithm. *Energies* **2018**, *11*, 3522. [CrossRef]
9. Sharp, C.; DuPont, B. Wave energy converter array optimization: A genetic algorithm approach and minimum separation distance study. *Ocean. Eng.* **2018**, *163*, 148–156. [CrossRef]
10. Previsic, M.; Karthikeyan, A.; Scruggs, J.T. A comparative study of model predictive control and optimal causal control for heaving point absorbers. *J. Mar. Sci. Eng.* **2021**, *9*, 805. [CrossRef]
11. Faedo, N.; Olaya, S.; Ringwood, J.V. Optimal control, MPC and MPC–like algorithms for wave energy systems: An overview. *IFAC J. Syst. Control* **2017**, *1*, 37–56. [CrossRef]
12. Haider, A.S.; Brekken, T.K.; Mccall, A.K. Real-Time nonlinear model predictive controller for multiple degrees of freedom wave energy converters with non–ideal power take-off. *J. Mar. Sci. Eng.* **2021**, *9*, 890. [CrossRef]
13. Jia, Y.; Meng, K.; Dong, L.; Liu, T.; Sun, C.; Dong, Z.Y. Economic model predictive control of a point absorber wave energy converter. *IEEE Trans. Sustain. Energy* **2021**, *12*, 578–586. [CrossRef]
14. Montoya, D.G.; Tedeschi, E.; Castellini, L.; Martins, T.D. Passive model predictive control on a two–body self–referenced point absorber wave energy converter. *Energies* **2021**, *14*, 1731. [CrossRef]
15. Haider, A.S.; Brekken, T.K.; Mccall, A.K. Application of real-time nonlinear model predictive control for wave energy conversion. *IET Renew. Power Gener.* **2021**, *15*, 3331–3340. [CrossRef]
16. Richter, M.; Magaña, M.E.; Sawodny, O.; Brekken, T.K. Nonlinear model predictive control of a point absorber wave energy converter. *IEEE Trans. Sustain. Energy* **2013**, *4*, 118–126. [CrossRef]
17. Oetinger, D.; Magaña, M.E.; Sawodny, O. Decentralized model predictive control for wave energy converter arrays. *IEEE Trans. Sustain. Energy* **2014**, *5*, 1099–1107. [CrossRef]
18. Oetinger, D.; Magaña, M.E.; Sawodny, O. Centralised model predictive controller design for wave energy converter arrays. *IET Renew. Power Gener.* **2015**, *9*, 142–153. [CrossRef]
19. Vicente, P.C.; Falcão, A.F.; Gato, L.M.; Justino, P.A. Dynamics of arrays of floating point-absorber wave energy converters with inter-body and bottom slack-mooring connections. *Appl. Ocean Res.* **2009**, *31*, 267–281. [CrossRef]
20. Li, G.; Belmont, M. Model predictive control of sea wave energy converters—Part I: A convex approach for the case of a single device. *Renew. Energy* **2014**, *69*, 453–463. [CrossRef]
21. Yu, Z.; Falnes, J. State-Space modelling of a vertical cylinder in heave. *Appl. Ocean Res.* **1995**, *17*, 265–275. [CrossRef]
22. Jama, M.A.; Wahyudie, A.; Noura, H. Robust predictive control for heaving wave energy converters. *Control. Eng. Pract.* **2018**, *77*, 138–149. [CrossRef]
23. Li, G.; Belmont, M. Model predictive control of sea wave energy converters—Part II: The case of an array of devices. *Renew. Energy* **2014**, *68*, 540–549. [CrossRef]
24. Coogan, S.D.; Arcak, M. A dissipativity approach to safety verification for interconnected systems. *IEEE Trans. Autom. Control* **2015**, *60*, 1722–1727. [CrossRef]
25. Falnes, J. Radiation impedance matrix and optimum power absorption for interacting oscillators in surface waves. *Appl. Ocean Res.* **1980**, *2*, 75–80. [CrossRef]
26. Product Specifications for the pb3. Available online: https://oceanpowertechnologies.com/product/ (accessed on 1 March 2021).
27. Viet, N.V.; Carpinteri, A.; Wang, Q. A novel heaving ocean wave energy harvester with a frequency tuning capability. *Arab. J. Sci. Eng.* **2019**, *44*, 5711–5722. [CrossRef]
28. Aranuvachapun, S. Parameters of JONSWAP spectral model for surface gravity waves—I. Monte Carlo simulation study. *Ocean Eng.* **1987**, *14*, 89–100. [CrossRef]

Journal of
Marine Science and Engineering

MDPI

Article

Harnessing of Different WECs to Harvest Wave Energy along the Galician Coast (NW Spain)

Beatriz Arguilé-Pérez [1,*], Américo Soares Ribeiro [1,2], Xurxo Costoya [1], Maite deCastro [1], Pablo Carracedo [3], João Miguel Dias [2], Liliana Rusu [4] and Moncho Gómez-Gesteira [1]

1 Environmental Physics Laboratory (EPhysLab), Centro de Investigación Mariña, Universidade de Vigo, Campus da Auga, 32004 Ourense, Spain; americosribeiro@ua.pt (A.S.R.); xurxocostoya@uvigo.es (X.C.); mdecastro@uvigo.es (M.d.); mggesteira@uvigo.es (M.G.-G.)
2 CESAM, Physics Department, University of Aveiro, 3810-193 Aveiro, Portugal; joao.dias@ua.pt
3 MeteoGalicia, 15707 Santiago de Compostela, A Coruña, Spain; pablo.enrique.carracedo.garcia@xunta.gal
4 Department of Mechanical Engineering, Faculty of Engineering, 'Dunarea de Jos' University of Galati, 111 Domneasca St., 800201 Galati, Romania; liliana.rusu@ugal.ro
* Correspondence: beatriz.arguile.perez@uvigo.es

Abstract: The wave power resource (WP) was calculated along the Galician coast (NW Spain) over the period 2014–2021 using high spatial resolution hourly data from the SWAN model. In addition, the electrical energy (P_E) that can be extracted for a particular wave energy converter (WEC) was analyzed for four different WECs (Oyster, Atargis, Aqua Buoy, and Pelamis). The performance of every WEC was also calculated attending to two parameters: the power load factor (ε) and the normalized capture width with respect to the WEC's geometry (efficiency). Results show that the WP resource is lower than 10 kWm^{-1} onshore, but it increases to about 50 kWm^{-1} offshore. Atargis obtained the highest P_E, and it is the most efficient device (ε ~40% and efficiency ~45%). Pelamis showed the lowest performance in offshore areas (ε ~15%, efficiency < 10%). A different type of WEC should be considered for every location along the coast depending on its size, performance parameters, and coexistence with other socio-economic activities and protected environmental areas.

Keywords: wave power; wave energy converters (WECs); Galician coast; SWAN

Citation: Arguilé-Pérez, B.; Ribeiro, A.S.; Costoya, X.; deCastro, M.; Carracedo, P.; Dias, J.M.; Rusu, L.; Gómez-Gesteira, M. Harnessing of Different WECs to Harvest Wave Energy along the Galician Coast (NW Spain). *J. Mar. Sci. Eng.* **2022**, *10*, 719. https://doi.org/10.3390/jmse10060719

Academic Editor: Luca Cavallaro

Received: 18 April 2022
Accepted: 21 May 2022
Published: 24 May 2022

1. Introduction and Studied Area

More than 81% of the world's primary energy consumption comes from fossil fuels [1], which have a clear influence on the increase in greenhouse gases and global temperature [2]. In addition, global energy demand continues to rise, and fossil resources are increasingly scarce [3,4], highlighting the urgency of the energy transition. At the United Nations Conference on Climate Change (COP26), held in Glasgow in the autumn of 2021, many countries agreed upon a series of political action guidelines. Among them, the 1.5 °C scenario was maintained, for which it is necessary to reduce greenhouse gas emissions by 45%—with respect to 2010 levels—at the end of the current decade and reach zero emissions in 2050 [5]. One of the United Nations 2030 Agenda for Sustainable Development goals is to "increase substantially the share of renewable energy in the global energy mix" to mitigate global warming and ensure energy supply [6]. To achieve the necessary success in developing renewable energy, it is essential to use all available renewable sources, including marine energy.

Ocean waves continually form on the ocean surface and can travel thousands of miles with minimal energy loss [7]. A higher and more stable density power [7,8] makes wave energy a clear alternative to other "more conventional" renewable energies. In fact, the theoretical potential of wave power is 29,000 TWh per year, which is enough to meet the world's electricity demand [9]. In addition, wave energy has the advantage of having more predictability than solar or wind power and is less dependent on environmental factors [10]. Unlike offshore wind power, the wave power resource is not yet being commercially exploited [9]. The survivability of the wave energy converters in the harsh marine environment, the lack of consensus on the choice of the most suitable device, and their economic costs have hindered their commercialization since the first patent in 1799 [11]. However, many of the wave energy converters can be scalable in size according to local wave characteristics to optimize their efficiency and make them economically profitable [10]. Another drawback is that wave energy devices have to coexist with other uses of the sea, such as shipping, fishing, or aquaculture. Nevertheless, these marine businesses—the so-called blue economy—can benefit from the installation of wave energy harvesting devices by not having to look for energy sources on land [9]. In addition, the installation of these devices in the vicinity of shellfishing areas could also protect them from waves by dissipating part of their energy.

The natural wave power (WP) resource depends on the sea state and is determined by the wave significant height (H_s) and the wave peak period (T_p). The WP resource is defined as the amount of energy per unit of time and length of the wave front (expressed in kWm^{-1}) transmitted in the direction of wave propagation [12]. Only a percentage of this resource will be transformed into electricity. This percentage depends on the wave energy converter (WEC) and its physical and technical characteristics (see [13–17] for more details). WECs can be classified attending to several factors. Depending on their operating principle, they can be oscillating water columns (OWS), oscillating bodies, wave-activated bodies, or cycloidal wave absorbers, among others (see [13,18–20] for more details). Regarding their optimal location, WECs are categorized into onshore, nearshore, and offshore devices [10,19–21]. They can also be classified as point absorbers, attenuators, and terminators according to their orientation with respect to the wavefront and their shape [20–23]. Attenuators and terminators have an elongated shape and are placed parallel and perpendicular to the direction of the wavefront, respectively, while point absorbers are similar to buoys. Attenuators and terminators must be larger than the predominant wavelength of the prevailing waves, while point absorbers have smaller dimensions than the main wavelength [21,24].

According to IRENA [25], in 2020, there was a total of 33 wave energy converters with a combined capacity of 2.3 MW distributed around the world. However, the European Commission aims to reach 1 GW of installed capacity by 2030 and 40 GW by 2050 [9]. Spain is the European country with the most R&D facilities in floating wind power and other marine energies [26,27]. For example, the experimental zone for the use of marine energy in Punta Langosteira (Galicia, Spain)—within the EnergyMare project—stands out. This site has the second-highest testbed wave energy density in the world, behind the south coast of Wales [27]. The Spanish government has set itself the goal of reaching 7 GW in floating wind power and between 40 and 60 MW of other marine energies—including wave power—by the end of this decade [27]. Galicia is the Spanish region with the highest wave energy potential (40–45 kWm^{-1}), followed by the Cantabrian Sea (30 kWm^{-1}) and the north of the Canary Islands (20 kWm^{-1}) [27].

There are several studies on the wave power resource in large areas such as the Mediterranean Sea [28], the Black Sea [29], or along the western coast of the Iberian Peninsula [7,12]. However, regional studies allow the identification of local phenomena and areas of greatest interest in terms of the available wave energy. These regional studies are of vital importance to promote studies at a local scale in various areas of the planet and achieve a global energy transition. Regional studies require a high or very high spatial resolution. Thus, the Galician shelf was previously analyzed with a spatial resolution of $0.25° \times 0.25°$ and with a higher resolution of 200 m × 200 m at the particular area between Cape San Adrián and Cape Ortegal [30]. In the same way, different locations in the north and west shelf of the Iberian Peninsula were analyzed with a nested grid of $0.5' \times 0.5'$ [31,32]. Apart from analyzing the WP resource, previous studies have analyzed the amount of energy that can be actually harnessed by particular WECs at different points of the Iberian Peninsula shelf [12,32,33]. In these previous studies, the performance of the

WECs is normally characterized by means of the capture width, which do not consider the dimensions of the different devices, making their comparison impossible.

The aim of this study is to analyze the wave power resource and the electrical power that could have been extracted from four WEC devices along the Galician coast between 2014 and 2021. Due to the narrowness of the continental shelf (Figure 1), the analysis of feasibility of the four WECs along the Galician coast makes it necessary to calculate the WP resource with a very high spatial resolution (a maximum of ~75 m) and also limits the type of device used. This very high spatial resolution makes it possible to calculate the wave power resource and the performance of these devices inside the estuaries. The performance of each WEC will be limited to the optimal depth range for the installation of the device and will be determined attending to two parameters: the power load factor (ε) and the normalized capture width according to the WEC's geometry (efficiency). At this point, it is important to mention that it was developed as a new metric that allows comparisons to be made between different WECs taking into account their dimensions.

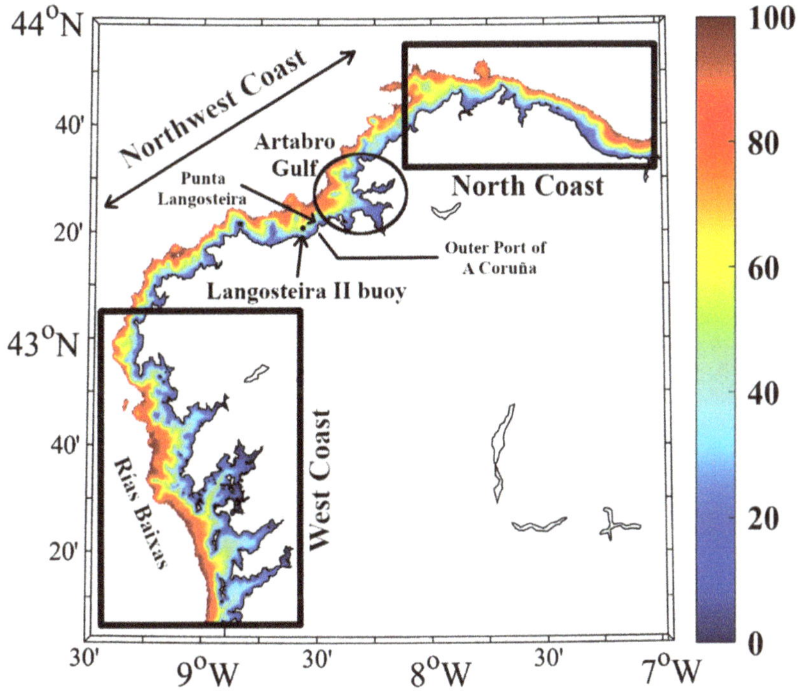

Figure 1. Study area. Colors represent the bathymetry (m).

2. Data and Methodology

2.1. Data

The wave parameters H_s and T_p were provided by MeteoGalicia, the regional administration in charge of weather forecast for Galicia, using simulations of the third-generation SWAN model developed by the Delft University of Technology [34–37]. SWAN solves the spectral action balance equations [34] (defined in [38]) and is the most adequate tool in order to simulate high-resolution coastal waves [34]. SWAN model uses an unstructured variable mesh, with a higher spatial resolution in coastal and estuarine areas —where it can reach 75 m resolution—than offshore [34]. Consequently, high-resolution wave data are obtained in the areas of interest without exceeding computational time. The SWAN model physics included bottom friction, shoaling, and depth-induced breaking, and the wave spectrum was discretized to a resolution of $\Delta f/f = 0.1$ between the frequency range of

0.0521–1 Hz and 36 directional bands. As boundary conditions, the implementation uses MeteoGalicia regional models for waves (WAVEWATCH III, with temporal and spatial resolution of 1 h and 2.5 arc minute, respectively), wind (WRF, with temporal and spatial resolution of 1 h and 4 km, respectively), and water level series (ROMS, with temporal and spatial resolution of 1 h and 4 km, respectively). These specifications are shown in Table 1.

Table 1. Main characteristics of the SWAN model set up in the present study.

Parameter	Value
Computational grid	
Spatial grid	Unstructured. Variable mesh step: 250 m−75 m
Frequency grid	Resolution: $\Delta f/f = 0.1$. Range: 0.0521 Hz–1 Hz
Directional grid	Resolution: 10°
Processes activated	
Local wave generation by wind	Yes (third-generation)
Limited depth processes	Yes (bottom friction, refraction, shoaling, depth-induced breaking)
Water level variations	Yes
Wave–current interactions	No
Forcing	
Boundary conditions	WaveWatch III. Temporal/Spatial resolution: 1 h/2.5 arc minutes
Wind	Meteogalicia WRF. Temporal/Spatial resolution: 1 h/4 km
Water level	Time series synthesis using the tidal solution of the MeteoGalicia ROMS model. Temporal/Spatial resolution: 1 h/4 km

Hourly time resolution data can be downloaded in daily folders from the THREDDS server (Thematic Realtime Environmental Distributed Data Service), http://mandeo.meteogalicia.gal/thredds/catalog/modelos/SWAN_HIST/galicia/catalog.html (accessed on 22 January 2022) [39]. Although each daily file contains data for a 4-day horizon, only the first 24 h of each file were selected to work in hindcast mode. Data series are available from 2014 onwards.

Simulation outputs were validated with Langosteira II buoy data, provided by Puertos del Estado (https://www.puertos.es/es-es/oceanografia/Paginas/portus.aspx (accessed on 22 January 2022), [40]), to assess the accuracy of the model results. This buoy is located in the Northwest Coast, at coordinates 8° 33′ W, 43°22′ N (Figure 1), being the only buoy that offers historical wave data on the Galician coast for the period considered in this study. Wave data have an hourly time resolution and are available from June 2013 to December 2021.

2.2. Methodology

Methods followed to validate the accuracy of SWAN model simulations to reproduce observed wave parameters (H_s and T_p) and to calculate the WP resource, and the WECs' efficiency are described below.

2.2.1. Validation

The validation process has been carried out by comparing H_s and T_p measured by the Langosteira II buoy and those calculated by the SWAN model at the point closest to the buoy for the period 2014–2021. This period was selected because it is the common period between both datasets. Four different statistics parameters have been used: normalized root mean square error (NRMSE), normalized bias error (NBias), Spearman's correlation coefficient (ρ_S), and the overlapping percentage (OP) between both series. RMSE is defined as the square root of the second sample moment of the differences between numerical and observed values [41,42]. As the magnitude of RMSE depends on the actual values of the series, the NRMSE defined in Equation (1) is a better option to compare two series of data.

$$\text{NRMSE}\ (\%) = \frac{100}{\frac{1}{2}\left(\overline{x_i^{num}} + \overline{x_i^{obs}}\right)} \cdot \sqrt{\frac{1}{N}\sum_{i=1}^{N}\left(x_i^{num} - x_i^{obs}\right)^2}, \tag{1}$$

where N is the total number of elements in both the data series, x_i^{num} refers to numerical values from the SWAN model, and x_i^{obs} denotes the observed data series retrieved from the buoy. Barred variables $\overline{x_i^{num}}$ and $\overline{x_i^{obs}}$ correspond to mean values.

The normalized bias error (in percentage) [41] was calculated as

$$\text{NBias}\ (\%) = \frac{100}{\frac{1}{2}\left(\overline{x_i^{num}} + \overline{x_i^{obs}}\right)} \cdot \frac{1}{N}\sum_{i=1}^{N}\left(x_i^{num} - x_i^{obs}\right). \tag{2}$$

The Spearman's correlation coefficient was computed as

$$\rho_S = \frac{\text{cov}\left(R\left(x_i^{num}\right), R\left(x_i^{obs}\right)\right)}{\sigma\left(R\left(x_i^{num}\right)\right) \cdot \sigma\left(R\left(x_i^{obs}\right)\right)}, \tag{3}$$

where $R(x_i)$ is the rank of the data series x_i, $\sigma(R(x_i))$ is its standard deviation, and $\text{cov}\left(R\left(x_i^{num}\right), R\left(x_i^{obs}\right)\right)$ is the covariance between the rank variables [43].

The overlapping percentage was calculated as

$$\text{OP}\ (\%) = 100 \cdot \sum_{i=1}^{n} \min\left(f_i\left(x^{num}\right), f_i\left(x^{obs}\right)\right), \tag{4}$$

where n is the number of bins in which series are classified, and $f_i(x)$ is the relative frequency of values in a given bin i. A number of 20 bins was used for H_s and 10 bins for T_p. An OP of 100% means that the model perfectly represents the observed data. The overlap test consists of calculating the OP between both series, and it has the advantage that the entire data distribution is considered. This method is based on the study of Perkins et al. [44] and has been used in previous studies [45].

2.2.2. Wave Power Resource

The WP resource is the wave power resource available in the natural environment. It is defined as the amount of wave energy flux per unit length of the wave front (expressed in kWm^{-1}) transmitted in the direction of wave propagation [12] and is represented by Equation (5):

$$\text{WP} = \frac{\rho g^2}{64\pi} H_S^2\, T_e, \tag{5}$$

where ρ is the density of seawater (considered here as $1025\ kgm^{-3}$), g is the gravitational acceleration, H_s is the significant wave height, and T_e is the energy period. The latter can be expressed in terms of wave peak period, T_p, as follows:

$$T_e = \alpha T_p. \tag{6}$$

Factor α varies with the shape of the wave spectrum. A value of $\alpha = 0.9$ was assumed in the present study. This supposition is equivalent to assuming a standard JONSWAP spectrum with a peak enhancement factor of $\gamma = 3.3$ [7].

2.2.3. Parameters to Analyze WEC Performance

Total electric power depends on both the natural WP resource available and the performance of the WEC in extracting it. Every device operates in a different range of depth, H_s and T_p. This means that if a sea state (H_s , T_p) does not belong to the WEC's operation range, no electric output will be generated. Operating intervals for H_s and T_p are included on the named "power matrix", which contains the electric power produced by the devices according to the sea state. In other words, each WEC obtains more energy with some sea states than with others so that each device is designed for certain wave conditions. The performance depends on this power matrix, which is provided by the manufacturer. Features such as WECs' geometric shape, size, and power take-off (PTO) parameters significantly influence the power matrix and thus the electric power output.

The expected average electric power (P_E, in kW) that can be extracted with a particular WEC is expressed by the Equation (7):

$$P_E = \frac{1}{100}\sum_{i=1}^{n_T}\sum_{j=1}^{n_H} p_{ij}P_{ij}, \tag{7}$$

where P_{ij} is the electric power obtained from an element ij of the power matrix of a particular WEC, p_{ij} is the probability of occurrence of a given sea state for this element of the power matrix (expressed in percentage), and n_T and n_H are the number of peak period and significant height bins considered, respectively. The methodology of [7,10,23,33,46] has been followed to compute P_E. Note that the power matrix is a matrix defined for a particular WEC, with specific geometric characteristics and a specific PTO principle. The same type of device with a different size, shape, or different PTO parameters would modify the power matrix and cause a change in the expected electrical power output. Additionally, if a sea state (H_s, T_p) is unlikely (small p_{ij}), the electrical power that can be extracted with that WEC will be smaller than if there is a more likely sea state.

Two parameters have been used to evaluate the performance of different devices to extract energy from waves: the power load factor (ε) and the efficiency. ε is defined as the relation between P_E and the maximum electric power that can produce a particular WEC (P_{max}), computed as the power matrix maximum value:

$$\varepsilon\ (\%) = 100\cdot\frac{P_E}{P_{max}}. \tag{8}$$

This dimensionless parameter reflects how far P_E is, which depends on the sea states at a particular location and the properties of the device, from the maximum power extractable from that device. Thus, a device well-fitted to the wave climate in the area would provide ε values close to 100%. Both P_E and P_{max} depend on the WEC size. Nevertheless, due to the fact that ε is defined as the ratio between the actual extracted power and the maximum extractable power, it can be used to compare devices with different geometries and sizes.

An additional parameter named efficiency was used to estimate the performance of the WECs in converting the wave energy into electricity. It is defined as the normalized capture width (C_w, in meters) with respect to the WEC's size [7,47]. In the literature, C_w is the width of the wave front that comprises the same amount of power as that absorbed by the WEC [47], and it is defined by Equation (9):

$$C_w = \frac{P_E}{WP}. \tag{9}$$

C_w is, like ε, one of the most common parameters to estimate WECs performances. However, as not all WECs have the same size, the new parameter efficiency has been introduced for this study. It is defined in the Equation (10):

$$\text{efficiency}\ (\%) = 100\cdot\frac{P_E}{WP\cdot L} \tag{10}$$

where and L is the WEC's length of the dimension by which the WEC captures the wave energy. The use of this parameter results in a dimensionless efficiency that allows the comparison among devices with different geometries and sizes.

Four different WECs have been considered: two point absorbers (Oyster and Aqua Buoy), an attenuator (Pelamis), and a terminator device (Atargis). Despite Oyster being classified on some occasions as an onshore terminator device [18,48], here, it is categorized as a point absorber due to its small size in comparison to the predominant wavelength. Although the development of the Oyster device was stopped a time ago [49], this device has been chosen as an example of a converter that operates at shallow depths since similar devices may be developed in the coming decades.

The technical properties of these devices can be found in [10,24,33,46,50–54] and summarized in Table 2. The different power matrices were obtained from [33,53].

Table 2. Type, maximum electric power, length opposite to the wave, and possible interval of depth for the installation of every analyzed WEC.

WEC	Type	P_{max} (kW)	L (m)	Depth (m)
Oyster	Point absorber	291	18	10–20
Atargis	Terminator	2530	60	40–100
Aqua Buoy	Point absorber	250	6	50–100
Pelamis	Attenuator	750	150	50–100

3. Results

3.1. Numerical Model Validation

The accuracy of the SWAN model to reproduce observed H_s and T_p measured by the Langosteira II buoy is shown in both Figure 2 and Table 3. Figure 2 represents the relative frequency of occurrence of a certain value of H_s (or T_p) contained in each bin. The red (blue) bars represent the relative frequency for the observed (numerical) data. The bars are plotted separately so that they can be properly differentiated. Blanks should not be interpreted as an absence of values, but rather the entire range of values on the x-axis is covered. Despite both data series being distributed over the same intervals, blue bars are drawn displaced with respect to the red ones for visualization purposes. Figure 2a shows that H_s data from MeteoGalicia are slightly underestimated for the smallest waves ($H_s < 1.575$ m) and overestimated for the largest ones. The comparison between the numerical and observed data is not clear for T_p. The SWAN model tends to overestimate T_p values in the intervals $(7, 9)$, $(9, 11)$, and $(13, 15)$, and underestimates in the other bins (Figure 2b).

Table 3. Normalized root mean square error (NRMSE), normalized bias error (NBias), Spearman correlation coefficient (ρ_S), and overlapping percentage (OP) for significant wave height (H_s) and peak period (T_p).

Variable	NRMSE (%)	NBias (%)	ρ_S	OP (%)
H_s	20.10	8.72	0.94	92.26
T_p	14.82	−1.15	0.83	91.70

The NRMSE, the NBias error, the Spearman correlation coefficient (ρ_S), and the overlapping percentage (OP) between both series are shown in Table 3. The NRMSE and the NBias error show smaller values for H_s and similar values for T_p than those of Bento et al. in [32]. Values of ρ_S close to unity and an OP around 90% were obtained for both variables, which allow considering that the numerical simulations adequately reproduce the reality [45].

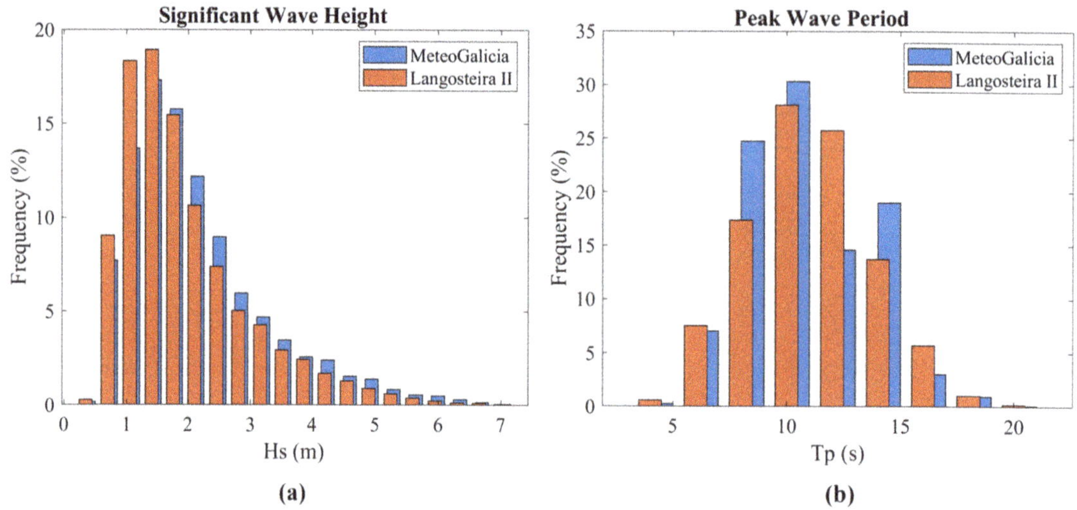

Figure 2. Histograms for (**a**) H_s and (**b**) T_p from the SWAN simulations (blue) and the Langosteira II buoy measures (red).

3.2. Wave Power Resource

First, the mean WP resource was calculated along the Galician coast from hourly data over the historical period (2014–2021). Figure 3 shows the WP resource for three regions: the West Coast (a), the Northwest Coast (b), and the North Coast (c). Overall, the WP resource is lower nearshore than offshore. Along the West Coast, WP increases from less than 10 kWm^{-1} inside the estuaries to around 50 kWm^{-1} at offshore (Figure 3a). As expected, the WP resource is related to the bathymetry—the shallowest areas are associated with the lowest WP resource. Spots of maximum WP resource (~60 kWm^{-1}) are found in submerged reef systems relatively far from the shore (Figure 3b,c). In the easternmost part of the North Coast, the resource is lower (around 20 kWm^{-1} at ~10 km from the coast, see Figure 3c) since waves usually come from NW, and that area is leeward.

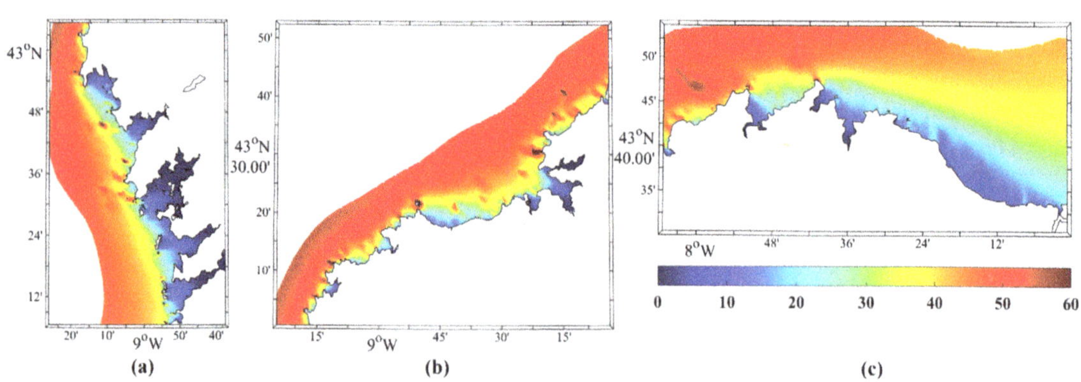

Figure 3. WP resource for (**a**) West Coast, (**b**) Northwest Coast, and (**c**) North Coast from 2014 to 2021.

3.3. Performance of WECs

Macroscopically, P_E patterns (Figure 4) are very similar for the three areas, decreasing landward, as previously observed for WP. The electric power that can be extracted from the Atargis device is greater compared to the other converters—an order of magnitude greater than Oyster and Pelamis and two orders greater than Aqua Buoy. These differences are justified by the disparities in the size of the devices, which result in the values of the WECs power matrices, and also by the fact that each device operates in a specific range of sea state (H_s, T_p). Similar values of P_E and their spatial distribution were found in [7] for the Aqua Buoy and Pelamis devices.

Along the West and Northwest Coast (Figure 5), the Atargis converter shows the most significant power load factor (ε ~45%), followed by the Oyster converter (~40% in shallow offshore areas) and Aqua Buoy with ε around 22%. Pelamis shows the most minor power load factor (~15%) in those regions. Note that the plotted area is different for each WEC, as each WEC works at different depth ranges. These results are consistent with those described in [7] for the Aqua Buoy and Pelamis devices, where P_E and ε show similar patterns. For example, along the North Coast the WECs performance decreases eastwards (right column). In this region, the power load factor for Atargis ranges between ~30% east of 8°36′ W and ~45% north of the Northwest Coast. It ranges between ~20% east of 8°36′ W and ~40% north of the Northwest Coast.

The efficiency percentage of WECs, represented in Figure 6, has a different distribution than the WP resource, P_E and ε, increasing landwards. Atargis shows the highest efficiency in the study area (more than 50% inside the estuaries and ~35% outside). For Oyster, efficiency around 35% is observed inside the estuaries (left column) and around 27% in the Artabro Gulf (middle column) and in the east part of the North Coast (right column). Aqua Buoy has a moderately homogeneous pattern of efficiency (~25%) in all regions. The Pelamis device shows a low efficiency (< 10%) to exploit the WP resource in all the regions. Taking into account that the capture width is the efficiency multiplied by the length of the WEC, the Aqua Buoy and Pelamis efficiencies are consistent with the capture width (C_w) of 1.5 m and 3 m obtained by [7] for Aqua Buoy and Pelamis, respectively, along the North Coast for the period (1979–2005).

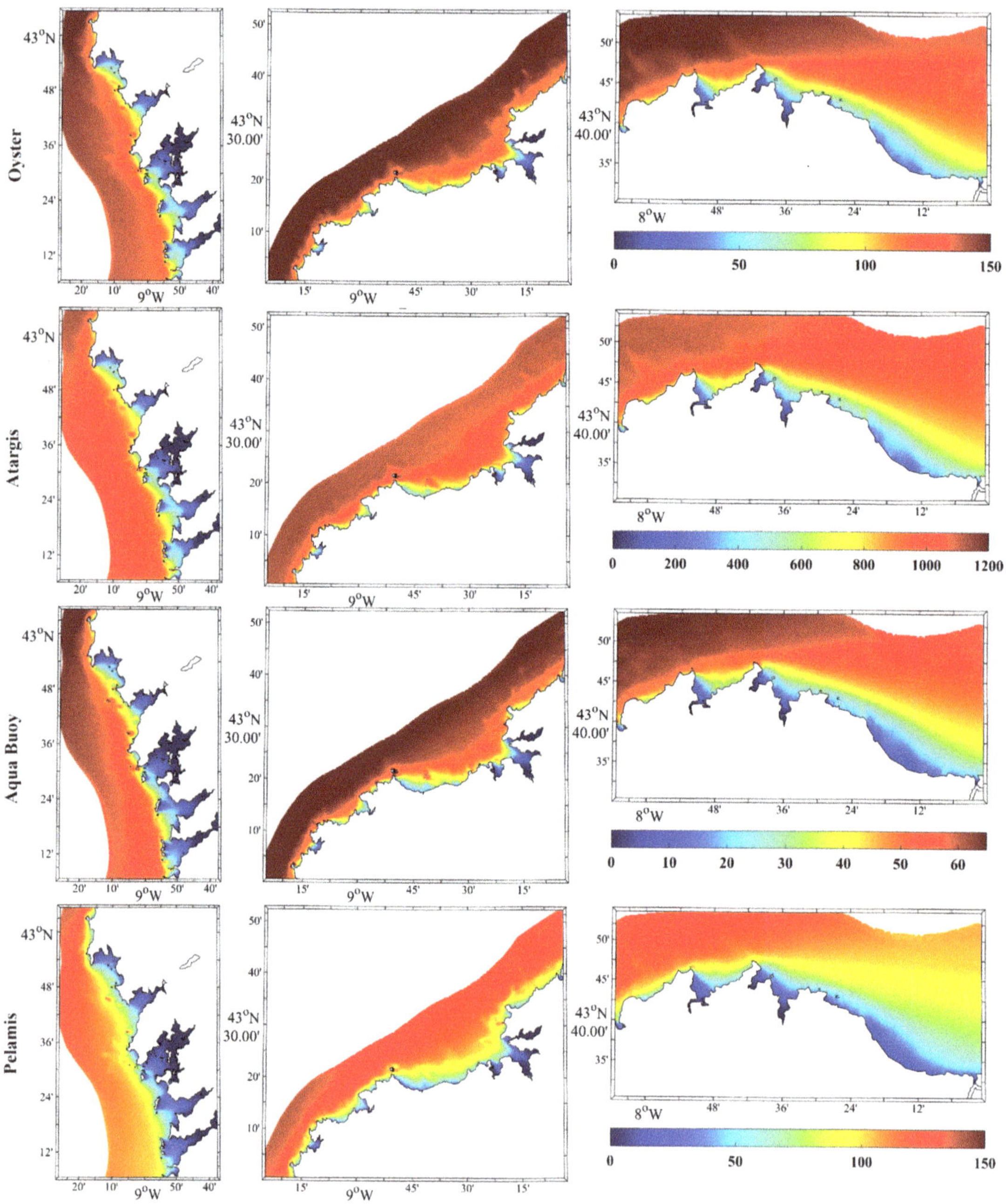

Figure 4. Electric power capacity (P_E, in kW) of the Oyster, Atargis, Aqua Buoy, and Pelamis devices for the West Coast (**left** column), the Northwest Coast (**middle** column), and the North Coast (**right** column) for the period (2014–2021).

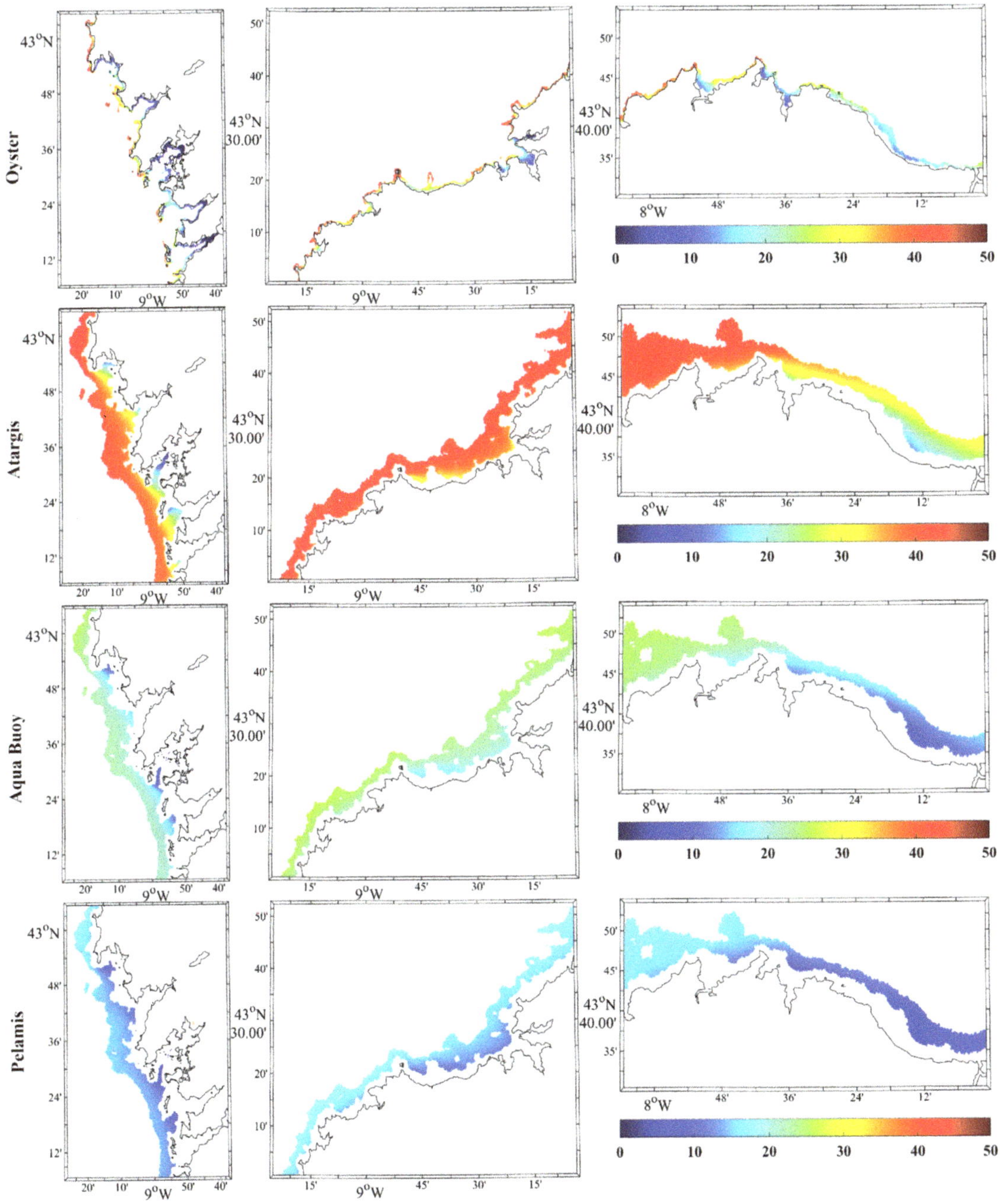

Figure 5. Power load factor (ε, in %) of the Oyster, Atargis, Aqua Buoy, and Pelamis devices for the West Coast (**left** column), the Northwest Coast (**middle** column), and the North Coast (**right** column) for the period (2014–2021).

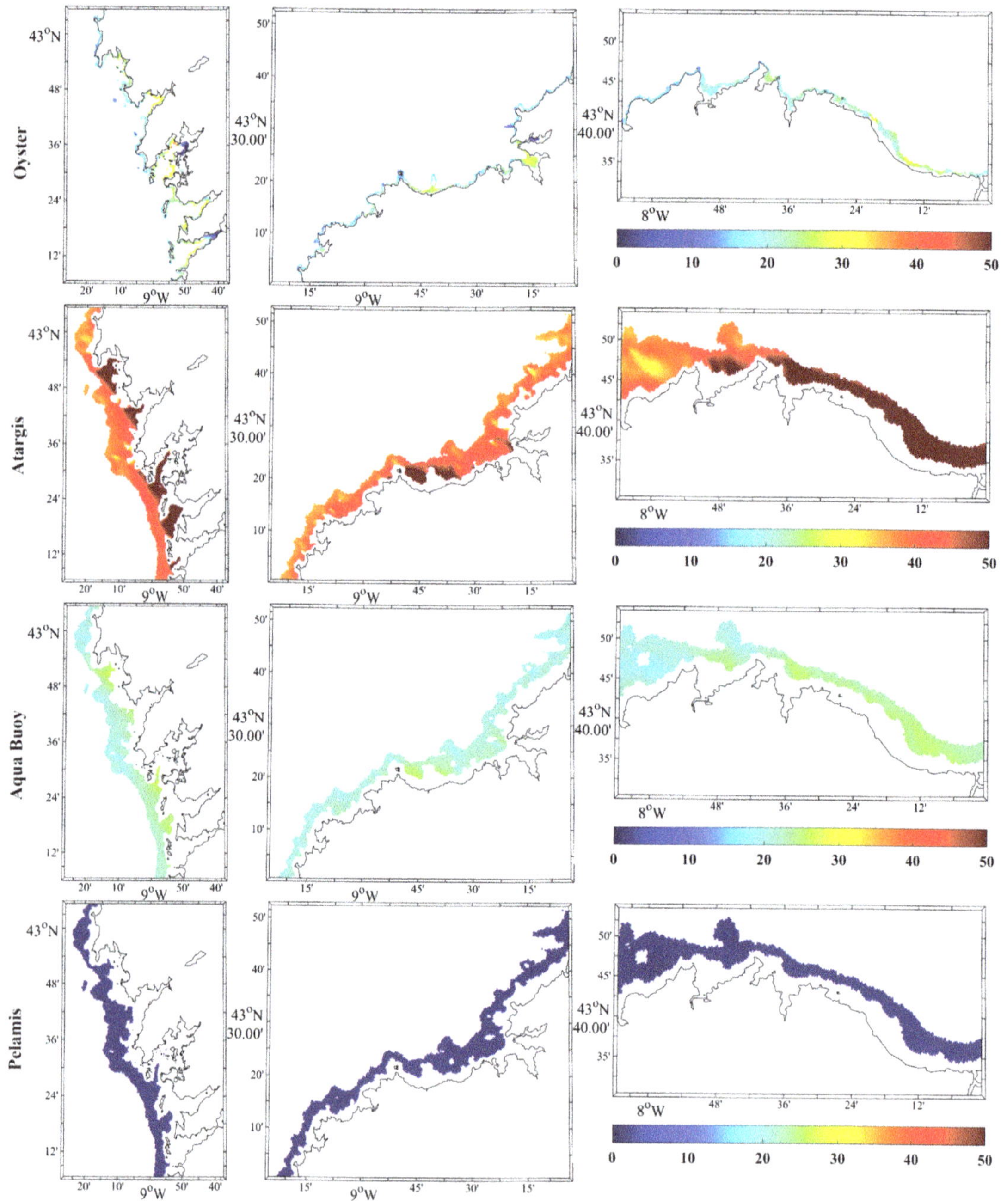

Figure 6. Efficiency (in %) of the Oyster, Atargis, Aqua Buoy, and Pelamis devices for the West Coast (**left** column), the Northwest (**middle** column), and the North Coast (**right** column) for the period (2014–2021).

4. Discussion

Although the implementation of renewable energies is a key factor in mitigating climate change, not all of them have reached the same level of maturity. Wave energy is an auspicious renewable energy due to its high stability, predictability, and power density. However, it has found many difficulties such as the lack of consensus between the best technology, the conflict with other socio-economic activities, and the high cost [11]. These uncertainties have delayed the development of this marine renewable energy. However, technological advances are expected to allow the installation of wave energy farms in the upcoming decades. As shown in this analysis, the use of high spatial resolution simulations containing wave data is crucial to discover the coastal areas with the best conditions for the future installation of wave energy farms. In addition, this approach is also helpful to know the WEC that best fits each area because it is possible to analyze its efficiency only in the depth range that can be installed. As a whole, this study highlights the great potential that Galicia has to become a wave energy-generating region.

Overall, the results have shown that the highest wave resource is found in the northernmost part of the West Coast, the Northwest Coast, and the westernmost part of the North Coast. This is mainly due to the fact that the swell has a north-western direction [34], and these areas are more exposed to waves, as previously pointed out in [30]. In the same way, the least WP resource is found in the easternmost part of the North Coast. This difference in values is possible because the swell is shielded by the northernmost part of Galicia. The spatial distribution of the WP resource and its values are very similar to those shown in [7,30,34] for other different periods.

The WECs use only part of the WP resource available to produce electricity. The electrical energy capable of producing the WEC depends on the power matrix and the probability of occurrence of the sea states. Looking at the power matrix for the four converters in [33], it is to be expected that the device capable of producing the highest electrical energy is Atargis since the elements in the power matrix are the highest. Pelamis has a high power output, especially for high waves (>6 m), while Aqua Buoy and Oyster have the smallest power elements. However, the probability of occurrence of a sea state (H_s, T_p) has to be also taken into account to estimate the P_E. Results have shown that the device capable of producing the most electricity in the three studied regions is undoubtedly Atargis. Oyster also gives high P_E values, especially along the Northwest Coast region (Figure 4). Installing Aqua Buoy seems not to be a good idea in terms of energy production because the lowest P_E has been obtained for that converter. The Pelamis device has not provided as much electrical power as expected when looking at the power matrix. This is because not as many optimal sea states have occurred for this converter. The P_E pattern observed in all WECs is very similar to the wave resource. As the P_E decreases towards the coast, it is possible to think that it is better to install wave energy farms away from the coast. However, installing offshore farms entails an increase in installation and maintenance costs, so a balance must be found between high energy production and proximity to the coast. The installation of a farm of Atargis devices seems to be the best option due to their P_E and the recommended depth. The Northwest Coast is likely to be the most suitable region for that installation.

The power load factor (ε) provides an approximation of how much the device is being availed because it compares the electrical energy obtained with the maximum energy it can produce. Atargis is the converter that showed the highest ε values in the study regions, followed by Oyster. Although Oyster and Pelamis had similar P_E, Pelamis shows the lowest ε due to the higher maximum electric power value (P_{max}).

The other parameter to estimate the performance of a WEC is efficiency. The efficiency represents the performance in harvesting the WP resource taking into account the dimensions of the WEC. The results have shown that the efficiency increases towards the coast. One possible explanation is that the WP resource is much larger than the observed offshore P_E because P_E depends on the WEC limitations and does not consider the most energetic wave states, whereas WP resource does. The similarity between the P_E and the WP resource is also observable onshore. Regarding efficiency, Atargis has shown the best results in the study regions although Oyster and Aqua Buoy are acceptable. Size is another aspect to consider, especially in areas with insufficient space available for a large device such as Atargis. The installation of several Aqua Buoys can be considered in this areas to increase energy production. Pelamis is the least efficient device in WP resource harvesting because of its dimensions, being much larger than Oyster (Table 2), although both show similar P_E. The high efficiency of Atargis is explained due to its small dimensions and large P_E.

Environmental and human factors must also be considered to choose the best device and installation area. Although the analysis of legal concerns is out of the scope of the present analysis, it is important to mention that Galicia has a large number of ports distributed throughout the coast [55] and places with environmental and fishing interest, such as the Atlantic Islands National Park of Galicia and the Os Miñarzos marine reserve in the Rías Baixas [56,57] or the Costa da Morte Bird Protection Zone (ZEPA in Spanish) in the Northwest Coast [56]. In fact, it can be observed that practically the entire Galician coast is bathed in Natural and National Parks, Sites of Community Importance (LIC, in Spanish), marine reserves of fishing interest, and ZEPAs and RAMSAR wetlands [56,58,59]. Fortunately, WECs are barely above the sea level [21], so they do not have to negatively influence the flight paths of birds.

Aqua Buoy and Pelamis have the advantage that they can be installed far from other important socio-economic activities and environmental places—usually located near the coast. However, the installation far from the coast can lead to inflate the installation and maintenance costs. Instead, its installation can be considered to attract offshore blue economy business. The installation of Oyster could be complicated because the most suited areas are coastal waters, where most of the environmental interest sites are located. Possible locations would be near artificial structures, e.g., ports where visual impact and conflict with other activities are less significant. A possible site could be near the Outer Port of A Coruña at Punta Langosteira. Installation near harbors can reduce maintenance costs [30]. Atargis has the advantage that it can be located somewhat further from the coast, avoiding conflict with protected areas. In addition, Atargis is installed under the surface, nullifying the visual impact. For this, Atargis seems to be the better option for implementing wave energy on the Galician coast. However, its size makes it difficult to install within the estuaries where the Aqua Buoy device appears to be one of the most efficient due to its small size. The Atargis impact on the seabed could also be studied in the future.

5. Conclusions

This study analyzed the wave energy resource in three regions of the Galician coast over the period 2014–2021 and the expected electrical energy output from four wave energy converters—Oyster, Atargis, Aqua Buoy, and Pelamis. Their performance was also investigated attending to their power load factor and efficiency. To fulfil this task, high-resolution significant wave height and peak period data were obtained from simulations of the third-generation SWAN model. These data were validated with the measurements of a buoy located on the Northwest Coast, obtaining a high overlapping percentage. The main findings of this study can be summarized as follows:

- The wave power resource, the expected electrical energy output, and the power load factor decrease landward, while the efficiency increases landward.
- The highest wave power resource is found in the northernmost part of the West Coast, the Northwest Coast, and the westernmost part of the North Coast.
- Atargis is the device that shows the highest expected electrical energy output, followed by Oyster. Aqua Buoy is the device with the lowest electrical energy output.
- Attending to the power load factor, Atargis is the best-availed device, followed by Oyster. Pelamis has obtained the lowest performance in the three regions.
- The parameter efficiency could substitute capture width in future studies in order to estimate the WEC performance because it allows making comparisons between different WECs regarding their dimensions.

- Looking at the efficiency, Atargis is definitely the most efficient device, while Oyster and Aqua Buoy are quite efficient in harvesting the wave power resource in the three regions. Pelamis has shown a low efficiency in the whole area.
- Atargis seems to be the best device to be installed in the Galician coast—especially in the Northwest Coast region—due to its expected electrical output, performance, its location under the surface, and optimum depth, avoiding areas of environmental interest.

The present study showed the Galicia's great potential to become a wave energy-generating region. Regarding the parameters considered to determine the performance of the different types of WECs, and taking into account the size and location of WECs, Atargis is the most suitable to be installed in the Artabro Gulf, where it shows high values of P_E, load capacity factor, and efficiency. In addition, it can be located somewhat further from the coast than the others, avoiding conflicts with protected areas. The Aqua Buoy is shown to be the most suitable to be installed within estuaries, where the available space is limited and must coexist with other economic activities, such as raft aquaculture, restricting the installation of large devices such as Atargis or Oyster. The Aqua Buoy has high efficiency due to its small size, and the power output can be increased by considering various devices. The installation of these devices in the vicinity of shellfishing areas could also protect them from waves by dissipating part of their energy. The appropriate locations for the Oyster device, which shows high efficiency and capacity factor for the entire area, are in the vicinity of harbors such as the Outer Port of A Coruña to reduce costs and avoid conflicts with protected and shellfishing areas. Finally, it should be mentioned that future studies should estimate the economic profitability of the devices to fulfil a more comprehensive assessment of their economic viability of the wave power in the study areas.

Author Contributions: Conceptualization, B.A.-P., X.C., M.d., and M.G.-G.; methodology, B.A.-P., X.C., M.d., A.S.R., J.M.D., P.C., L.R., and M.G.-G.; software, B.A.-P., X.C., and M.G.-G.; validation, B.A.-P.; formal analysis, B.A.-P. and L.R.; investigation, P.C. and M.G.-G.; resources, M.d., P.C., and M.G.-G.; data curation, B.A.-P. and P.C.; writing—original draft preparation, B.A.-P.; writing—review and editing, M.d., X.C., A.S.R., J.M.D., P.C., L.R., and M.G.-G.; visualization, B.A.-P., X.C., M.d., A.S.R., J.M.D., L.R., and M.G.-G.; supervision, M.G.-G.; project administration, M.d.; funding acquisition, M.d. All authors have read and agreed to the published version of the manuscript.

Funding: This research was funded by the Spanish Government through a Juan de la Cierva Postdoctoral Fellowship (IJC2020-043745-I). This work was partially supported by Xunta de Galicia under project ED431C 2021/44 (Grupos de Referencia Competitiva) and Ministry of Science and Innovation of the Government of Spain under the project SURVIWEC PID2020-113245RB-I00. We acknowledge financial support to CESAM by FCT/MCTES (UIDP/50017/2020+UIDB/50017/2020+LA/P/0094/2020), through national funds

Institutional Review Board Statement: Not applicable.

Informed Consent Statement: Not applicable.

Data Availability Statement: Publicly available datasets were analyzed in this study. This data can be found here: http://mandeo.meteogalicia.gal/thredds/catalog/modelos/SWAN_HIST/galicia/catalog.html and https://www.puertos.es/es-es/oceanografia/Paginas/portus.aspx (accesed on 22 January 2022).

Acknowledgments: The authors thank the MeteoGalicia and Puertos del Estado organizations for the free distribution of wave data.

Conflicts of Interest: The authors declare no conflict of interest. The funders had no role in the design of the study; in the collection, analyses, or interpretation of data; in the writing of the manuscript, or in the decision to publish the results.

References

1. IEA. World Energy Balances: Overview. Available online: https://www.iea.org/reports/world-energy-balances-overview (accessed on 22 January 2022).
2. Masson-Delmotte, V.; Zhai, P.; Pirani, A.; Connors, S.L.; Péan, C.; Berger, S.; Caud, N.; Chen, Y.; Goldfarb, L.; Gomis, M.I.; et al. *Climate Change 2021: The Physical Science Basis. Contribution of Working Group I to the Sixth Assessment Report of the Intergovernmental Panel on Climate Change*; Cambridge University Press: Cambridge, UK; IPCC: New York, NY, USA, 2021. [CrossRef]
3. Oliveira-Pinto, S.; Stokkermans, J. Assessment of the Potential of Different Floating Solar Technologies—Overview and Analysis of Different Case Studies. *Energy Convers. Manag.* **2020**, *211*, 112747. [CrossRef]
4. Ritchie, H.; Roser, M. Fossil Fuels. Available online: https://ourworldindata.org/fossil-fuels?country= (accessed on 22 January 2022).
5. Iberdrola. COP26: Iberdrola En La Cumbre Del Clima 2021. Available online: https://www.iberdrola.com/sostenibilidad/contra-cambio-climatico/cop26 (accessed on 22 January 2022).
6. General Assembly. Resolution Adopted by the General Assembly on 6 July 2017; United Nations, A/RES/71/313. 2017. Available online: https://ggim.un.org/documents/a_res_71_313.pdf (accessed on 22 January 2022).
7. Ribeiro, A.S.; deCastro, M.; Rusu, L.; Bernardino, M.; Dias, J.M.; Gomez-Gesteira, M. Evaluating the Future Efficiency of Wave Energy Converters along the NW Coast of the Iberian Peninsula. *Energies* **2020**, *13*, 3563. [CrossRef]
8. Lavidas, G.; Blok, K. Shifting Wave Energy Perceptions: The Case for Wave Energy Converter (WEC) Feasibility at Milder Resources. *Renew. Energy* **2021**, *170*, 1143–1155. [CrossRef]
9. IRENA. *Offshore Renewables: An Action Agenda for Deployment*; International Renewable Energy Agency: Abu Dhabi, United Arab Emirates, 2021; ISBN 978-92-9260-349-6. Available online: https://www.irena.org/-/media/Files/IRENA/Agency/Publication/2021/Jul/IRENA_G20_Offshore_renewables_2021.pdf (accessed on 22 January 2022).
10. Rusu, E. Evaluation of the Wave Energy Conversion Efficiency in Various Coastal Environments. *Energies* **2014**, *7*, 4002–4018. [CrossRef]
11. Aderinto, T.; Li, H. Review on Power Performance and Efficiency of Wave Energy Converters. *Energies* **2019**, *12*, 4329. [CrossRef]
12. Mota, P.; Pinto, J. Wave Energy Potential along the Western Portuguese Coast. *Renew. Energy* **2014**, *71*, 8–17. [CrossRef]
13. Kim, S.-J.; Koo, W.; Kim, M.-H. The Effects of Geometrical Buoy Shape with Nonlinear Froude-Krylov Force on a Heaving Buoy Point Absorber. *Int. J. Nav. Archit. Ocean Eng.* **2021**, *13*, 86–101. [CrossRef]
14. Wang, L.; Zhao, T.; Lin, M.; Li, H. Towards Realistic Power Performance and Techno-Economic Performance of Wave Power Farms: The Impact of Control Strategies and Wave Climates. *Ocean Eng.* **2022**, *248*, 110754. [CrossRef]
15. Ciappi, L.; Cheli, L.; Simonetti, I.; Bianchini, A.; Talluri, L.; Cappietti, L.; Manfrida, G. Wave-to-Wire Models of Wells and Impulse Turbines for Oscillating Water Column Wave Energy Converters Operating in the Mediterranean Sea. *Energy* **2022**, *238*, 121585. [CrossRef]
16. Wang, L.; Ringwood, J.V. Control-Informed Ballast and Geometric Optimisation of a Three-Body Hinge-Barge Wave Energy Converter Using Two-Layer Optimisation. *Renew. Energy* **2021**, *171*, 1159–1170. [CrossRef]
17. Mohtat, A.; Yim, S.; Osborne, A. Energy Content Characterization of Water Waves Using Linear and Nonlinear Spectral Analysis. *J. Offshore Mech. Arct. Eng.* **2022**, *144*, 011203. [CrossRef]
18. Zhang, Y.; Zhao, Y.; Sun, W.; Li, J. Ocean Wave Energy Converters: Technical Principle, Device Realization, and Performance Evaluation. *Renew. Sustain. Energy Rev.* **2021**, *141*, 110764. [CrossRef]
19. Lehmann, M.; Karimpour, F.; Goudey, C.A.; Jacobson, P.T.; Alam, M.-R. Ocean Wave Energy in the United States: Current Status and Future Perspectives. *Renew. Sustain. Energy Rev.* **2017**, *74*, 1300–1313. [CrossRef]
20. Mofor, L.; Goldsmith, J.; Jones, F. Ocean Energy: Technology Readiness, Patents, Deployment Status and Outlook. *Abu Dhabi* **2014**, 27. Available online: https://www.irena.org/publications/2014/Aug/Ocean-Energy-Technologies-Patents-Deployment-Status-and-Outlook (accessed on 22 January 2022).
21. Pecher, A.; Kofoed, J.P. *Handbook of Ocean Wave Energy*; Springer Nature: Berlin, Germany, 2017.
22. Olmo, B. *Explotación Del Potencial de Energía Del Oleaje En Función Del Rango de Trabajo de Prototipos Captadores*; Universitat Politècnica de Catalunya: Barcelona, Spain, 2009. Available online: https://upcommons.upc.edu/handle/2099.1/8720 (accessed on 22 January 2022).
23. Rusu, L.; Onea, F. The Performance of Some State-of-the-Art Wave Energy Converters in Locations with the Worldwide Highest Wave Power. *Renew. Sustain. Energy Rev.* **2017**, *75*, 1348–1362. [CrossRef]
24. Bozzi, S.; Archetti, R.; Passoni, G. Wave Electricity Production in Italian Offshore: A Preliminary Investigation. *Renew. Energy* **2014**, *62*, 407–416. [CrossRef]
25. La Camera, F. IRENA Chief: Europe Is 'the Frontrunner' on Tidal and Wave Energy. 2020. Available online: https://www.euractiv.com/section/energy/interview/irena-chief-europe-is-the-frontrunner-on-tidal-and-wave-energy/ (accessed on 22 January 2022).
26. IDAE. Eólica Marina y Energías Del Mar En España. Available online: https://www.idae.es/tecnologias/energias-renovables/uso-electrico/eolica/eolica-marina/eolica-marina-y-energias-del-mar (accessed on 22 January 2022).
27. MITECO. *Hoja de Ruta Para El Desarrollo de La Eólica Marina y de Las Energías del Mar en España*; MITECO: Madrid, Spain, 2021. Available online: https://www.miteco.gob.es/es/ministerio/planes-estrategias/desarrollo-eolica-marina-energias/eshreolicamarina-pdfaccesiblev5_tcm30-534163.pdf (accessed on 22 January 2022).

28. Ferrari, F.; Besio, G.; Cassola, F.; Mazzino, A. Optimized Wind and Wave Energy Resource Assessment and Offshore Exploitability in the Mediterranean Sea. *Energy* **2020**, *190*, 116447. [CrossRef]
29. Rusu, L. Evaluation of the near Future Wave Energy Resources in the Black Sea under Two Climate Scenarios. *Renew. Energy* **2019**, *142*, 137–146. [CrossRef]
30. Iglesias, G.; López, M.; Carballo, R.; Castro, A.; Fraguela, J.A.; Frigaard, P. Wave Energy Potential in Galicia (NW Spain). *Renew. Energy* **2009**, *34*, 2323–2333. [CrossRef]
31. Silva, D.; Bento, A.R.; Martinho, P.; Guedes Soares, C. High Resolution Local Wave Energy Modelling in the Iberian Peninsula. *Energy* **2015**, *91*, 1099–1112. [CrossRef]
32. Bento, A.R.; Martinho, P.; Soares, C.G. Wave Energy Assessement for Northern Spain from a 33-Year Hindcast. *Renew. Energy* **2018**, *127*, 322–333. [CrossRef]
33. Silva, D.; Rusu, E.; Soares, C.G. Evaluation of Various Technologies for Wave Energy Conversion in the Portuguese Nearshore. *Energies* **2013**, *6*, 1344–1364. [CrossRef]
34. Carmeáns Rodríguez, M.; Suárez Rey, R.; Arean Varela, N.; Suárez Bilbao, M.; Carracedo García, P.; Gómez Hombre, B. *Atlas de Oleaje de Galicia. Caracterización Del Oleaje Costero Con Alta Resolución*; MeteoGalicia: Santiago de Compostela, Spain, 2014. Available online: https://www.meteogalicia.gal/datosred/infoweb/meteo/proxectos/energymare/Atlas_Ondas_Galicia.pdf (accessed on 22 January 2022).
35. SWAN. Available online: https://www.tudelft.nl/en/ceg/about-faculty/departments/hydraulic-engineering/sections/environmental-fluid-mechanics/research/swan (accessed on 22 January 2022).
36. Welcome to the SWAN Home Page. Available online: https://swanmodel.sourceforge.io/ (accessed on 22 January 2022).
37. Features of SWAN. Available online: https://swanmodel.sourceforge.io/features/features.htm (accessed on 22 January 2022).
38. Spectral Action Balance Equation. Available online: https://swanmodel.sourceforge.io/online_doc/swantech/node12.html (accessed on 22 January 2022).
39. THREDDS Data Server. Available online: http://mandeo.meteogalicia.gal/thredds/catalog/modelos/SWAN_HIST/galicia/catalog.html (accessed on 22 January 2022).
40. Puertos del Estado Prediccion de Oleaje, Nivel Del Mar; Boyas y Mareografos. Available online: https://www.puertos.es/es-es/oceanografia/Paginas/portus.aspx (accessed on 22 January 2022).
41. Des, M.; Martínez, B.; deCastro, M.; Viejo, R.M.; Sousa, M.C.; Gómez-Gesteira, M. The Impact of Climate Change on the Geographical Distribution of Habitat-Forming Macroalgae in the Rías Baixas. *Mar. Environ. Res.* **2020**, *161*, 105074. [CrossRef]
42. Costoya, X.; Rocha, A.; Carvalho, D. Using Bias-Correction to Improve Future Projections of Offshore Wind Energy Resource: A Case Study on the Iberian Peninsula. *Appl. Energy* **2020**, *262*, 114562. [CrossRef]
43. Kumar, A.; Abirami, S. Aspect-Based Opinion Ranking Framework for Product Reviews Using a Spearman's Rank Correlation Coefficient Method. *Inf. Sci.* **2018**, *460–461*, 23–41. [CrossRef]
44. Perkins, S.E.; Pitman, A.J.; Holbrook, N.J.; McAneney, J. Evaluation of the AR4 Climate Models' Simulated Daily Maximum Temperature, Minimum Temperature, and Precipitation over Australia Using Probability Density Functions. *J. Clim.* **2007**, *20*, 4356–4376. [CrossRef]
45. Costoya, X.; Decastro, M.; Santos, F.; Sousa, M.; Gómez-Gesteira, M. Projections of Wind Energy Resources in the Caribbean for the 21st Century. *Energy* **2019**, *178*, 356–367. [CrossRef]
46. Rusu, L.; Onea, F. Assessment of the Performances of Various Wave Energy Converters along the European Continental Coasts. *Energy* **2015**, *82*, 889–904. [CrossRef]
47. Price, A.A.E.; Dent, C.J.; Wallace, A.R. On the Capture Width of Wave Energy Converters. *Appl. Ocean Res.* **2009**, *31*, 251–259. [CrossRef]
48. Choupin, O.; Andutta, F.P.; Etemad-Shahidi, A.; Tomlinson, R. A Decision-Making Process for Wave Energy Converter and Location Pairing. *Renew. Sustain. Energy Rev.* **2021**, *147*, 111225. [CrossRef]
49. Curto, D.; Franzitta, V.; Guercio, A. Sea Wave Energy. A Review of the Current Technologies and Perspectives. *Energies* **2021**, *14*, 6604. [CrossRef]
50. Henry, A.; Doherty, K.; Cameron, L.; Whittaker, T.; Doherty, R. Advances in the Design of the Oyster Wave Energy Converter. In Proceedings of the Marine Renewable and Offshore Wind Energy Conference, Royal Institution of Naval Architects, London, UK, 21–23 July 2010.
51. Farkas, A.; Degiuli, N.; Martić, I. Assessment of Offshore Wave Energy Potential in the Croatian Part of the Adriatic Sea and Comparison with Wind Energy Potential. *Energies* **2019**, *12*, 2357. [CrossRef]
52. Frandsen, J.; Doblaré, M.; Rodriguez, P.; Reyes, M. Technical Assessment of the Pelamis Wave Energy Converter Concept. 2012. Available online: https://www.researchgate.net/publication/317014924_Technical_assessment_of_the_Pelamis_wave_energy_converter_concept (accessed on 22 January 2022).
53. Ermakov, A.; Ringwood, J.V. Rotors for Wave Energy Conversion—Practice and Possibilities. *IET Renew. Power Gener.* **2021**, *15*, 3091–3108. [CrossRef]
54. Siegel, S.G. Numerical Benchmarking Study of a Cycloidal Wave Energy Converter. *Renew. Energy* **2019**, *134*, 390–405. [CrossRef]
55. Localizador de Portos EPPG. Available online: https://portosdegalicia.gal/es/web/portos-de-galicia/locportos (accessed on 23 January 2022).
56. Intecmar. Available online: http://ww3.intecmar.gal/intecmar/ (accessed on 23 January 2022).

57. Reserva Marina de Interés Pesquero. Available online: https://www.marsostenible.com/reserva-marina-de-interes-pesquero/ (accessed on 23 January 2022).
58. Zona Especial Protección Para Las Aves—ZEPA. Available online: https://www.indemares.es/areas-marinas/zona-especial-proteccion-para-las-aves-zepa (accessed on 23 January 2022).
59. Portos de Galicia. *Delimitación de Los Espacios y Usos Portuarios del Puerto de BRENS-CEE*; Portos de Galicia: Spain, 2012. Available online: https://portosdegalicia.gal/documents/10627/36363/DEUP+Brens+-+Cee_Memoria_1.0.pdf (accessed on 22 January 2022).

Journal of
*Marine Science
and Engineering*

MDPI

Article

Characteristics of the Wind and Wave Climate along the European Seas Focusing on the Main Maritime Routes

Ana-Maria Chirosca and Liliana Rusu *

Department of Mechanical Engineering, Faculty of Engineering, "Dunarea de Jos" University of Galati, 47 Domneasca Street, 800008 Galati, Romania; ana.chirosca@ugal.ro
* Correspondence: liliana.rusu@ugal.ro

Abstract: European seas have a strong economic role both in terms of transport and tourism. Providing more knowledge, regarding the mean and extreme values of the wind and sea state conditions in the areas characterized by high maritime traffic, helps to improve navigational safety. From this perspective, six zones with high maritime traffic are studied. ERA5 database, a state-of-the-art global reanalysis dataset provided by ECMWF (European Centre for Medium-Range Weather Forecasts), is used to assess the average values and the percentiles for the wind speed and the main wave parameters in the target areas considering the period 2001–2020. The main European routes and the extreme conditions along them as well as the areas characterized by high values of wind speed and high waves were also identified. A more comprehensive picture of the expected dynamics of the environmental matrix along the most significant shipping routes is useful because in this way the most dangerous areas could be avoided by ships for the safety of passengers and transported goods.

Keywords: European seas; sea state conditions; wind climate; wave climate; statistical analysis

Citation: Chirosca, A.-M.; Rusu, L. Characteristics of the Wind and Wave Climate along the European Seas Focusing on the Main Maritime Routes. *J. Mar. Sci. Eng.* **2022**, *10*, 75. https://doi.org/10.3390/jmse10010075

Academic Editor: José-Santos López-Gutiérrez

Received: 19 December 2021
Accepted: 4 January 2022
Published: 7 January 2022

1. Introduction

The shipbuilding industry, especially the maritime traffic of both goods and people, has a considerable impact on the economy and technological evolution in the European sea basin. Even though the pandemic situation due to the Coronavirus virus (COVID-19) affected passenger transport during this period, freight transport continued to increase due to high market demand. For these reasons, it is necessary to investigate how maritime transport is carried out and what are the problems it faces, both in order to optimize it so that we can benefit from its advantages at maximum levels, but also to counteract various accidents at sea. For Europe, the transport of people, as well as goods and services, is a major sector that contributes to economic development and is constantly evolving in order to improve transport capacity, ship speed, port management time, safety, and security conditions.

One of the main factors that affect maritime transport is weather conditions, especially wind and wave parameters. Unfortunately, these parameters are constantly changing, and therefore have been the subject of several studies [1–3], in order to investigate how it behaves over the years and how climatic conditions evolve, but also to be able to perform predictions for their future behavior [4]. Another important aspect is the knowledge of the wind and wave climate that helps to reduce the number of accidents in areas prone to severe weather conditions, but also in coastal areas where due to congestion and limited space for various maneuvers, as well as limited water depth, the accident rate is higher [1].

At the same time, the study of wave and wind characteristics is of great interest for the offshore wind platform industry, which has begun to increase in number, especially in the North Sea area, where an offshore power grid is developing [5,6]. Climate change effects are reflected depending on the characteristics of each area, as well as the activities performed in each sea. Therefore, a detailed study of the present climate is needed to identify the changes in future climatic conditions [3], especially on the most frequented routes, in order to be able to counteract unfavorable navigation situations.

The European seas have various wind and wave climate characteristics. Thus, some of them are open seas and they have a direct connection with the Atlantic Ocean, and their wind and wave climate is strongly affected by the weather conditions in the ocean, while other seas are semi-closed basins where the wave climate is affected by local weather conditions. The Black Sea and the Baltic Sea, being semi-closed seas, are characterized by not very aggressive climatic conditions. In the Black Sea, only the eastern area is prone to the greatest swelling [7], but it is nevertheless an area suitable for maritime activities and operations [8]. Compared to the rest of the European seas, the Mediterranean Sea is most prone to severe climate change [9–11]. More information regarding the wind and wave climate in the Mediterranean Sea is presented in various studies, e.g., [12–14]. In the Baltic Sea and the North Sea, such conditions are met only in the autumn and winter seasons [15,16].

The study of wind speed conditions is of great interest to the North Sea due to commercial interest, both due to the transport of goods and for the offshore industry [17], but a study of waves is not negligible due to the sea connection with North Atlantic Ocean [18]. Also, the wave climate and the changes observed in the last years affect the fatigue life of the offshore structure deployed in this sea [19]. The Norwegian Sea is also influenced by the North Atlantic Ocean and is facing several problems due to global warming and climate changes [20–22]. Thus, a decrease in mean wave height is projected in North Atlantic [23] and across most of the European coasts, with an increase in the annual maximum and 99th percentile wave height [24]. However, in some areas, an increase in waves is observed (e.g., north of Scotland) partly caused by the reduction in sea ice that induced an increased fetch for northerly winds [24]. One of the most important transport routes is located on the English Channel. The Strait of Dover separates the North Sea from the English Channel and is of great importance both in terms of transport ships and ferries [25,26].

Considering the ones presented above, one objective of this study is to analyze the importance of the maritime traffic in the European seas, in terms of the total number of ships registered in ports, the number of passengers and the number of goods transported. The analysis is performed for a 20-year time interval (2001–2020). Taking into consideration the strong economic role of the European seas in terms of transport and tourism, it was considered of interest to carry out a study about the wind and wave climate that influence the safety and efficiency of navigation along the main routes crossing these seas. Thus, this paper presents also a study regarding the mean and extreme values of the wind and sea state conditions in the areas characterized by high maritime traffic during the same period (2001–2020).

The latest reanalysis data provided by the European Center for Medium-Range Weather Forecasts (ECMWF) were considered for this study, more details on the data used are given in the following sections where the wind and wave climate in target areas is analyzed in detail, together with studies regarding the extreme wind and wave conditions along the maritime routes.

2. Marine Traffic

Throughout the years 2001–2020, only two periods in which maritime transport suffered were identified, namely in the 2008–2009 period due to the financial crisis, and in the 2019–2020 period, due to the coronavirus pandemic (COVID-19). However, as the transport of goods and people both in Europe and internationally is a key factor for economic development but also for accessing goods globally, maritime traffic has recovered quickly after the end of these periods of crisis.

Maritime transport significantly influences the development of the European economy, due to the large number of passengers who choose this method of transport, as well as the level of goods transported, because it is the most preferred method of transport due to low prices and large quantities that can be shipped in only one trip. Due to the increased cargo volume transported by vessels, maritime shipping is among the most affordable ways to move goods.

Figure 1 shows an overview of the density of maritime traffic in European countries in 2019, for all months of the year (Figure 1a). To compare the differences in traffic between the months of the year (and also between various seasons), Figure 1b shows the density map for June (summer season), and Figure 1c shows the density map for December (winter season). As expected, during the summer maritime traffic is much higher than in the winter, and the main reason is the increased number of passenger ships.

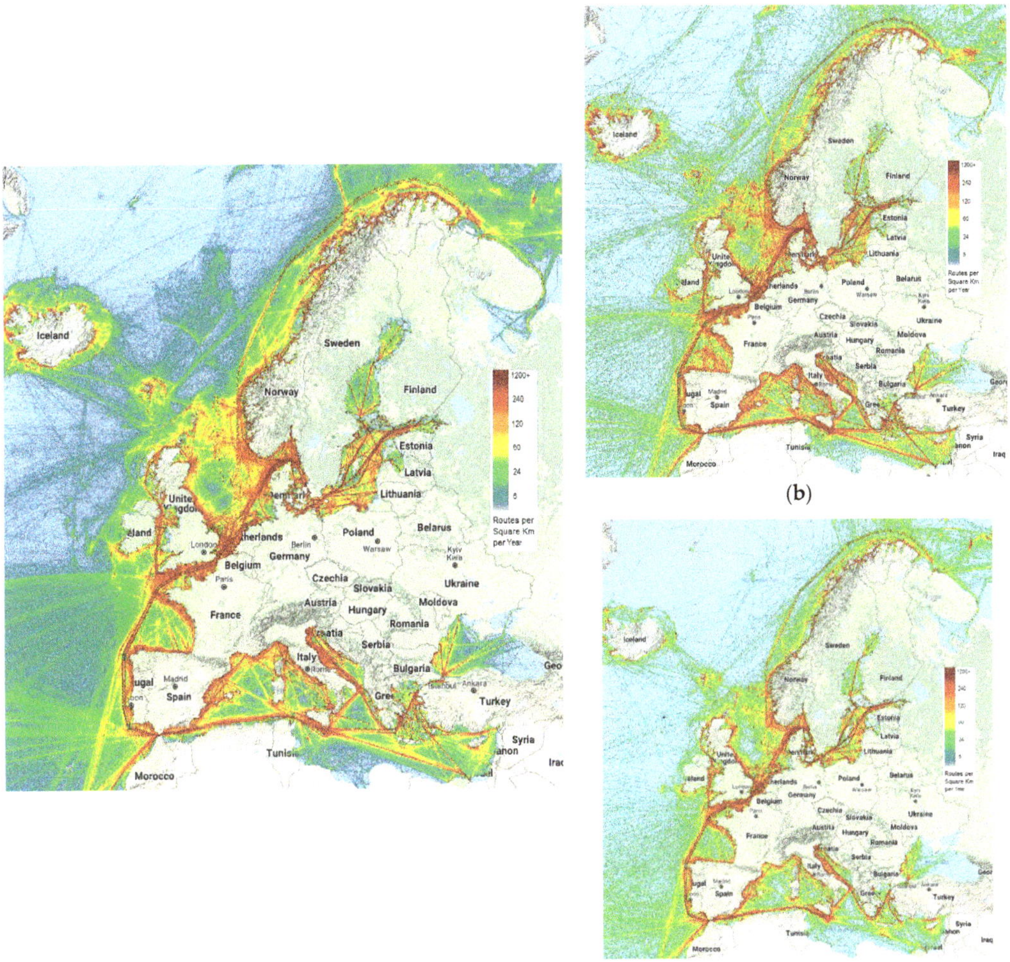

Figure 1. European Union vessel density map: (**a**) During the year 2019; (**b**) for July 2019; (**c**) for December 2019. [27].

The European Marine Observation and Data Network (EMODnet, [27]) platform was used to generate these maps that have been made from ship reporting data of the Automatic Identification System (AIS). As we can see, the maritime traffic is quite developed in Europe, especially in coastal regions and we find almost all types of existing ships.

From the maps presented in Figure 1, it was observed that for various reasons (e.g., weather conditions, economical, geographical location, etc.), the maritime traffic has a higher density along certain routes. Furthermore, using the information provided by the World Meteorological Organization (WMO) through the interactive map of the Ship Observations Team (SOT, https://www.ocean-ops.org/sot/, accessed on 12 December 2021) together with data from the Global Sipping Routes from the ArcGIS platform (https://www.arcgis.com/apps/mapviewer/index.html, accessed on 12 December 2021) the tracks of the core and secondary European routes were identified and they are presented in Figure 2. The seas crossed by these routes and their lengths are indicated in Table 1.

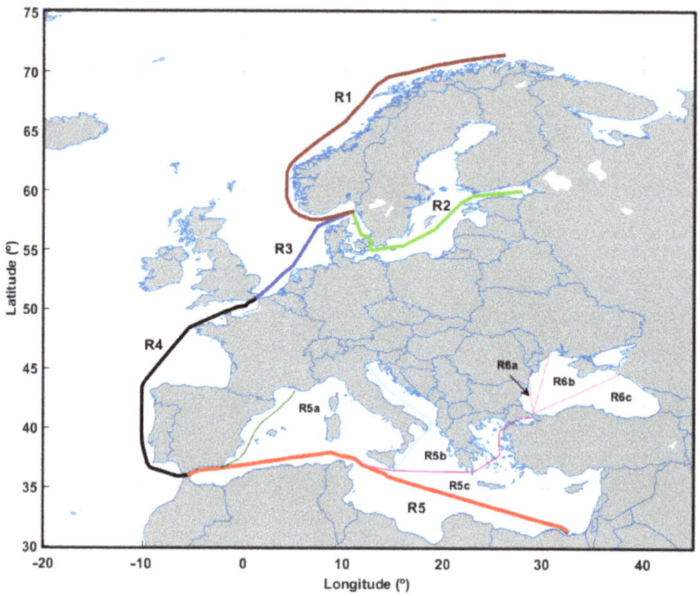

Figure 2. Tracks of the European routes (the main routes are represented with thicker lines, while the secondary routes are marked with thinner lines).

Table 1. Characteristics of the European routes.

Routes		Length (nmi)	Seas Crossed by the Route
Main routes	R1	1280	Norwegian Sea
	R2	814	Baltic Sea
	R3	585	North Sea; Northern English Cannel
	R4	1266	English Cannel; Bay of Biscay; North Atlantic Ocean
	R5	1911	Mediterranean Sea
Secondary routes	R5a	487	Western Mediterranean Sea; Gulf of Lion
	R5b	710	Ionian Sea; Adriatic Sea
	R5c	985	Sicilian Channel; Ionian Sea; Sea of Crete; Aegean Sea; Marmara Sea
	R6a	178	Western Black Sea
	R6b	323	Western Black Sea
	R6c	495	Black Sea

The route R1 departs from the Skagerrak strait that connects the North Sea and the Kattegat Sea, following the Norwegian Trough and then the Scandinavian Peninsula coastline. From the same point starts also the route R2 that passes Copenhagen and reaches the Gulf of Finland, and the route R3 that crosses the North Sea until the Strait of Dover. Route R4 is one of the intense routes, connecting the British Channel (from the Strait of Dover) with the Strait of Gibraltar and then through route R5 with the Suez Canal. In the Mediterranean Sea, in addition to the main route R5, three secondary routes have been identified: route R5a follows the south of the Iberian Peninsula until Marseilles, route R5b departs from the southern coast of Sicily, crosses the Adriatic Sea, and arrives at Venice, and route R5c connects the main route R5 with the Bosporus Strait. In the Black Sea, three secondary routes are defined, all of them depart from Bosporus Strait and connect the main ports of this basin (R6a—Constanta port, R6b—Odesa port, R6c—Novorossiysk) with the main European routes. Apart from the route R4 and route R5, which are two of the busiest sea routes in the world, the connection between the route R1 and the route R2, represented by the Skagerrak Strait, is of great interest because it is the gateway between the Baltic Sea and the North Sea.

In recent years, there has been a trend to improve the characteristics of transport vessels and port infrastructure [28]. Especially in the case of container ships, it is desired to improve the loading capacity of this type of ship, its travel speed as well as to increase the travel distance. At the same time, the development of port infrastructure is also necessary to make it possible for high-capacity ships to be introduced on regular routes and used over shorter distances.

Based on data provided by Eurostat, the statistical office of the European Union [29] which provides statistical data on Europe up to date, the maritime traffic for the period 2001–2020 was investigated. The total number of ships registered for the period 2001–2020 period in European countries (Figure 3) was 44,539,687 vessels, and during the 20 years studied we can observe a slight upward trend from the beginning of the period to the end of the period, with small exceptions in a few years.

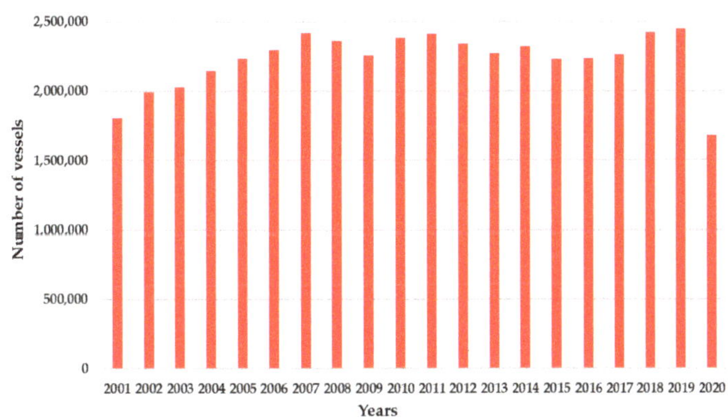

Figure 3. Total number of vessels in the main ports of European countries for the period 2001–2020.

Reported annually, the highest number of ships registered in the main ports was in 2019, when a total number of 2,447,421 ships were recorded, and the lowest number of ships was recorded in 2020 when they were registered with 30% fewer ships due to the crisis caused by the COVID-19 virus.

A major factor in shipping is the routes on which goods or people travel. Some routes are popular throughout the year, such as those arriving in the port of Rotterdam, which is the most important port for shipping and ranks first in terms of the gross weight of goods handled in each port [30]. Other routes, although they have races throughout the year, become much more popular in the summer season, due to cruise passengers or tourists, such as routes in the Mediterranean Sea area.

From the point of view of the ranking of the countries in which the most vessels were registered, Italy takes the lead with 20.44% of the total number of ships, followed by Greece with 19.73% and Denmark with 15.38%, as we can see in Figure 4.

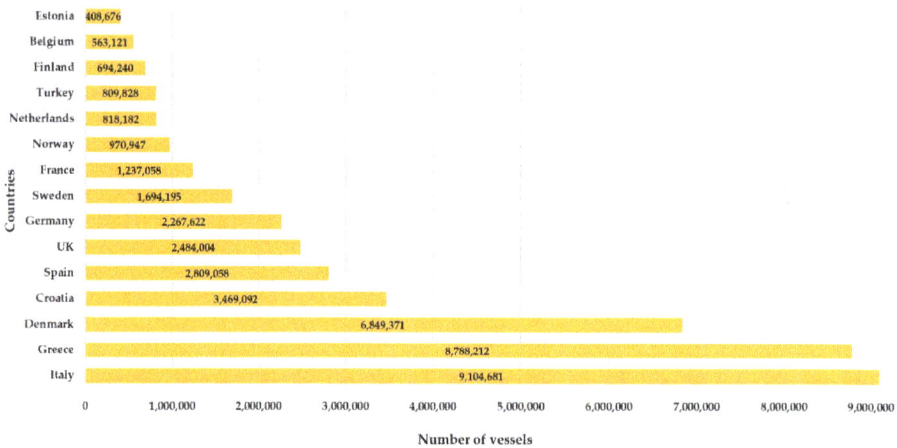

Figure 4. Top 15 European countries regarding the number of vessels registered in the period 2001–2020.

During the years 2001–2020, 83,002,314 thousand tons gross weight of goods were transited (Figure 5). Inbound goods were 10% more than outbound goods out of total goods transited. As we can see in Figure 6, for the studied period, most goods were registered in the Netherlands (12.58%) and the United Kingdom (12.35%) and Italy (11.55%) in the 2nd and 3rd place, respectively, and followed very closely by Spain (9.92%).

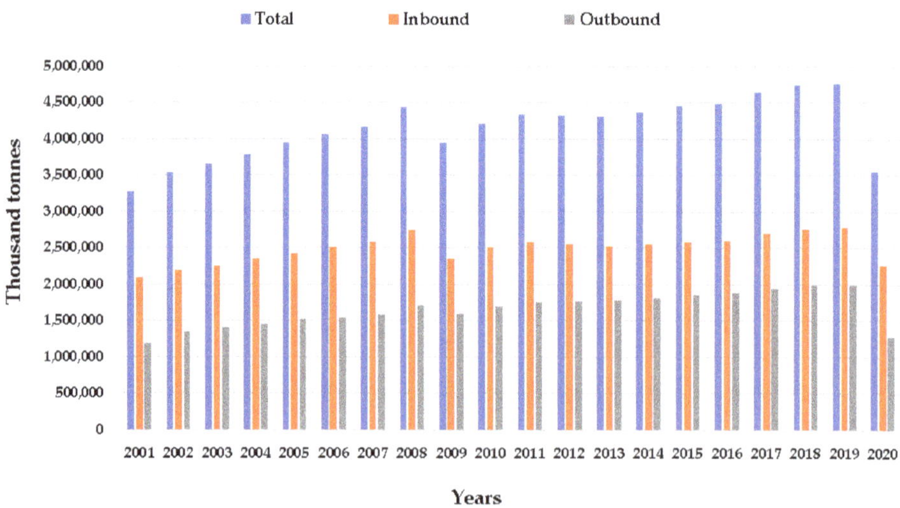

Figure 5. The gross weight of goods handled for the period 2001–2020.

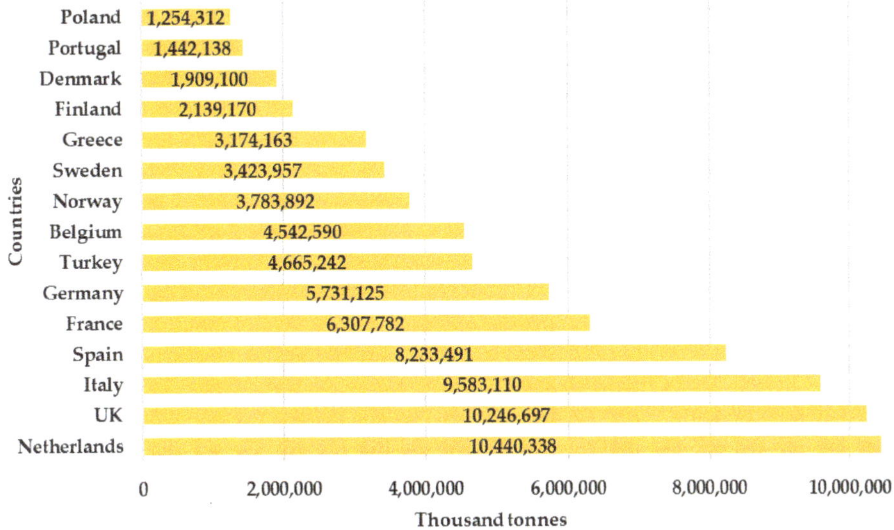

Figure 6. Top 15 European countries regarding the number of goods registered in the period 2001–2020.

After cargo ships, the second most predominant type of ship, in the basin of the European seas, is that of passenger ships. A total of 5,501,141 thousand passengers excluding cruise passengers was registered in the European sea area during the period 2001–2020 (Figure 7).

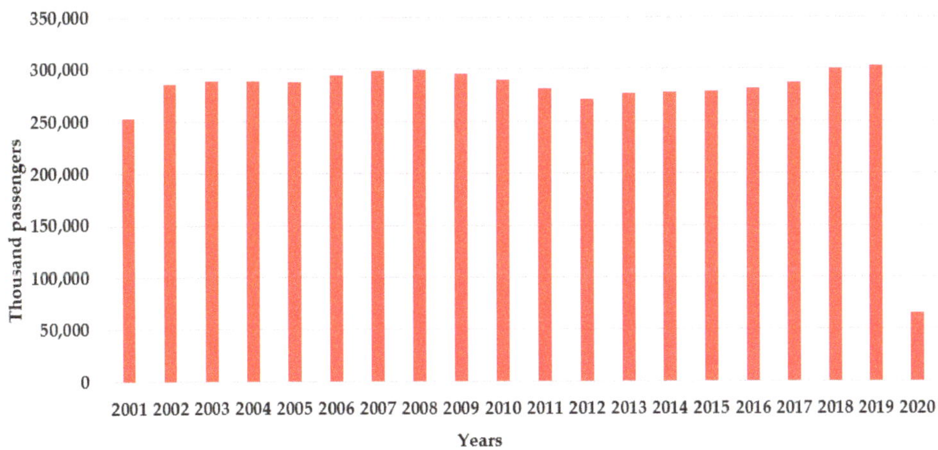

Figure 7. Total passengers transported to/from main ports of European countries for the period 2001–2020.

The countries with the most passengers recorded were (Figure 8) Italy, Greece and Denmark. The pandemic crisis' impact was seen most in passenger traffic, decreasing considerably by about 80% compared to 2019.

The most popular routes for the transport of goods and people cross areas where climatic conditions are constantly changing, and therefore a study on wind and wave climate is required along them, together with the assessment of the extreme condition. In the following section analyses of the wind and wave climate will be performed in the European seas and along the maritime routes defined in Figure 2.

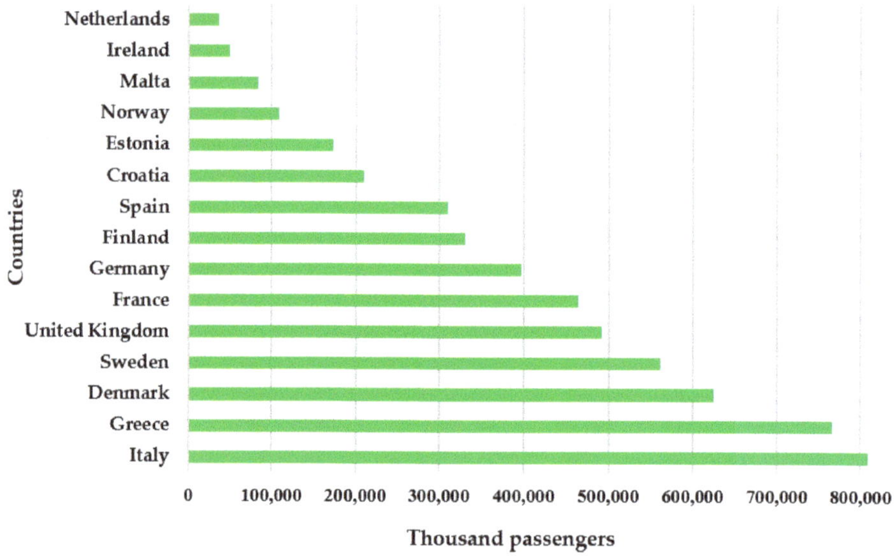

Figure 8. Top 15 countries of European countries regarding the number of passengers for the period 2001–2020.

3. Data Used in the Climate Analysis

The study of climatic conditions for wind and waves was performed using data extracted from the ERA5 database [31,32]. ERA5 data provided by the Copernicus Climate Change Service (C3S) at ECMWF represents a new generation reanalysis for global climate and weather in recent decades. This database has a significantly improved horizontal resolution compared to ERA-Interim [33] (previous reanalysis data provided by ECMWF) and is based on the Cy41r2 Integrated Forecasting System (IFS). Another series of improvements over ERA-Interim is information on variation in quality over space and time, improved representation of tropical cyclones, better precipitation over land in the deep tropics, more consistent sea surface temperature and sea ice, and many others [32].

ERA5 provides hourly estimates for several atmospheric parameters, both for land and sea dust, and covers the period from 1979 to the present. For this study, every 3 h wind data were extracted on single levels for the period 2001–2020, as well as the main wave parameters (significant height of combined wind waves and swell-Hs and mean zero-crossing wave period-Tm). The spatial resolution of the wind fields is 0.25°, while the wave parameters are provided with a resolution of 0.5°, both in latitude and longitude.

Six zones with high maritime traffic were considered to study the wind and wave climate (see Figure 9), and three of them are semi-enclosed seas (Mediterranean Sea, Black Sea, and Baltic Sea). Three other zones are marked in Figure 9, and they cover the Norwegian and North seas and a zone hereinafter referred to as the English Channel, which includes the channel with the same name and also the Celtic Sea. In these zones are some of the most important ports in Europe. The areas of the semi-enclosed seas were not marked as in the case of the other three target areas because their basins are clearly delimited on the map by the surrounding coasts. The characteristics of the geographical domains considered to analyze wind and wave climate are presented in Table 2, together with some characteristics of the seas covered by them.

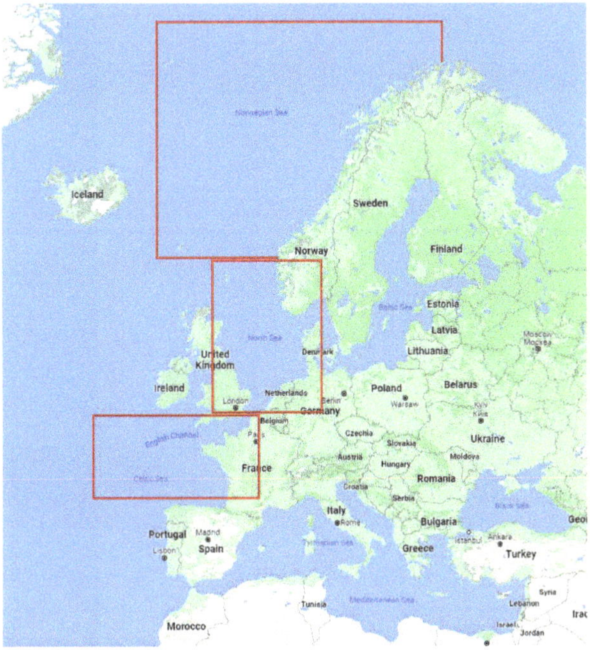

Figure 9. The geographical zones investigated in the study (figure processed from Google Maps).

Table 2. Main characteristics of the geographical zones considered in this study.

Zone	Coordinates	Surface Area	Average Depth	Maximum Depth
Black Sea	27° E–42° E/41° N–47° N	436,402 km^2	1253 m	2212 m
Mediterranean Sea	5° W–35° E/30° N–40° N	2,500,000 km^2	1500 m	5267 m
English Channel	15° W–10° E/45° N–51° N,	375,000 km^2	63 m	174 m
North Sea	5° W–10° E/51° N–60° N,	570,000 km^2	95 m	700 m
Baltic Sea	12° E–32° E/53° N–65° N,	377,000 km^2	55 m	459 m
Norwegian Sea	10° W–25° E/60.5° N–74° N	1,380,000 km^2	2000 m	3970 m

4. Analysis of Wind Climate

4.1. Wind Climate in the European Seas

To study wind speed (V), the 10 m u-component of wind and 10 m v-components of wind were extracted from the ERA5 database for the considered period, and the magnitude of wind speed at 10 m above the sea level was calculated as follows:

$$V = \sqrt{u^2 + v^2}\ [m/s]$$

The 3-h resolution time series of wind speed at each point of the wind field, over the 20 years studied, were averaged. The mean values of the wind speed for the period 2001–2020 are presented in Figure 10. The maximum mean value of wind speed in the region of Europe is around 9.81 m/s and as expected it is located on the western side of the area, toward the North Atlantic Ocean. It should be mentioned here that in all following maps, where a climatic analysis is represented, the position of the maximum value of the field is marked with a black circle.

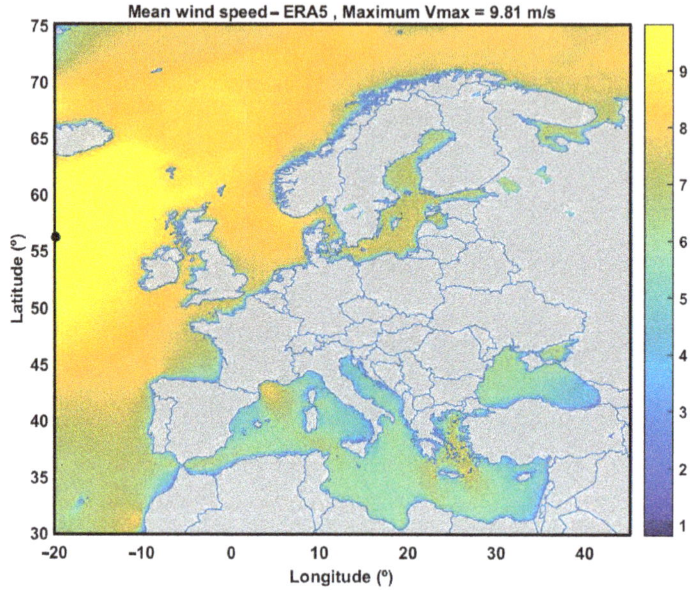

Figure 10. Mean wind speed for the Europe area.

In the north-western European seas, namely the Norwegian Sea, the North Sea, and England Channel, in general, the highest averages of wind speed are found, where the maximum mean values reach even 9 m/s. For these reasons, the North Sea area is popular in terms of installing wind platforms, due to the favorable wind conditions, but also low water depth [34]. Also, on the northern coast of Brittany, high winds are found, mainly due to the channeling effect of the English Channel [35]. The Norwegian Sea is also characterized by high wind speeds [36], and projections of future wind speeds show that these high values will be maintained [37].

Lower mean values were found in the Black Sea area, but this area also has some regions with very good offshore wind energy potential as shown in various studies [38–40]. Moreover, the projections of the future wind conditions show an increase in wind speeds in the coastal environment of the Black Sea, especially on the western side of the basin [41].

The mean values of the wind speed in the Mediterranean Sea are comparable with those encountered in the Black Sea. However, there are some areas where the average values are higher, reaching maximum values ranging between 8–8.5 m/s. As observed in various studies [42,43], these areas are the Gulf of Lion and the Aegean Sea.

For the seasonal variation analysis of the mean wind speed, the usual quarterly partition for the months of the year was considered (DJF: December-January-February, MAM: March-April-May, JJA: June-July-August, SON: September-October-November) and the results are shown in Figure 11 (the same colorbar is used for all panels). One unified colormap that covers the range of values across all four maps will help us to better observe the seasonal variability of the mean wind speed.

Figure 11 clearly indicates the seasonal evolution of the mean wind speed, the windiest season is the winter (DJF), while the calmest wind conditions are usually in the summer months (JJA). In winter the mean wind speed reaches values around 10 m/s near to the western European coasts and along the routes crossing these areas. The mean wind speed field in intermediate seasons (spring and autumn) present similar patterns, with slightly higher values in autumn, while in summer the mean wind speed drops by about 40% the values seen in winter. An unusual feature appears in the Aegean Sea in July and August when the Etesian winds prevail over the area, with dry and relatively cool air masses originating from the region of Russia and the Caspian Sea [42].

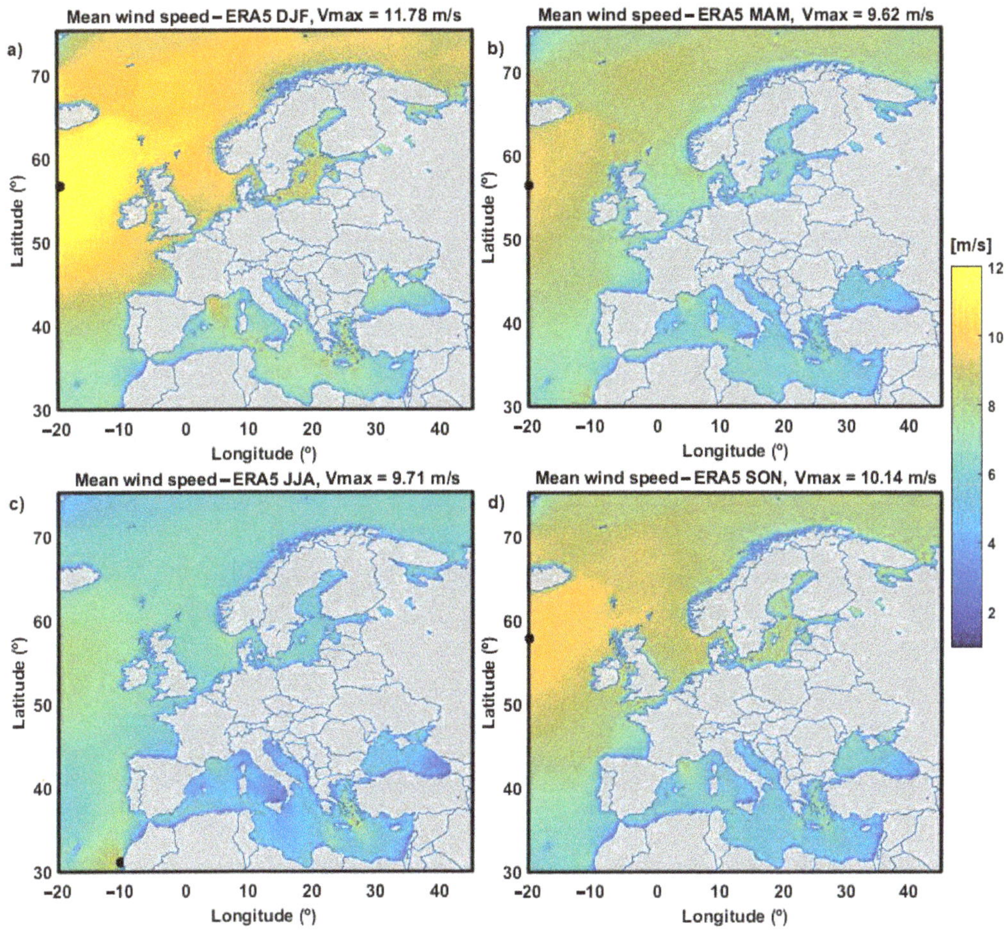

Figure 11. Seasonal variability of the mean wind speed over the 20-year time interval (from 2001 to 2020) based on ERA5 data: (**a**) winter (DJF), (**b**) spring (MAM), (**c**) summer (JJA), (**d**) autumn (SON).

An image of the statistical distribution of a certain parameter analyzed over a time interval is provided through the computation of the percentiles. The 50th, 75th, 90th, and 95th percentiles of the wind speed based on ERA5 data over the 20-year interval considered are also computed using the Matlab function dedicated to this and the results are illustrated as maps. This allows for the analysis of the data in terms of percentages and an image of the extreme values encountered in the studied areas. Thus, the 95th percentile indicates the value below which 95% of the data are found, and above which are only 5%. The 50th percentile represents the median of the distribution (or second quartile Q2), while the 75th percentile is the third quartile (Q3). The results obtained for the 20-years period considered in this study are presented as maps for entire Europe in Figure 12.

In general, the maps for the 50th and 75th percentiles have a similar distribution, with wind speeds not exceeding 8.5 m/s and 11 m/s, respectively, near western European coasts, while the maximum values (9.61 m/s and 12.75 m/s) are located far to the coast, on the western side of the Europe zone. In the case of the 90th and 95th percentiles, the higher wind magnitudes are moving to the northern part, as also indicated by the location of the maximum values.

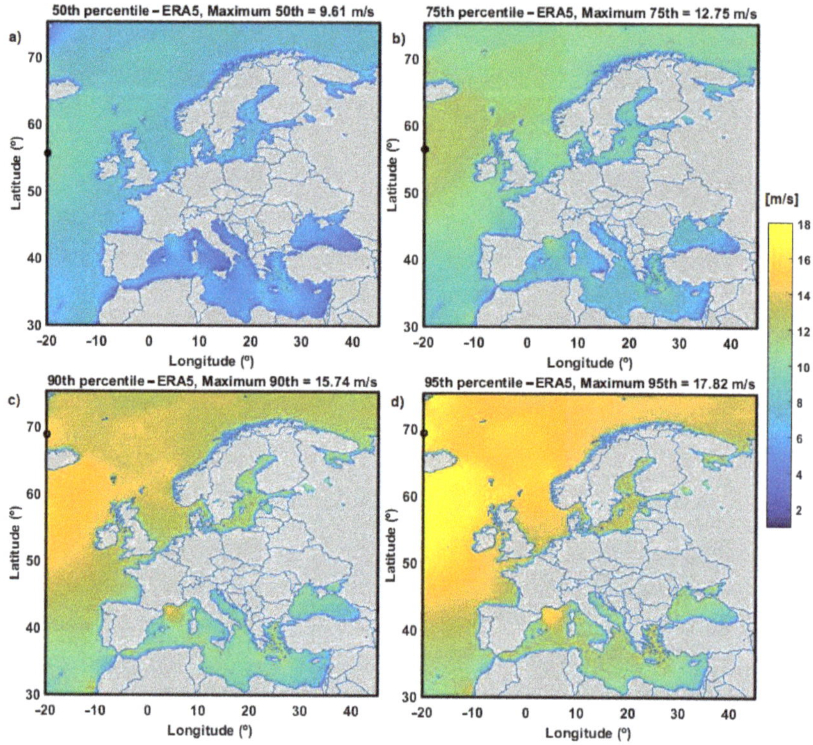

Figure 12. Statistical investigation on wind speed: (**a**) 50th percentile; (**b**) 75th percentile; (**c**) 90th percentile; (**d**) 95th percentile.

4.2. Extreme Wind Conditions along the Maritime Routes

The extreme wind conditions at 10 m above the sea level, represented by the wind speed 95th percentiles, were extracted along the maritime routes and the results are presented in Figure 13. The maximum value of the 95th percentile on each route was computed and they are given in Table 3.

Table 3. Maximum wind speed 95th percentile on each route considered in this study.

Routes		Maximum wind Speed 95th Percentile (m/s)
Main routes	R1	15.98
	R2	13.59
	R3	14.51
	R4	14.43
	R5	12.75
Secondary routes	R5a	15.92
	R5b	11.93
	R5c	13.34
	R6a	11.39
	R6b	11.60
	R6c	11.32

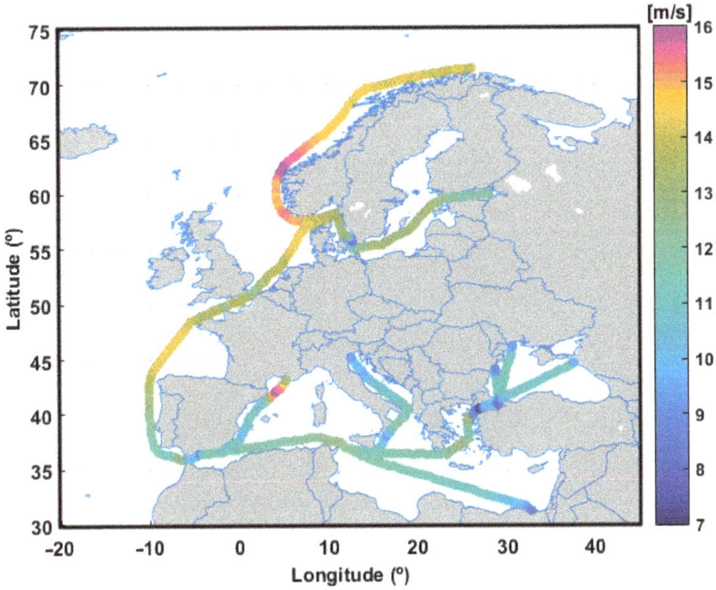

Figure 13. Wind speed 95th percentile along the European maritime routes.

As can be observed from Table 3 and Figure 13, the highest values (above 15 m/s) of the wind speed 95th percentile are on routes R1 and R5a (in Gulf of Lion). On route R3 and the northern part of route R4 maximum values around 14.5 m/s can be found.

5. Analysis of Wave Climate

5.1. Wave Climate in the European Seas

Regarding the sea state conditions, they were analyzed from the perspective of the two main wave parameters (Hs and Tm) which are also important in the analysis of ship dynamics and influence the operability of vessels. As in the case of wind analyses, for the Europe zone, the mean values of both wave parameters were calculated (Figure 14). An analysis of the mean fields for both parameters was also made for each zone. The importance of knowing the wave climate along the maritime routes has been mentioned in various studies, e.g., [1,44].

For the Europe zone, the mean values computed using data covering the twenty years have a maximum value of 3.48 m for Hs and 6.85 s for Tm. As expected, the lowest value of the maximum mean Hs was found in the Black Sea (0.97 m) and the highest maximum mean Hs in the English Channel (3.14 m) being located in the northwest corner of the area, followed by the Norwegian Sea, North Sea, Mediterranean Sea and Baltic Sea. The same order is also maintained in the case of Tm.

The maximum value of the Hs mean is higher with about 30% than the 95th percentile of the field in the North Sea, while in the other seas and whole of Europe these are higher with about 4%. This high difference shows that there are some extreme values in certain areas compared with the behavior of the whole basin. The maps indicate that at the northern part of the North Sea basin the waves are affected by the swell coming from North Atlantic Ocean, and for this reason the highest values of the Hs are encountered here, while in the rest of the basin the Hs means do not exceed 2.3 m. In the case of the mean wave period, the highest mean values are found in the English Channel zone, being influenced by the swell affecting the Celtic Sea waves.

Given that in the northern hemisphere there is a greater seasonal variability of the wave climate than in the southern hemisphere [45–47], the next step was to perform an analysis of the seasonality of the wave climate in the target areas. and the results obtained for both wave parameters are presented in Figures 15 and 16. A seasonal analysis of the maximum percentiles of the mean fields encountered in each sea was also performed only for Hs and the results are presented in Figure 17.

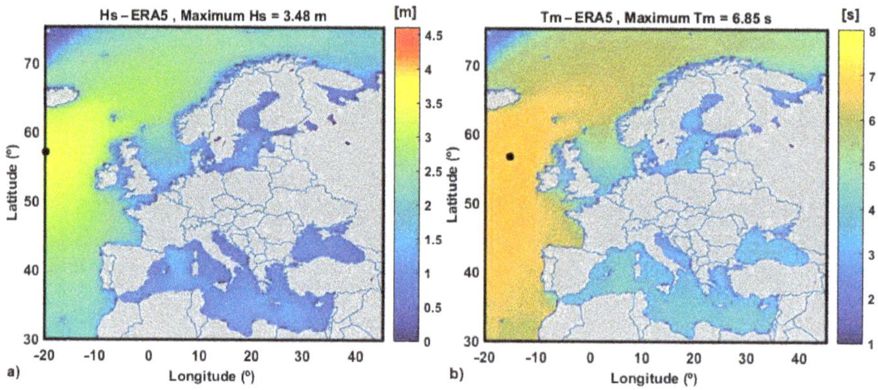

Figure 14. Mean wave conditions for the Europe region in the period 2001–2020: (**a**) Wave heights; (**b**) Wave periods.

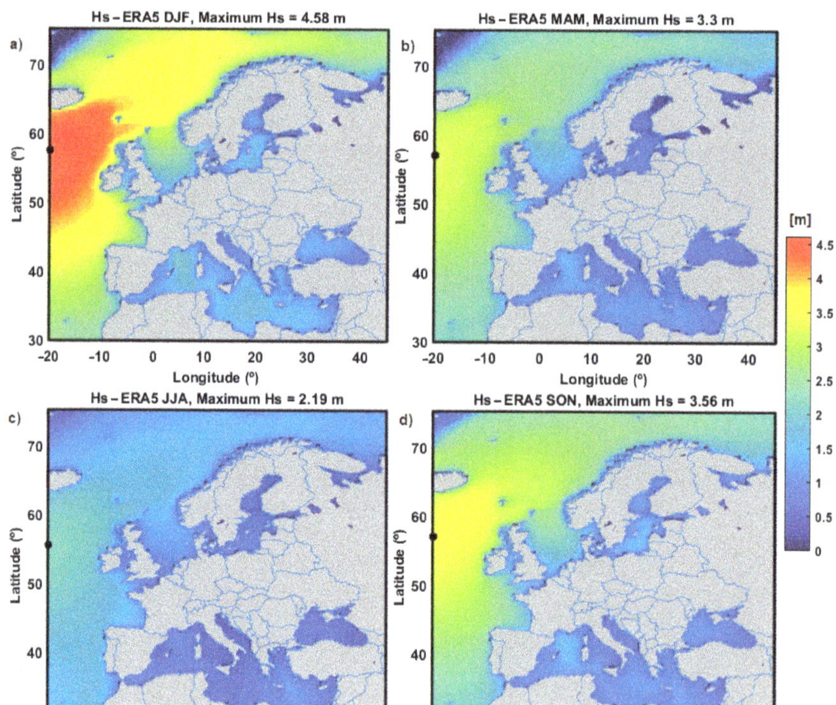

Figure 15. Seasonal variability of the mean significant wave height over the 20-year time interval (from 2001 to 2020) based on ERA5 data: (**a**) winter (DJF), (**b**) spring (MAM), (**c**) summer (JJA), (**d**) autumn (SON).

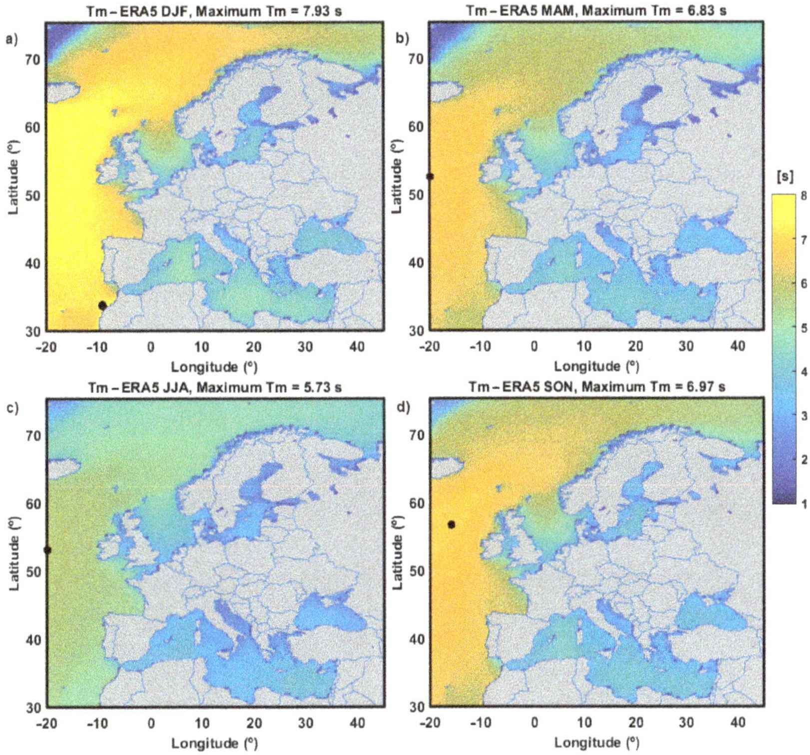

Figure 16. Seasonal variability of the mean wave period over the 20-year time interval (from 2001 to 2020) based on ERA5 data: (**a**) winter (DJF), (**b**) spring (MAM), (**c**) summer (JJA), (**d**) autumn (SON).

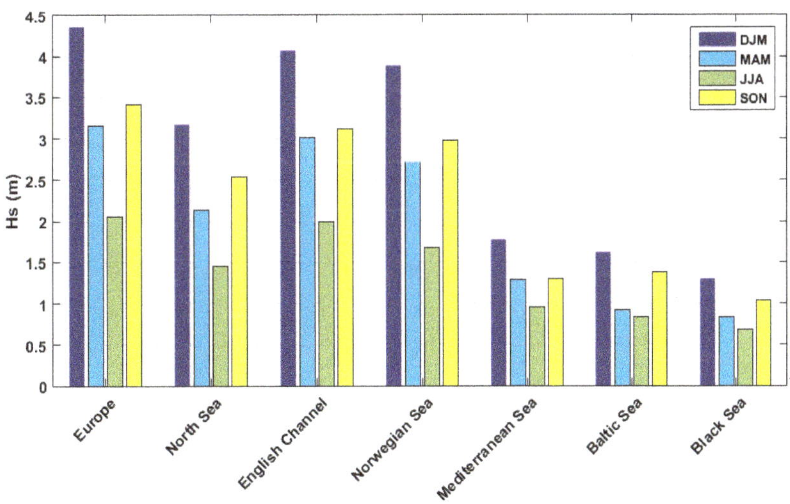

Figure 17. Seasonal characteristics 95th percentile of the mean Hs fields for each area considered.

From Figures 15 and 17, it can be observed that the maximum value of Hs means and percentiles in summer (JJA) are half those encountered in winter (DJF), except in the North Sea where the difference between winter and summer values are even greater. In the transition seasons (spring and autumn) there are more balanced maximum values, a little higher in autumn.

The most severe wave conditions are encountered in winter when the highest means of the wave parameters are in the English Channel (3.14 m for Hs and 6.74 s for Tm), and the calmest sea states are in the Black Sea where the maximum values are 0.97 m and 3.62 s, respectively.

As in the case of wind speed analysis, the 50th, 75th, 90th, and 95th percentiles of Hs were computed using the 20-year time series of data available in each point of the ERA5 grid covering the Europe zone. The results are presented in Figure 18, while in Table 4 the maximum values of percentiles for each zone are given. It should be taken into consideration that, in each defined area, the analysis was made considering both information on the sea state conditions near to the coast, and also from the offshore. In order to identify the differences between these areas (nearshore vs. offshore), which sometimes can be high (see for example the Norwegian Sea zone), the figures in which the maps are presented must be also consulted.

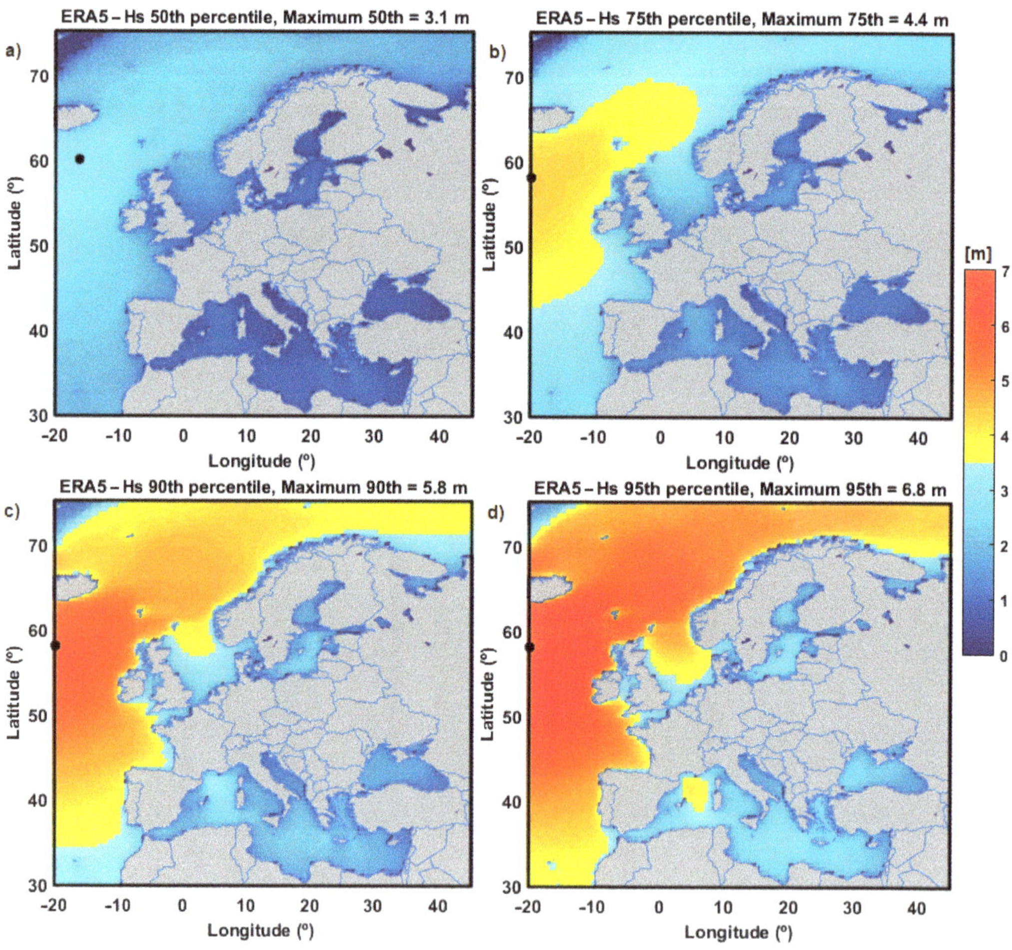

Figure 18. Statistical investigation on wave height conditions: (**a**) 50th percentile; (**b**) 75th percentile; (**c**) 90th percentile; (**d**) 95th percentile.

Table 4. Maximum values of percentiles computed for Hs in each area considered.

Zone	Max 95th (m)	Max 90th (m)	Max 75th (m)	Max 50th (m)
Europe	6.8	5.8	4.4	3.1
North Sea	5.9	5.1	3.8	2.7
English Channel	6.4	5.4	4.0	2.9
Norwegian Sea	5.6	4.8	3.7	2.6
Mediterranean Sea	4.0	3.2	2	1.1
Baltic Sea	2.9	2.4	1.6	1.0
Black Sea	2.3	1.9	1.2	0.8

From Figure 18 it can be observed very clear that in the Mediterranean Sea the highest values of percentiles (especially 95th and 90th) are in the Gulf of Lion and the Aegean Sea. Also, the western part of the Black Sea and southeastern part of the Baltic Sea present the highest values of these percentiles. It is obvious that in the other zones the highest values are in the vicinity of the Atlantic Ocean.

5.2. Extreme Wave Conditions along the Maritime Routes

As in the case of the extreme wind analysis, significant wave height values along the European maritime routes considered in this study were extracted and the results are presented in Figure 19. Table 5 shows the maximum value of the 95th percentile computed on each route.

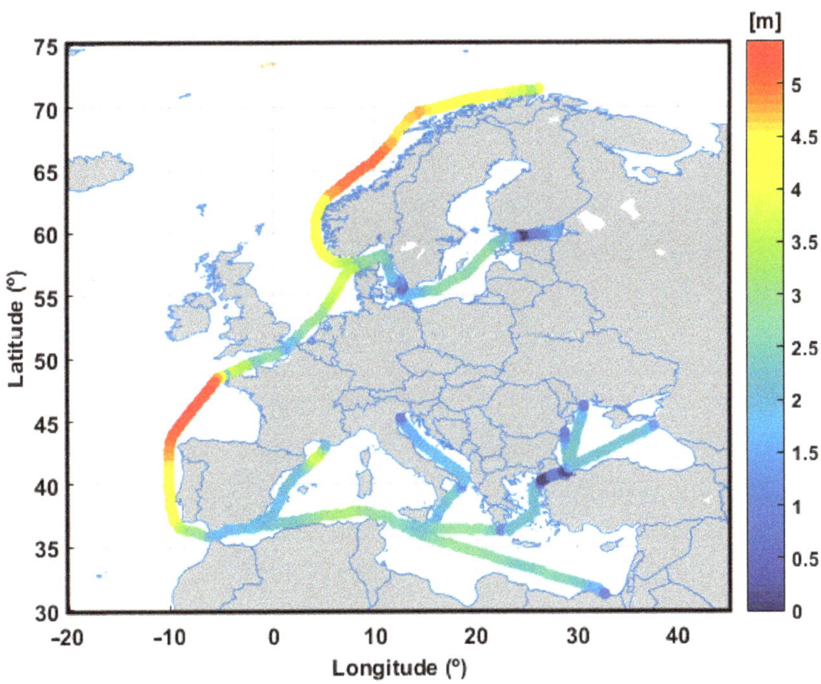

Figure 19. Hs 95th percentile along the European maritime routes.

Table 5. Maximum Hs 95th percentile on each route considered in this study.

Routes		Maximum Hs 95th Percentile (m)
Main routes	R1	5.14
	R2	2.80
	R3	3.67
	R4	5.31
	R5	3.19
Secondary routes	R5a	3.68
	R5b	2.35
	R5c	2.80
	R6a	2.29
	R6b	2.31
	R6c	2.23

As could be expected, the highest values (above 4.5 m) of Hs 95th percentile are on the North Atlantic coasts, on routes R1 (Norwegian Sea) and R4 (south of English Channel), respectively. Although on route R3 high values of wind speed 95th percentile are met, the maximum value for Hs 95th percentile does not reach 4 m.

6. Conclusions

Currently, maritime transport covers about 90% of world trade, with a strong impact on the economies of the countries in the area where it can take place. During the studied period, a tendency to improve the characteristics of the ships in terms of the amount transported, the reduction of the amount of fuel consumed, and the implementation of methods to reduce pollution were observed. Even with these expensive measures, sea transport remains the most popular and cheapest method of transport.

On the other hand, not only is the transport of goods on an upward trend, but also the marine transportation of people is increasing day by day. The number of regular passengers is doubled during the holidays due to tourists who are becoming more and more interested in traveling along the coastal area. Italy, Greece, and Denmark have the busiest ports in terms of the number of registration vessels, and most quantities of goods were registered in the Netherland and the United Kingdom. Although the Netherlands is in first place in the transport of goods, when it comes to passengers it is in 15th place in Europe, at a fairly large difference from Italy and other countries.

In order to be able to continuously support the development of maritime traffic, the climatic conditions in this area, as well as their extreme events are necessary to be known. Therefore, based on the data extracted from ERA5 database, a series of analyses were performed, to evaluate the mean values, 95th, 90th, 75th, and 50th percentiles, as well as seasonal variations of wind and wave conditions.

Following the study carried out based on the data covering a 20-year period, the areas with the most severe conditions were identified. The most important parameters for the analysis and operability of ships, in terms of climatic conditions, are the significant wave height and the wave period. Both in terms of wind and wave conditions, the North Sea zone and the English Channel region present the highest values due to some areas with severe conditions (northern part of the North Sea zone and the Celtic Sea). The calmest zones seem to be the Black and Baltic seas. However, even in these basins extreme events in the winter season that are dangerous for maritime traffic can appear [7,48–50].

Regarding the maritime routes in the European area, severe conditions are met on the routes that pass through the Norwegian Sea (R1) and on the route that passes through the Gulf of Lion (R5a). Moreover, the seasonal variation is an important factor in analyzing wind conditions, and as expected, the winter season is the windiest season and summer is the calmest, with values lower by about 40%. The influence of the North Atlantic Ocean on the North Sea produces extreme conditions at the entrance in this basin, and for this reason, here we encounter high values of significant wave height. In terms of seasonal variation, the same pattern as wind conditions is encountered.

Awareness of wave and wind climate in this region has several benefits, the most important of which would be avoiding hazardous areas and thus reducing accidents, preventing and preparing for crossing different passages, and exploiting them by installing wind and wave farms for electricity production based on renewable sources.

Author Contributions: Conceptualization, A.-M.C. and L.R.; methodology, L.R.; software simulations, validation, analysis and visualization, A.-M.C.; writing—original draft preparation, A.-M.C.; writing final version, L.R.; supervision, L.R. All authors have read and agreed to the published version of the manuscript.

Funding: The work of the first author is supported by the project ANTREPRENORDOC, in the framework of Human Resources Development Operational Programme 2014–2020, financed from the European Social Fund under the contract number 36355/23.05.2019 HRD OP/380/6/13—SMIS Code: 123847. The work of the second author was carried out in the framework of the research project DREAM (Dynamics of the REsources and technological Advance in harvesting Marine renewable energy), supported by the Romanian Executive Agency for Higher Education, Research, Development and Innovation Funding – UEFISCDI, grant number PN-III-P4-ID-PCE-2020-0008.

Institutional Review Board Statement: Not applicable.

Informed Consent Statement: Not applicable.

Data Availability Statement: Not applicable.

Acknowledgments: The data used in this study are openly available. ERA5 data used in this study were obtained from the ECMWF data server.

Conflicts of Interest: The authors declare no conflict of interest.

References

1. Vettor, R.; Soares, C.G. Characterization of the expected weather conditions in the main European coastal traffic routes. *Ocean. Eng.* **2017**, *140*, 244–257. [CrossRef]
2. Chirosca, A.; Rusu, L. Sea state characteristics and the maritime traffic in the European Seas. In Proceedings of the 20th International Multidisciplinary Scientific GeoConference SGEM, Albena, Bulgaria, 18–24 August 2020; Volume 20, pp. 863–870. [CrossRef]
3. Rusu, E. A 30-year projection of the future wind energy resources in the coastal environment of the Black Sea. *Renew. Energy* **2019**, *139*, 228–234. [CrossRef]
4. Lobeto, H.; Menendez, M.; Losada, I.J. Future behavior of wind wave extremes due to climate change. *Sci. Rep.* **2021**, *11*, 7869. [CrossRef] [PubMed]
5. Lavidas, G.; Polinder, H. North sea wave database (NSWD) and the need for reliable resource data: A 38 year database for metocean and wave energy assessments. *Atmosphere* **2019**, *10*, 551. [CrossRef]
6. Novo-Corti, I.; Gonzalez-Laxe, F. The integrated offshore grid in Europe: Exploring challenges for regional energy governance. *Energy Res. Soc. Sci.* **2019**, *52*, 55–67.
7. Divinsky, B.V.; Fomin, V.V.; Kosyan, R.D.; Rather, Y.D. Extreme wind waves in the Black Sea. *Oceanologia* **2020**, *62*, 23–30. [CrossRef]
8. Onea, F.; Rusu, L. Long-Term Analysis of the Black Sea Weather Windows. *J. Mar. Sci. Eng.* **2019**, *7*, 303. [CrossRef]
9. Pastor, F.; Valiente, J.A.; Khodayar, S. A Warming Mediterranean: 38 Years of Increasing Sea Surface Temperature. *Remote Sens.* **2020**, *12*, 2687. [CrossRef]
10. Castells, M.; Martínez de Osés, F.X.; Usabiaga, J.J. Analysis of the Prevailing Weather Conditions Criteria to Evaluate the Adoption of a Future ECA in the Mediterranean Sea. In *Marine Navigation and Safety of Sea Transportation*; Weintrit, A., Ed.; CRC Press Taylor & Francis Group: Boca Raton, FL, USA, 2013; pp. 161–165. [CrossRef]
11. Lejeusne, C.; Chevaldonne, P.; Pergent-Martini, C.; Boudouresque, C.F.; Perez, T. Climate change effects on a miniature ocean: The highly diverse, highly impacted Mediterranean Sea. *Trends Ecol. Evol.* **2010**, *25*, 250–260. [CrossRef]
12. Ganea, D.; Amortila, V.; Mereuta, E.; Rusu, E. A joint evaluation of the wind and wave energy resources close to the Greek Islands. *Sustainability* **2017**, *9*, 1025. [CrossRef]
13. Barbariol, F.; Davison, S.; Falcieri, F.M.; Ferretti, R.; Ricchi, A.; Sclavo, M.; Benetazzo, A. Wind Waves in the Mediterranean Sea: An ERA5 Reanalysis Wind-Based Climatology. *Front. Mar. Sci.* **2021**, *8*, 760614. [CrossRef]

14. Onea, F.; Ruiz, A.; Rusu, E. An Evaluation of the Wind Energy Resources along the Spanish Continental Nearshore. *Energies* **2020**, *13*, 3986. [CrossRef]
15. Lensu, M.; Goerlandt, F. Big maritime data for the Baltic Sea with a focus on the winter navigation system. *Mar. Policy* **2019**, *104*, 53–65. [CrossRef]
16. Bonaduce, A.; Stanevra, J.; Beherens, A.; Bidlot, J.; Wilcke, R.A.I. Wave Climate Change in the North Sea and Baltic Sea. *J. Mar. Sci. Eng.* **2019**, *7*, 166. [CrossRef]
17. Suselj, K.; Sood, A.; Heinemann, D. North Sea near-surface wind climate and its relation to the large-scale circulation patterns. *Theor. Appl. Climatol* **2010**, *99*, 403–419. [CrossRef]
18. Bell, R.J.; Gray, S.L.; Jones, O.P. North Atlantic storm driving of extreme wave heights in the North Sea. *J. Geophys. Res. Oceans* **2017**, *122*, 3253–3268. [CrossRef]
19. Mosquera, I.A.; Sagrilo, L.V.; Videiro, P.M.; Sousa, F. Assessing the impact of climate change on the fatigue life of offshore structures. *J. Offshore Mech. Arct. Eng.* **2021**, 1–35. [CrossRef]
20. Mork, K.A.; Skagseth, O.; Soiland, H. Recent Warming and Freshening of Norwegian Sea Observed by Argo Data. *J. Clim.* **2019**, *32*, 36–3705. [CrossRef]
21. Dickson, B.; Østerhus, S. One hundred years in the Norwegian Sea. *Nor. Geogr. Tidsskr.-Nor. J. Geogr.* **2007**, *61*, 56–75. [CrossRef]
22. Simonsen, M.; Gossling, S.; Walnum, J.H. Cruise ship emissions in Norwegian waters: A geographical analysis. *J. Transp. Geogr.* **2019**, *78*, 87–97. [CrossRef]
23. Lemos, G.; Menendez, M.; Semedo, A.; Miranda, P.M.; Hemer, M. On the decreases in North Atlantic significant wave heights from climate projections. *Clim. Dyn.* **2021**, *57*, 2301–2324. [CrossRef]
24. Bricheno, L.M.; Wolf, J. Future wave conditions of Europe, in response to high-end climate change scenarios. *J. Geophys. Res. Oceans* **2018**, *123*, 8762–8791. [CrossRef]
25. Rodrigue, J.P. Maritime transport. In *The International Encyclopedia of Geography: People, the Earth, Environment and Technology*; Richardson, D., Castree, N., Goodchild, M.M., Kobayashi, A., Liu, W., Marston, R.A., Eds.; John Wiley & Sons, Ltd.: Hoboken, NJ, USA, 2017; Volume 15, pp. 1–7. [CrossRef]
26. Squire, D. The Hazards of Navigating The Dover Strait (Pas-de-Calais) Traffic Separation Scheme. *J. Navig.* **2003**, *56*, 195–210. [CrossRef]
27. European Marine Observation and Data Network (EMODnet). Available online: https://www.emodnet-humanactivities.eu/ (accessed on 12 December 2021).
28. Dedecca, J.G.; Hakvoort, R.A.; Herder, P.M. Maritime Transport and Trade: The Impact of European Transport Policy. An Overview of Maritime Freight Transport Patterns. *Eur. Res. Stud.* **2009**, *XII*, 131–147.
29. Eurostat Database. Available online: https://ec.europa.eu/eurostat/data/database (accessed on 10 July 2021).
30. Chirosca, A.; Rusu, L. Statistical analysis of the types of ships that have crossed the European ports in the last decade. In Proceedings of the 20th International Multidisciplinary Scientific GeoConference SGEM, Albena, Bulgaria, 18–24 August 2020; Volume 20, pp. 249–256. [CrossRef]
31. ECMWF Reanalysis v5 (ERA5). Available online: https://www.ecmwf.int/en/forecasts/dataset/ecmwf-reanalysis-v5 (accessed on 12 July 2021).
32. Hersbach, H.; Bell, B.; Berrisford, P. The ERA5 global reanalysis. *QJR Meteorol Soc.* **2020**, *146*, 1999–2049. [CrossRef]
33. Dee, D.P.; Uppala, S.M.; Simmons, A.J.; Berrisford, P.; Poli, P.; Kobayashi, S.; Andrae, U.; Balmaseda, A.M.; Balsamo, G.; Bauer, P.; et al. The ERA-Interim reanalysis: Configuration and performance of the data assimilation system. *Q. J. R. Meteorol. Soc.* **2011**, *137*, 553–597. [CrossRef]
34. Buatois, A.; Gibescu, M.; Rawn, B.G.; Van der Meijden, A.M. Analysis of North Sea offshore wind power variability. *Resources* **2014**, *3*, 454–470. [CrossRef]
35. Bentamy, A.; Croize-Fillon, D. Spatial and temporal characteristics of wind and wind power off the coasts of Brittany. *Renew. Energy* **2014**, *66*, 670–679. [CrossRef]
36. Karagali, I.; Badger, M.; Hahmann, A.N.; Peña, A.; Hasager, C.B.; Sempreviva, M.A. Spatial and temporal variability of winds in the Northern European Seas. *Renew. Energy* **2013**, *57*, 200–210. [CrossRef]
37. Moemken, J.; Reyers, M.; Feldmann, H.; Pinto, J.G. Future changes of wind speed and wind energy potentials in EURO-CORDEX ensemble simulations. *J. Geophys. Res. Atmos.* **2018**, *123*, 6373–6389. [CrossRef]
38. Diaconita, A.; Andrei, G.; Rusu, L. New insights into the wind energy potential of the west Black Sea area based on the North Sea wind farms model. *Energy Rep.* **2021**, *7*, 112–118. [CrossRef]
39. Akpınar, A.; Bingölbali, B.; van Vledder, G.P. Wind and wave characteristics in the Black Sea based on the SWAN wave model forced with the CFSR winds. *Ocean Eng.* **2016**, *126*, 276–298. [CrossRef]
40. Rusu, L. The wave and wind power potential in the western Black Sea. *Renew. Energy* **2019**, *139*, 1146–1158. [CrossRef]
41. Koletsis, I.; Kotroni, V.; Lagouvardos, K.; Soukissian, T. Assessment of offshore wind speed and power potential over the Mediterranean and the Black Seas under future climate changes. *Renew. Sustain. Energy Rev.* **2016**, *60*, 234–245. [CrossRef]
42. Soukissian, T.; Karathanasi, F.; Axaopoulos, P.; Voukouvalas, E.; Kotroni, V. Offshore wind climate analysis and variability in the Mediterranean Sea. *Int. J. Climatol.* **2018**, *38*, 384–402. [CrossRef]
43. Menendez, M.D.C.R.; García-Díez, M.; Fita, L.; Fernández, J.; Méndez, F.J.; Gutiérrez, J.M. High-resolution sea wind hindcasts over the Mediterranean area. *Clim. Dyn.* **2014**, *42*, 1857–1872. [CrossRef]

44. Vettor, R.; Soares, C.G. Assessment of the storm avoidance effect on the wave climate along the main North Atlantic routes. *J. Navig.* **2016**, *69*, 127–144. [CrossRef]
45. Echevarria, E.R.; Hemer, M.A.; Holbrook, N.J. Seasonal variability of the global spectral wind wave climate. *J. Geophys. Res. Oceans* **2019**, *124*, 2924–2939. [CrossRef]
46. Rusu, L.; Rusu, E. Evaluation of the worldwide wave energy distribution based on ERA5 data and altimeter measurements. *Energies* **2021**, *14*, 394. [CrossRef]
47. Takbash, A.; Young, I.R. Long-Term and Seasonal Trends in Global Wave Height Extremes Derived from ERA-5 Reanalysis Data. *J. Mar. Sci. Eng.* **2020**, *8*, 1015. [CrossRef]
48. Rutgersson, A.; Kjellström, E.; Haapala, J.; Stendel, M.; Danilovich, I.; Drews, M.; Jylhä, K.; Kujala, P.; Larsén, X.G.; Halsnæs, K.; et al. Natural hazards and extreme events in the Baltic Sea region. *Earth Syst. Dyn. Discuss.* **2021**, 1–80, preprint. [CrossRef]
49. Bernardino, M.; Rusu, L.; Soares, C.G. Evaluation of extreme storm waves in the Black Sea. *J. Oper. Oceanogr.* **2021**, *14*, 114–128. [CrossRef]
50. Soomere, T.; Räämet, A. Spatial patterns of the wave climate in the Baltic Proper and the Gulf of Finland. *Oceanologia* **2011**, *53*, 335–371. [CrossRef]

Journal of
Marine Science and Engineering

MDPI

Article

Site Selection of Offshore Solar Farm Deployment in the Aegean Sea, Greece

Dimitra G. Vagiona [1], George Tzekakis [2], Eva Loukogeorgaki [2,*] and Nikolaos Karanikolas [1]

[1] Department of Spatial Planning and Development, Aristotle University of Thessaloniki, 54124 Thessaloniki, Greece; dimvag@plandevel.auth.gr (D.G.V.); karanik@plandevel.auth.gr (N.K.)
[2] Department of Civil Engineering, Aristotle University of Thessaloniki, 54124 Thessaloniki, Greece; georgetzek@gmail.com
* Correspondence: eloukog@civil.auth.gr; Tel.: +30-2310-995951

Abstract: Offshore solar energy presents a new opportunity for low-carbon energy transition. In this research, we identify and rank suitable Offshore Solar Farm (OSF) sites in the Aegean Sea, Greece, considering various constraints and assessment criteria. The methodology includes two distinct phases. In the first phase, Geographic Information Systems (GIS) are used to spatially depict both incompatible and compatible marine areas for OSF deployment, while in the second phase, two models based on different combinations of multi-criteria decision-making methods are deployed to hierarchically rank the eligible areas for OSF deployment. The first model (Objective Model—OM) attributes weights to assessment criteria using an entropy-based weight method, while the second model (Subjective Model—SM) utilizes the pairwise comparison of the Analytical Hierarchy Process (AHP) method. Both models use TOPSIS (Technique for Order of Preference by Similarity to Ideal Solution) to prioritize the suitable OSF sites. The results indicate the existence of nine suitable OSF marine areas in the Greek marine environment (total surface area of 17.25 km^2) and a different ranking of these sites depending upon the deployed model (OM or SM). The present approach provides useful guidelines for OSF site selection in Greece as well as in other countries.

Keywords: solar farm siting; assessment criteria; entropy weight method; AHP; TOPSIS

check for updates

Citation: Vagiona, D.G.; Tzekakis, G.; Loukogeorgaki, E.; Karanikolas, N. Site Selection of Offshore Solar Farm Deployment in the Aegean Sea, Greece. *J. Mar. Sci. Eng.* **2022**, *10*, 224. https://doi.org/10.3390/jmse10020224

Academic Editor: Liliana Rusu

Received: 14 January 2022
Accepted: 5 February 2022
Published: 8 February 2022

Publisher's Note: MDPI stays neutral with regard to jurisdictional claims in published maps and institutional affiliations.

1. Introduction

The scarcity of habitable land, combined with rising energy consumption and the environmental consequences of fossil fuels, is forcing the development of offshore renewable energy projects [1]. Offshore wind, wave, and tidal energy are the main forms of renewable energy in the marine environment, and intensive research is being conducted to develop technologies in these fields [2]. However, a renewable energy form that has received little attention in the marine environment so far is solar energy [3]. Floating photovoltaic systems are required to exploit and harvest this resource in the oceans and seas. Although the use of this technology in the marine environment is relatively new, several applications of floating photovoltaic farms have emerged in lakes and reservoirs around the world [4].

Photovoltaic technology converts solar radiation into electricity without emitting pollutants or negatively impacting the environment. Furthermore, offshore solar power plants offer two major technical advantages: (i) sun-tracking around a vertical axis, which simplifies concentrator system requirements and avoids shading between collector rows, and (ii) unlimited cooling water availability, which can improve thermodynamic cycle efficiency [5]. In addition, this type of renewable energy solution is characterized by the limited need for (land) space and cost (efficient use of space) [6], as well as by the feasibility of large-scale implementations that face less public opposition compared to analogous land-based projects [7].

Although many studies investigating offshore wind farm siting can be found in the recent literature [8–14], the Offshore Solar Farm (OSF) siting applications are missing. This research gap is addressed in the present paper which aims to investigate the suitability of the Aegean Sea in Greece to deploy OSF projects and to identify and prioritize the most appropriate sites, considering various constraints and several assessment criteria. More specifically, this study addresses two research questions: (i) which criteria should be used to assess OSF siting deployment and (ii) how ranking results may change due to different criteria weighting methods.

Geographic Information Systems (GIS) in combination with Multi-Criteria Decision-Making (MCDM) methods have been adopted in the literature to comprehensively address renewable energy siting problems. The identification of suitable locations for solar farm deployment is a difficult task involving several criteria that can influence decision making [15–17]. The use of MCDM methods is suggested in the solution of these problems [17–19]. MCDM methods can be used either in the criteria weighting and/or in the evaluation of available alternatives. Considering the MCDM methods that have been widely applied, Ilbahar et al. [20] noted that there is a distinct upward trend in the utilization of the Analytical Hierarchy Process (AHP) for renewable energy exploitation, followed by ELimination Et Choice Translating REality (ELECTRE), Technique for Order Preference by Similarity to Ideal Solution (TOPSIS) and other MCDM methods such as VIKOR (VIseKriterijumska Optimizacija I Kompromisno Resenje), PROMETHEE (Preference Ranking Organization METHod for Enrichment of Evaluations), data envelopment analysis, conjoint analysis, etc. Most recent publications of the two frequently used MCDM methods on onshore solar farm siting problems include AHP (e.g., [21–26]) and TOPSIS (e.g., [27,28]).

This study proposes a framework for OSF deployment that is divided into two phases. GIS is used in the first phase (Phase I) to spatially depict both incompatible and compatible marine areas for OSF deployment, while in the second phase (Phase II), two models are used to hierarchically rank the suitable sites for OSF deployment, based on different combinations of MCDM methods. The first model (Objective Model—OM) assigns weights to assessment criteria using an Entropy-based Weight Method (EWM), whereas the second model (Subjective Model—SM) uses the pairwise comparison process of AHP. To prioritize the suitable marine areas, both models employ TOPSIS. The correlations between the two models are examined, using the Kendall rank correlation coefficient.

To the authors' knowledge, this is the first study that investigates OSF siting. Both exclusion and assessment criteria are selected and applied for the first time in solving OSF siting problems. Moreover, a significant advantage of the present work is that it uses both objective and subjective weighting methods to provide the appropriate weight for each assessment criterion. The use of EWM avoids the interference of human factors in the weighting of assessment criteria and thereby could improve the objectivity of the assessment results. In addition, the use of both objective and subjective weighting methods in the criteria weighting and the comparison of their results, in a way, address the rank reversal phenomenon that AHP has been widely criticized for and is at the core of many debates in MCDM methods. However, a limitation of the present study could be considered the selection of exclusion criteria, as on the one hand, there is no relevant national legislation concerning OSF deployment and, on the other, marine spatial planning is yet missing in Greece.

The remainder of this paper is organized as follows. Section 2 describes the exclusion and assessment criteria as well as the MCDM methods used in this study. In Section 3, the results are presented and discussed, while, in Section 4, the main conclusions are drawn.

2. Materials and Methods

In order to identify and prioritize the most appropriate marine areas for the deployment of OSF projects in the Aegean Sea, Greece, the methodological approach of Figure 1 is developed and applied. The approach consists of two phases and considers various constraints and several assessment criteria.

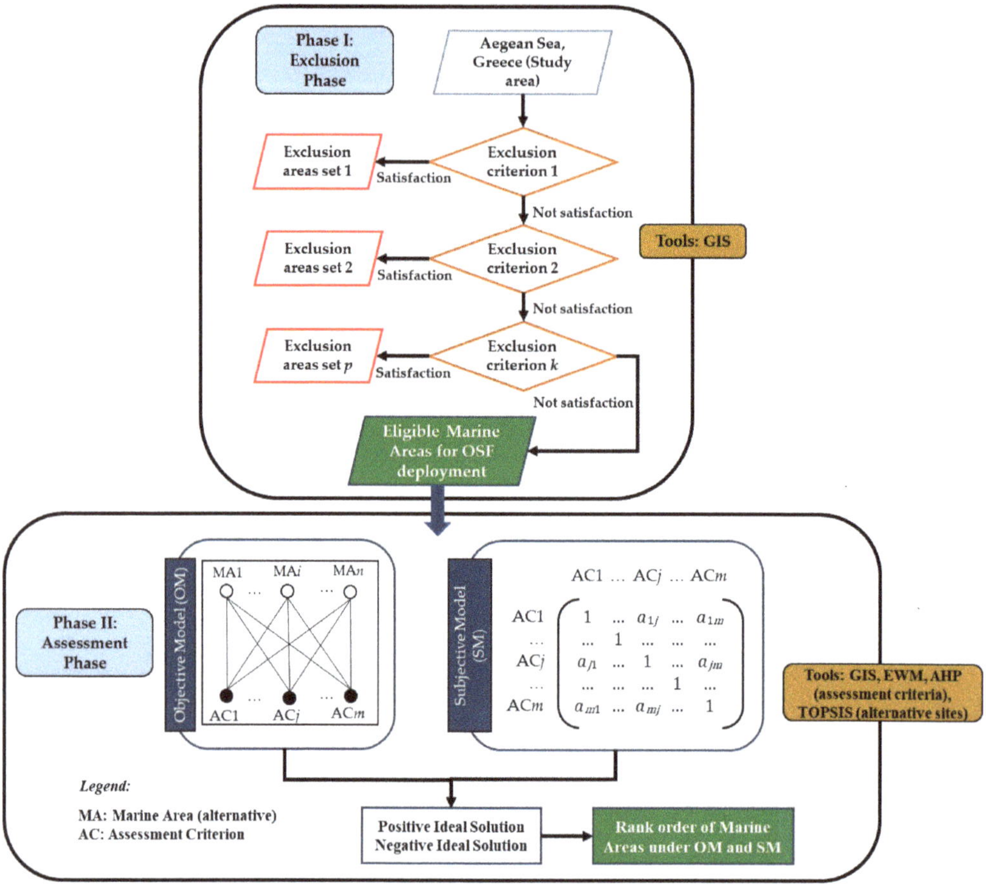

Figure 1. Methodological approach for site selection of OSFs in the Aegean Sea, Greece.

In Phase I, suitable marine areas for OSF deployment in the study area are identified. This is accomplished by creating thematic maps of exclusion criteria in a GIS environment. Following that, in Phase II, two multi-criteria decision models are used to evaluate these marine areas/alternatives. The first model (OM) includes EWM combined with the TOPSIS method to prioritize the potential marine areas, while the second model (SM) uses a combination of the AHP and the TOPSIS method.

2.1. Exclusion Criteria

GIS aids in the implementation of Phase I of the proposed methodological approach (Figure 1). Thematic layers that represent and define the study area, as well as sets of exclusion criteria and relevant restriction zones, where OSF implementation is not possible, are produced. The aforementioned criteria along with their imposed limitations are shown in Table 1. The values are derived either from the Greek Specific Framework for Spatial Planning and Sustainable Development for Renewable Energy Sources (SFSPSD/RES) [29] or scientific publications on siting issues of various renewable energy sources (e.g., offshore wind and wave, onshore solar).

Table 1. Exclusion criteria and restriction zones.

Exclusion Criteria	Unsuitable Areas
Areas to be licensed for Exploration and Exploitation of Hydrocarbons (AEEH)	Occupied areas [30]
Military Exercise Areas (MEA)	Occupied areas (e.g., [31,32])
Ports and Shipping Routes (PSR)	<1 km buffer from sea route [32], >100 km from deep water ports, and >50 km from small piers [33]
Protected Areas (PA)	<1 km (e.g., [11,34])
Aquaculture Zones (AZ)	Occupied areas (e.g., [31,35])
Distance from Shore (DS)	<10 km [29]
Areas where Offshore Renewable Energy Projects (AOREP) have been already installed or planned to be installed	Occupied areas [30]
Water Depth (WD)	>100 m based on [36]
Site Area Limitations (SAL)	<0.3 and >7 km^2

2.2. Assessment Criteria

The assessment criteria are defined, as in the case of the exclusion criteria, through literature review (e.g., [11,34,37–39]) on various renewable energy sources (e.g., offshore wind and wave, onshore solar). Water depth (AC1), distance from shore (AC2), main voltage at a maximum distance of 100 km from the site area (AC3), distance from ports (AC4), serving population (AC5), solar radiation (AC6), and installation site area (AC7) are considered among the most important assessment criteria, which have been used in this study, and are described in the following paragraphs.

Water depth (AC1): Floating photovoltaic applications are currently limited to inland water bodies such as lakes or hydroelectric dam reservoirs and there is no commercially available technology yet available that can be employed in open seas [40]. Like other offshore renewable energy technologies (wind and wave), the construction costs increase with water depth due to mooring, anchoring, and cabling costs [33]. The areas with shallower water depth are considered preferable as they provide technical solutions with reduced construction and maintenance costs.

Distance from shore (AC2): This criterion is selected for technical and aesthetic reasons. On the one hand, the proximity to the shore is an important criterion for the reduction in the costs associated with the installation's connection, while on the other hand, siting solar energy installations in proximity to the shoreline can cause visual and landscape impacts to tourist activities. Visibility from the shore is frequently a planning constraint, so a minimum distance should be defined as representative of the planning preferences. For that reason, the distance from the shore is also used as an exclusion criterion in this study (Table 1). The five categories for this AC2 in decreasing preference order are: 11–25, 26–50, 51–100, 101–150, and 151–200 km.

Main voltage at a maximum distance of 100 km from the site area (AC3): To export the electricity generated, a grid connection point close to the proposed project location with sufficient capacity is required. The proximity of an eligible marine area for OSF deployment to a local electrical grid with high voltage capacity improves its suitability. Three classes of grid capacity (400, 150, and 66 kV) are selected based on the available capacity of the Greek local grid and the existing conditions in the study area.

Distance from ports (AC4): The areas with the shortest distances from ports are preferred as they result in lower installation costs [30]. The four categories for this criterion in decreasing preference order are: ≤50, 51–70, 71–90, and >90 km.

Serving population (AC5): The population that could be served in terms of coverage of energy needs is crucial both for the economic sustainability of the project and its social acceptance. The study area is grouped into four zones (North Aegean, Cyclades, Eastern Aegean, and South Aegean) and the permanent populations of the islands in these zones are aggregated. The assessment of this criterion is performed based on the position of each eligible marine area in the aforementioned zones. The most preferable zones are those that include the highest number of permanent residents.

Solar radiation (AC6): In many studies related to onshore solar farm siting, the total solar radiation incident on a horizontal surface (i.e., global horizontal irradiance) is regarded as an extremely important assessment criterion, e.g., [16,41]. The intensity of a solar PV system's radiation determines the size of its electrical output. Areas with high solar potential contribute significantly to the project's efficiency and economic feasibility. Therefore, the evaluation of the potential OSF sites in the Aegean Sea is based on the following three categories, in increasing preference order: 1601–1700, 1701–1800, and 1801–1900 kWh/m^2.

Installation site area (AC7): Larger sites allow for greater flexibility in terms of the exact installation point based on the conditions, the size of the project, and the number of systems to be installed [42]. The size and scale of floating solar projects are expected to grow further as technologies become more mature. Marine sites with a large area are considered preferable for OSF deployment.

It is noted that all the necessary data describing the above assessment criteria, as well as the exclusion criteria of Table 1, are obtained from specific sources as follows: (i) areas to be licensed for exploration and exploitation of hydrocarbons from [43], (ii) military exercise areas and water depth from [44], (iii) ports and shipping routes from [45,46], (iv) aquaculture zones from [47], (v) distance from shore, protected areas, and distance from ports from [45], (vi) areas where offshore renewable energy projects have been already installed or planned to be installed from [48], (vii) existing high-voltage electricity grid from [49], (viii) serving population from [50], and (ix) solar radiation from [51].

2.3. *Multi-Criteria Decision Making Methods*

2.3.1. Entropy Weighted Method (EWM)

Entropy was originally a concept in thermodynamics, and it was used to calculate the disorder of a system, namely, the degree of its confusion [52]. EWM is an important information weight model that eliminates the influence of human factors on the weight of indicators, thereby improving the objectivity of the overall evaluation results [53]. EWM consists of four steps.

In Step 1, the initial assessment matrix is defined, including the numerical value x_{ij} of each i-th, $i = 1, \ldots, n$, alternative for each assessment criterion ACj, $j = 1, \ldots, m$. From this initial assessment matrix, and as the assessment criteria are expressed in different units, a new normalized decision matrix is calculated in Step 2 to retrieve the values on a common basis. The normalized rating r_{ij} is calculated using Equation (1) below:

$$r_{ij} = \frac{x_{ij}}{\sum_{i=1}^{n} x_{ij}} \tag{1}$$

where n is the total number of alternatives.

Successively, in Step 3, the entropy value E_j, $j = 1, \ldots, m$, for each j-th assessment criterion is calculated by deploying Equation (2), as defined in [54].

$$E_j = -\frac{\sum_{i=1}^{n} r_{ij} ln(r_{ij})}{ln(n)} \tag{2}$$

Finally, in Step 4, the entropy weight w_j, $j = 1, \ldots, m$, for each j-th assessment criterion is calculated as follows [53,55]:

$$w_j = \frac{(1 - E_j)}{\sum_{j=1}^{m} (1 - E_j)} \tag{3}$$

where m is the number of assessment criteria.

For a given j-th criterion, the lower the entropy value, E_j, is, the greater the degree of diversity among alternative values within this criterion. As a result, the corresponding criterion provides more useful decision information for the decision-making problem at hand, and the criterion would have a higher importance weight within the decision procedure.

2.3.2. Analytic Hierarchy Process (AHP)

AHP was initiated by Professor Thomas L. Saaty back in the 1970s [56]. Its process entails decomposing a problem into a hierarchy with a goal at the top, criteria at the second level of the hierarchy, and alternatives at the bottom of the hierarchy. In AHP, each factor is compared as a binary value at each level of the hierarchy using pairwise comparisons, and the relative values are assessed in accordance with the level of importance among themselves to each other based on Saaty's fundamental scale (Table 2).

Table 2. Saaty's fundamental scale [57].

Intensity of Importance on an Absolute Scale	Definition	Reasoning
1	Equal importance	Two activities contribute equally to the goal
3	Moderate importance of one over another	One activity is preferred over another based on experience and judgment
5	Essential or strong importance	One activity is clearly superior to another based on experience and judgment
7	Very strong importance	An activity is strongly preferred, and its dominance is evident in practice
9	Extreme importance	The evidence favoring one activity over another is of the highest possible order of affirmation
2, 4, 6, 8	Intermediate values	When a compromise is required

In this study, we use the AHP method to derive the relative weights of ACj, $j = 1, \ldots, m$. The judgement matrix **A** with elements denoting the decision maker's preference of one criterion over another based on Saaty's scale is formed as follows:

$$\mathbf{A} = \begin{bmatrix} 1.0 & \cdots & a_{1j} & \cdots & a_{1m} \\ & & \cdots & & \\ a_{j1} & \cdots & 1.0 & \cdots & a_{jm} \\ & & \cdots & & \\ a_{m1} & \cdots & a_{mj} & \cdots & 1.0 \end{bmatrix} \quad (4)$$

The relative weights of the compared criteria ACj, $j = 1, \ldots, m$ are calculated by normalizing matrix **A** into a new matrix, where the elements of each column are divided by the sum of the elements of the same column. The row average of the new normalized matrix is then used to compute the relative weights of the criteria.

The degree of inconsistency of comparison matrices is expressed by the Consistency Index (CI) and the Consistency Ratio (CR), which are given by Equations (5) and (6), respectively:

$$CI = \frac{\lambda_{max} - m}{m - 1} \quad (5)$$

$$CR = \frac{CI}{RI} \tag{6}$$

In Equation (5), λ_{max} is the maximum eigenvalue of the m x m comparison matrix and m is the size of this matrix. If $CR < 0.10$, the degree of consistency is considered satisfactory and acceptable [56].

2.3.3. TOPSIS

TOPSIS is a straightforward and computationally efficient MCDM technique for selecting the best solution from a set of alternatives. It follows a series of steps outlined in [58], with references to [59]. The method's central idea is that the chosen solution should be as close as possible to the positive ideal solution while remaining as far away as possible from the negative ideal solution [60]. Alternative priority order can be achieved based on the comparison of the relative distance. The steps of the TOPSIS method are described below.

In Step 1, Equation (7) is employed to normalize the decision matrix, where R_{ij} is the TOPSIS normalized rating.

$$R_{ij} = \frac{x_{ij}}{\sqrt{\sum_{i=1}^{n} x_{ij}^2}} \tag{7}$$

Then, in Step 2, the weighted normalized value v_{ij}, $i = 1, \ldots, n$, $j = 1, \ldots, m$, is calculated as follows:

$$v_{ij} = w_j * R_{ij} \tag{8}$$

where w_j is the weight of the j-th assessment criterion and $\sum_{j=1}^{m} w_j = 1$.

Successively, in Step 3, we determine the ideal A^+ and the negative ideal solution A^- using Equations (9) and (10), respectively [61].

$$A^+ = \{v_1^+, \ldots, v_m^+\} = \{(max\, v_{ij} | j \in J'), (\, minv_{ij} | j \in J'')\} \tag{9}$$

$$A^- = \{v_1^-, \ldots, v_m^-\} = \{(min\, v_{ij} | j \in J'), (\, maxv_{ij} | j \in J'')\} \tag{10}$$

where J' is associated with benefit criteria and J'' is associated with non-benefit (cost) criteria.

In Step 4, we calculate the Euclidean distance of each alternative from the optimal ideal (S_i^+) and the negative ideal choice (S_i^-) by deploying Equations (11) and (12), respectively.

$$S_i^+ = \sqrt{\sum_{j=1}^{m} \left(v_{ij} - v_j^+\right)^2} \tag{11}$$

$$S_i^- = \sqrt{\sum_{j=1}^{m} \left(v_{ij} - v_j^-\right)^2} \tag{12}$$

The closeness coefficient C_i^+ of each i-th alternative to the optimal ideal and the negative ideal solution is next calculated in Step 5 as follows:

$$C_i^+ = \frac{S_i^-}{S_i^+ + S_i^-} \tag{13}$$

Finally, in Step 6, the ranking order of the alternatives based on the relative closeness coefficient C_i^+ are determined.

3. Results and Discussion

3.1. *Determination of Eilgible Marine Areas for OSF Deployment*

Using the exclusion criteria and the restriction zones of Table 1, several thematic maps have been created in GIS. Figure 2 shows indicatively the thematic maps of the exclusion criteria related to Protected Areas (PA), Aquaculture Zones (AZ), and Distance from Shore (DS).

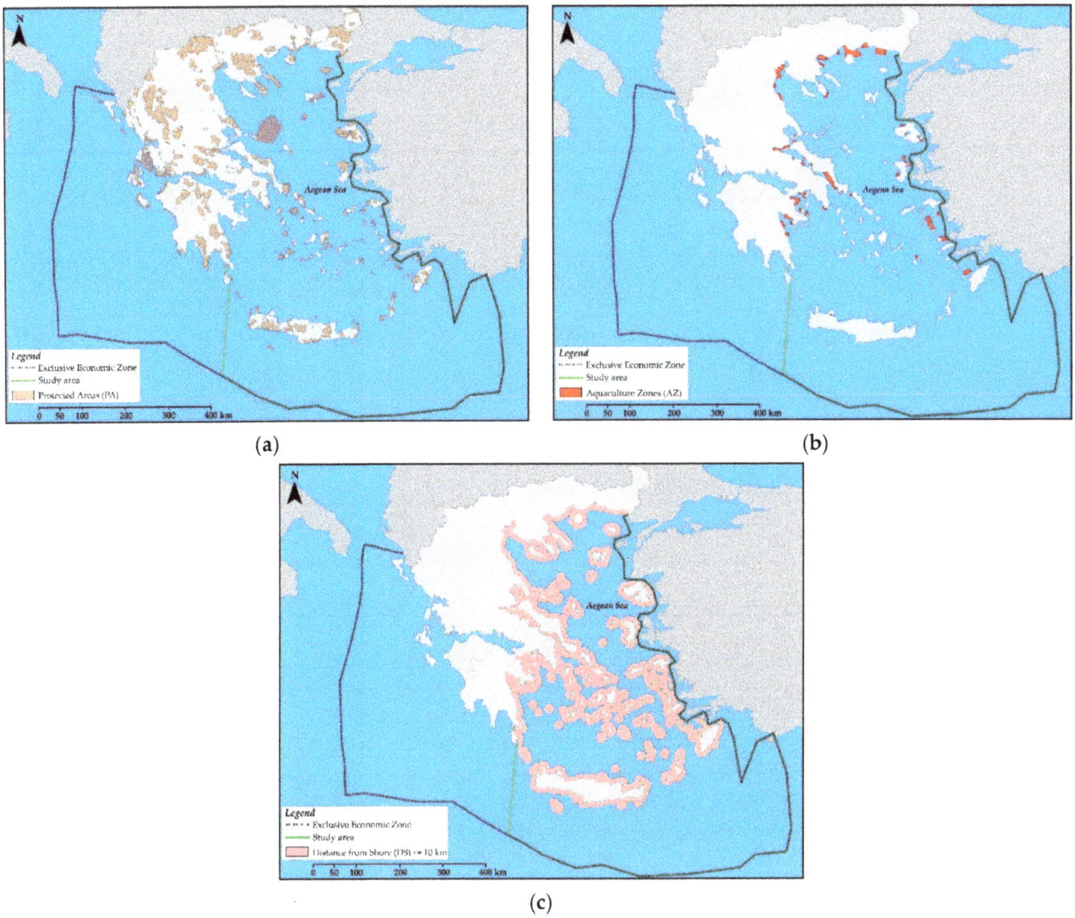

Figure 2. Thematic maps of exclusion criteria related to: (**a**) Protected Areas (PA), (**b**) Aquaculture Zones (AZ), (**c**) Distance from Shore (DS).

By superimposing the thematic maps of all exclusion criteria, marine areas unsuitable for OSF siting are determined and, accordingly, marine sites eligible for the deployment of OSFs in the study area are identified (Phase I). Specifically, the corresponding results have indicated nine (9) eligible Marine Areas (MAs) in the Aegean Sea, Greece, which are shown in Figure 3 and are considered for further assessment and evaluation (Phase II). Six MAs (MA4–MA9) are located in the North Aegean, one (MA2) in the central Aegean, while the remaining two are located East of Crete (MA1) and offshore (East) of Euboea (MA3). The size of these areas is cited in Table 3.

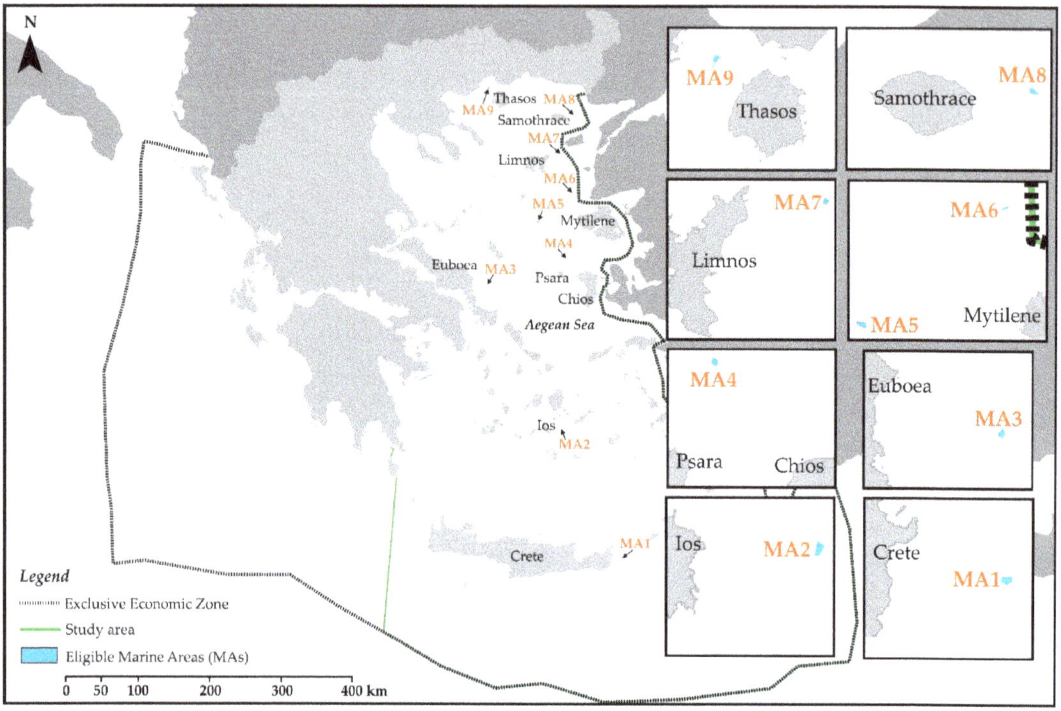

Figure 3. Eligible areas for OSF siting in the Aegean Sea, Greece.

Table 3. Initial assessment matrix for OM.

	AC1 (m)	AC2 (km)	AC3 (kV)	AC4 (km)	AC5 (Population)	AC6 (kWh/m^2)	AC7 (km^2)
MA1	100	11–25	150	≤50	686,969	1801–1900	0.973
MA2	100	11–25	150	≤50	119,887	1801–1900	1.071
MA3	100	11–25	150	≤50	176,264	1701–1800	1.112
MA4	100	11–25	66	51–70	176,264	1801–1900	1.322
MA5	100	26–50	66	≤50	176,264	1701–1800	4.885
MA6	100	26–50	66	51–70	176,264	1601–1700	1.669
MA7	50	11–25	66	51–70	176,264	1601–1700	0.974
MA8	50	11–25	400	≤50	176,264	1601–1700	1.615
MA9	50	11–25	400	≤50	176,264	1601–1700	3.628

3.2. Assessment and Ranking of Eligible Marine Areas

3.2.1. Weights of Assessment Criteria

The weights of the assessment criteria defined in Section 2.2 are calculated through EWM (Objective Model—OM) and AHP (Subjective Model—SM). EWM is the weighting method used herein to measure value dispersion in the examined decision-making problem, while the AHP method uses the pairwise comparison of the assessment criteria to quantify their relative weights. There is no predefined procedure or rules to perform pairwise comparisons and assign weights to the assessment criteria. Thus, AHP is a subjective process that, in most cases, depends on either the researchers' decision or the expertise of relevant stakeholders and policymakers. In the current study, pairwise comparisons are performed based on the authors' expertise [11,30,62] and their understanding of the study area's local conditions and constraints.

Table 3 provides the initial assessment matrix used in OM, including the numerical values x_{ij} of each eligible MAi, $i = 1, \ldots, 9$, for each ACj, $j = 1, \ldots, 7$, while Table 4 presents the pairwise comparison matrix of the seven assessment criteria used in SM. From Table 3, it is obvious that AC3, AC7, and AC5 show the highest degree of discrimination and grade discrimination, while the opposite holds true for AC4 and AC2. Accordingly, the entropy values of AC3, AC7, and AC5 (calculated equal to 0.881, 0.911, and 0.916, respectively) are smaller compared to those of AC4 and AC2 (equal to 0.996 and 0.998, respectively). Regarding Table 4, it is noted that the consistency ratio of the pairwise comparison matrix is $CR = 0.02$, meaning that the results are consistent and acceptable.

Table 4. Pairwise comparison matrix of assessment criteria for SM.

	AC1	AC2	AC3	AC4	AC5	AC6	AC7
AC1	1	4	1/2	1/2	2	1/4	4
AC2	1/4	1	1/5	1/5	1/3	1/7	1
AC3	2	5	1	1	3	1/3	5
AC4	2	5	1	1	3	1/3	5
AC5	1/2	3	1/3	1/3	1	1/5	3
AC6	4	7	3	3	5	1	7
AC7	1/4	1	1/5	1/5	1/3	1/7	1

The relative weights of the assessment criteria are quantified through EWM and AHP in OM and SM, respectively, and the corresponding results are presented in Figure 4. Starting with OM, it can be seen that AC3, AC7, and AC5 have the largest importance weights (32.46, 24.13, and 22.86%, respectively). This, in turn, indicates that the main voltage at a maximum distance of 100 km from the site area, the installation site area, and the serving population are, respectively, the three most important assessment criteria for determining the preference order of the OSF siting in the Aegean Sea, Greece, when OM is deployed. On the other hand, the distance from ports (AC4) and from the shore (AC2) have the smallest weights (1.07 and 0.49%, respectively) and, therefore, contribute slightly to the relevant decision-making process. All the above are in absolute accordance with the discussion made above regarding the entropy values and the degree/grade of discrimination of ACj, $j = 2, \ldots, 5$, and 7.

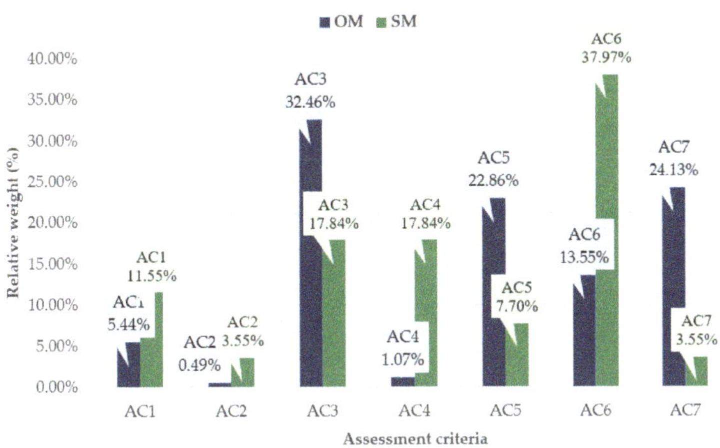

Figure 4. Relative weights of assessment criteria (AC1~AC7) in OM and SM.

In the case of SM, the relative weight of AC6 has the greatest value (37.97%) indicating that the solar radiation is the most important criterion for ranking MAs in the Aegean Sea, Greece. This result is in line with several studies on onshore solar farm siting that have highlighted that the total solar radiation is the assessment criterion with the greatest weighting factor [16,18,41]. The main voltage at a maximum distance of 100 km from the site area (AC3) and the distance from ports (AC4) are the next two most important criteria, having the same relative weight (17.84%). It should be noted that the total weight of the above three assessment criteria equals 73.65%. Consequently, the decision upon the sustainability of MAs for OSF deployment in the case of SM strongly depends on the availability of solar energy sources as well as on technical/economic factors. The priority weights of the remaining four assessment criteria in decreasing order are as follows: AC1 (11.55%), AC5 (7.70%), AC7 (3.55%), AC2 (3.55%).

3.2.2. Ranking of Eligible Marine Areas

For both models (OM and SM), eligible MAs for OSF deployment are prioritized using TOPSIS. The distance of every feasible solution (MA1–MA9) from the ideal solution (Equation (11)) and the negative ideal solution (Equation (12)) is obtained, and each MA is ranked by the relative degree of approximation (Equation (13)). The corresponding results are presented in Tables 5 and 6 for OM and SM, respectively, while comparison of the ranks between the two different methods is graphically presented in the radar chart of Figure 5.

Table 5. Distance of each MA from the ideal and the negative ideal solution and final ranking in OM.

	S_i^+	S_i^-	C_i^+	**Ranking**
MA1	0.1865	0.1661	0.4712	3
MA2	0.2403	0.0611	0.2027	5
MA3	0.2309	0.0507	0.1799	6
MA4	0.2525	0.0476	0.1586	7
MA5	0.2213	0.1382	0.3844	4
MA6	0.2506	0.0286	0.1025	8
MA7	0.2620	0.0186	0.0663	9
MA8	0.1847	0.1728	0.4834	2
MA9	0.1521	0.1945	0.5611	1

Table 6. Distance of each MA from the ideal and the negative ideal solution and final ranking in SM.

	S_i^+	S_i^-	C_i^+	**Ranking**
MA1	0.0779	0.1344	0.6330	1
MA2	0.0936	0.1239	0.5696	2
MA3	0.1093	0.0654	0.3745	6
MA4	0.1086	0.1228	0.5307	3
MA5	0.1242	0.0642	0.3410	7
MA6	0.1628	0.0174	0.0967	9
MA7	0.1617	0.0279	0.1472	8
MA8	0.1324	0.0964	0.4214	5
MA9	0.1315	0.0973	0.4253	4

The ranking results are different between the two models. In the case of OM (Table 5), the first three most preferable sites for the OSF deployment in the Aegean Sea, Greece, are MA9, MA8, and MA1 located, respectively, near Thasos, Samothrace (North Aegean), and Crete (Figure 3). Regarding the first two top choices, the existence of the highest (400 kV) capacity grid within a maximum distance of 100 km (AC3) from MA9–MA8, as well as the benefit of these two sites to serve a large population (AC5) and provide large installation area (AC7) contribute mainly to this ranking. MA1 corresponds to the third top choice due to the potential of this site to serve a large population (AC5). As for SM (Table 6), the first three top choices correspond to MA1, MA2, and MA4 offshore of Crete, Ios (Central Aegean), and Psara (North Aegean, near Chios), respectively (Figure 3). The large solar radiation values (AC6) in those three MAs contribute mainly to this ranking. For both models, the two least preferable sites correspond to MA6 and MA7 near Mytilene and

Limnos, respectively (Figure 3). This is mainly attributed to the small values of AC3 (main voltage at a maximum distance of 100 km from the site area) and AC6 (solar radiation) for both these MAs.

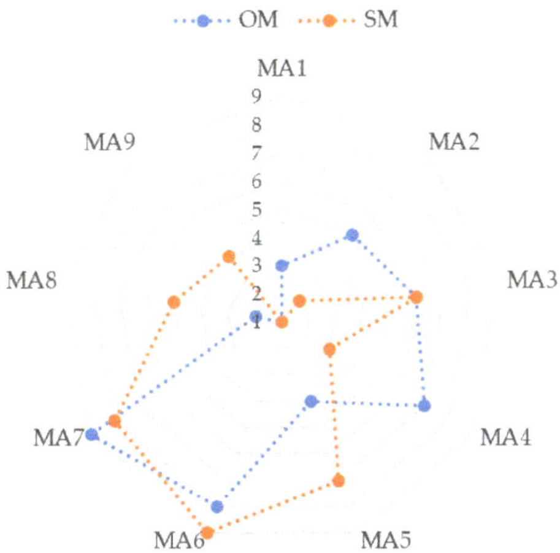

Figure 5. Radar chart for MAs ranking in terms of OM and SM.

The correlations in rankings between OM and SM are further examined, using Kendall rank correlation coefficient (Kendall's τ coefficient) and the correlation value (0.39) reveals a low agreement between rankings.

4. Conclusions

Given the world's growing interest in sustainable energy development and the vast and clean source of energy available for long-term exploitation, the current paper develops and presents a methodological framework for identifying the most appropriate marine areas in the Aegean Sea, Greece, for OSF siting.

Through the application of certain exclusion criteria and the use of GIS, nine (9) eligible MAs for the siting of OSFs in the study area are identified. Two different multi-criteria models (OM and SM), based on different weighting methods (EWM and AHP, respectively), are deployed to evaluate seven selected assessment criteria. The nine MAs are evaluated using TOPSIS according to the value of the relative degree of approximation. The main conclusions of the current research are summarized as follows:

1. Seven (7) assessment criteria are selected based on selected renewable energy resources literature (e.g., onshore solar and offshore wind and wave).
2. OM and SM give different relative weights to the assessment criteria and consequently different ranking of eligible MAs.
3. The solar radiation assessment criterion obtained the largest relative weight (37.97%) in the case of SM. This result is in line with several studies that consider solar radiation as the assessment criterion with the greatest weighting factor [16,18,41].
4. The offshore area (MA9) located near Thasos in North Aegean (size equal to 3.628 km^2) presents the most suitable site for OSF deployment based on OM. This is attributed to the proximity of MA9 with the grid of the highest capacity as well as to the potential of the specific site to serve a large population and provide an extended installation area.
5. The offshore area (MA1) located near Crete (size equal to 0.973 km^2) presents the most suitable site for OSF deployment based on SM. This is mainly attributed to the large value of solar radiation in this area.
6. AHP is one of the most suitable, easily applicable, and flexible MCDM methods for solving energy sector problems [63,64]. This method is recommended when experts

in the field can perform the pairwise comparisons. Therefore, in this study, the results obtained by SM could be considered precise and reliable.

7. Entropy method is used when a decision maker is non-existent and relatively subjective weights cannot be obtained. Although the results of EWM are considered reliable and effective according to the traditional literature, the engineering practice supports that the EWM's weighted result does not always accurately reflect the index's information amount and importance [53]. This conclusion is also confirmed by the results of our study.
8. As the offshore solar industry develops, the technical characteristics and spatial requirements might change, which, in turn, might make other sites more feasible. However, the methodological framework proposed in this study provides a starting point for investigating where OSFs could be installed.

Offshore solar energy could be a viable option for making many coastal communities, islands, and isolated locations more sustainable. This investigation provides a logical scientific methodological approach that could be used to rank the site suitability of OSF technology and can be used efficiently in various renewable energy projects. In addition, this paper contributes to the fulfillment of one of the main goals of the European Green Deal related to the decarbonization of the EU's energy system for reaching climate objectives. One of the key principles includes the deployment of a power sector based largely on renewable energy resources [65] and the presented methodology contributes to this direction.

The present investigation could be extended in order to include additional physical parameters as assessment siting criteria, as for example: (i) wind and wave conditions, which are critical for ensuring the structural reliability of OSF systems [66], and (ii) the water temperature, which contributes to increased efficiency (up to 10%) due to the water-cooling effect [67]. On the other hand, the assessment of the electricity production of OSF systems for different solar technologies might provide useful insights in future considerations. Finally, the subject of co-locating different marine renewable energy systems and, more specifically, the combination of offshore solar technologies with offshore wind turbines could be investigated as future work, as it can yield to sustainable solutions and contribute to launching the commercial feasibility of OSFs.

Author Contributions: Conceptualization, D.G.V., G.T., E.L. and N.K.; methodology, D.G.V., G.T., E.L. and N.K.; software, G.T. and N.K. formal analysis, D.G.V., G.T. and E.L.; investigation, G.T.; data curation, D.G.V., G.T. and N.K.; writing—original draft preparation, D.G.V. and E.L.; writing—review and editing, D.G.V., G.T., E.L. and N.K.; visualization, D.G.V., G.T., E.L. and N.K.; supervision, D.G.V. and E.L. All authors have read and agreed to the published version of the manuscript.

Funding: This research received no external funding.

Institutional Review Board Statement: Not applicable.

Informed Consent Statement: Not applicable.

Data Availability Statement: Not applicable.

Conflicts of Interest: The authors declare no conflict of interest.

Abbreviations

AC, Assessment criteria; AEEH, Areas to be licensed for Exploration and Exploitation of Hydrocarbons; AHP, Analytical Hierarchy Process; AOREP, Areas where Offshore Renewable Energy Projects; AZ, Aquaculture Zones; CI, Consistency Index; CR, Consistency Ratio; DS, Distance from Shore; ELECTRE, ELimination Et Choice Translating Reality; EWM, Entropy-based Weight Method; GIS, Geographic Information Systems; MA(s), Marine Area(s); MCDM, Multi-Criteria Decision Making; MEA, Military Exercise Areas; OM, Objective Model; OSF(s), Offshore Solar Farm(s); SM, Subjective Model; PA, Protected Areas; PROMETHEE, Preference Ranking Organization METHod for Enrichment of Evaluations; PSR, Ports and Shipping Routes; SAL, Site Area Limitations; SF-SPSD/RES, Specific Framework for Spatial Planning and Sustainable Development for Renewable

Energy Source; SM, Subjective Model; TOPSIS, Technique for Order Preference by Similarity to Ideal Solution; VIKOR, VIseKriterijumska Optimizacija I Kompromisno Resenje; WD, Water Depth.

References

1. López, M.; Rodríguez, N.; Iglesias, G. Combined Floating Offshore Wind and Solar PV. *J. Mar. Sci. Eng.* **2020**, *8*, 576. [CrossRef]
2. Taveira-Pinto, F.F.; Iglesias, G.; Rosa-Santos, P.; Deng, Z.D. Preface to Special Topic: Marine Renewable Energy. *J. Renew. Sustain. Energy* **2015**, *7*, 061601. [CrossRef]
3. Kumar, V.; Shrivastava, R.L.; Untawale, S.P. Solar Energy: Review of Potential Green & Clean Energy for Coastal and Offshore Applications. *Aquat. Procedia* **2015**, *4*, 473–480. [CrossRef]
4. Santafé, M.R.; Ferrer Gisbert, P.S.; Sánchez Romero, F.J.; Torregrosa Soler, J.B.; Ferrán Gozálvez, J.J.; Ferrer Gisbert, C.M. Implementation of a photovoltaic floating cover for irrigation reservoirs. *J. Clean Prod.* **2014**, *66*, 568–570. [CrossRef]
5. Diendorfer, C.; Haider, M.; Lauermann, M. Performance Analysis of Offshore Solar Power Plants. *Energy Procedia* **2014**, *49*, 2462–2471. [CrossRef]
6. Cagle, A.E.; Armstrong, A.; Exley, G.; Grodsky, S.M.; Macknick, J.; Sherwin, J.; Hernandez, R.R. The Land Sparing, Water Surface Use Efficiency, and Water Surface Transformation of Floating Photovoltaic Solar Energy Installations. *Sustainability* **2020**, *12*, 8154. [CrossRef]
7. de Lima, R.L.P.; Paxinou, K.C.; Boogaard, F.; Akkerman, O.; Lin, F.Y. In-Situ Water Quality Observations under a Large-Scale Floating Solar Farm Using Sensors and Underwater Drones. *Sustainability* **2021**, *13*, 6421. [CrossRef]
8. Kim, C.K.; Jang, S.; Kim, T.Y. Site selection for offshore wind farms in the southwest coast of South Korea. *Renew. Energy* **2018**, *120*, 151–162. [CrossRef]
9. Mahdy, M.; Bahaj, A.S. Multi criteria decision analysis for offshore wind energy potential in Egypt. *Renew. Energy* **2018**, *118*, 278–289. [CrossRef]
10. Schallenberg-Rodríguez, J.; Montesdeoca, N.G. Spatial planning to estimate the offshore wind energy potential in coastal regions and islands. Practical case: The Canary Islands. *Energy* **2018**, *143*, 91–103. [CrossRef]
11. Vagiona, D.G.; Kamilakis, M. Sustainable site selection for offshore wind farms in the South Aegean—Greece. *Sustainability* **2019**, *10*, 749. [CrossRef]
12. Wu, B.; Yip, T.L.; Xie, L.; Wang, Y. A fuzzy-MADM based approach for site selection of offshore wind farm in busy waterways in China. *Ocean. Eng.* **2018**, *168*, 121–132. [CrossRef]
13. Argin, M.; Yerci, V.; Erdogan, N.; Kucuksari, S.; Cali, U. Exploring the offshore wind energy potential of Turkey based on multicriteria site selection. *Energy Strategy Rev.* **2019**, *23*, 33–46. [CrossRef]
14. Deveci, M.; Özcan, E.; John, R.; Covrig, C.F.; Pamucar, D. A study on offshore wind farm siting criteria using a novel interval-valued fuzzy-rough based Delphi method. *J. Environ. Manag.* **2020**, *270*, 110916. [CrossRef]
15. Fang, H.; Li, J.; Song, W. Sustainable site selection for photovoltaic power plant: An integrated approach based on prospect theory. *Energy Convers. Manag.* **2018**, *174*, 755–768. [CrossRef]
16. Doorga, J.R.; Rughooputh, S.D.; Boojhawon, R. Multi-criteria GIS-based modelling technique for identifying potential solar farm sites: A case study in Mauritius. *Renew. Energy* **2019**, *133*, 1201–1219. [CrossRef]
17. Rigo, P.D.; Rediske, G.; Rosa, C.B.; Gastaldo, N.G.; Michels, L.; Neuenfeldt Júnior, A.L.; Siluk, J.C.M. Renewable energy problems: Exploring the methods to support the decision-making process. *Sustainability* **2020**, *12*, 10195. [CrossRef]
18. Al Garni, H.Z.; Awasthi, A. Solar PV power plant site selection using a GIS-AHP based approach with application in Saudi Arabia. *Appl. Energy* **2017**, *206*, 1225–1240. [CrossRef]
19. Wang, C.N.; Nguyen, V.T.; Thai, H.T.N.; Duong, D.H. Multi-Criteria Decision Making (MCDM) approaches for solar power plant location selection in Viet Nam. *Energies* **2018**, *11*, 1504. [CrossRef]
20. Ilbahar, E.; Cebi, S.; Kahraman, C. A state-of-the-art review on multi-attribute renewable energy decision making. *Energy Strategy Rev.* **2019**, *25*, 18–33. [CrossRef]
21. Keeley, A.R.; Matsumoto, K. Relative significance of determinants of foreign direct investment in wind and solar energy in developing countries—AHP analysis. *Energy Policy* **2018**, *123*, 337–348. [CrossRef]
22. Ozdemir, S.; Sahin, G. Multi-criteria decision-making in the location selection for a solar PV power plant using AHP. *Measurement* **2018**, *129*, 218–226. [CrossRef]
23. Colak, H.E.; Memisoglu, T.; Gercek, Y. Optimal site selection for solar photovoltaic (PV) power plants using GIS and AHP: A case study of Malatya Province, Turkey. *Renew. Energy* **2020**, *149*, 565–576. [CrossRef]
24. Ruiz, H.; Sunarso, A.; Ibrahim-Bathis, K.; Murti, S.; Budiarto, I. GIS-AHP Multi Criteria Decision Analysis for the optimal location of solar energy plants at Indonesia. *Energy Rep.* **2020**, *6*, 3249–3263. [CrossRef]
25. Prieto-Amparán, J.A.; Pinedo-Alvarez, A.; Morales-Nieto, C.R.; Valles-Aragón, M.C.; Álvarez-Holguín, A.; Villarreal-Guerrero, F. A regional GIS-assisted multi-criteria evaluation of site-suitability for the development of solar farms. *Land* **2021**, *10*, 217. [CrossRef]
26. Günen, M.A. A comprehensive framework based on GIS-AHP for the installation of solar PV farms in Kahramanmaraş, Turkey. *Renew. Energy* **2021**, *178*, 212–225. [CrossRef]

27. Alhuyi Nazari, M.; Aslani, A.; Ghasempour, R. Analysis of solar farm site selection based on TOPSIS approach. *Int. J. Soc. Ecol Sustain. Dev.* **2019**, *9*, 12–25. [CrossRef]
28. Ali Sadat, S.; Vakilalroaya Fini, M.; Hashemi-Dezaki, H.; Nazififard, M. Barrier analysis of solar PV energy development in the context of Iran using fuzzy AHP-TOPSIS method. *Sustain. Energy Technol. Assess.* **2021**, *47*, 101549. [CrossRef]
29. Ministry of Environment, Energy and Climate Change (MEECC). Specific Framework for Spatial Planning and Sustainable Development for Renewable Energy Sources (SFSPSD/RES). JMD 49828/2008 OGHE B′ 2464/3-12-08. 2008. Available online: https://ypen.gov.gr/wp-content/uploads/2020/11/FEK2464B_2008_RES.pdf (accessed on 13 January 2022).
30. Vasileiou, M.; Loukogeorgaki, E.; Vagiona, D.G. GIS-based multi-criteria decision analysis for site selection of hybrid offshore wind and wave energy systems in Greece. *Renew. Sustain. Energy Rev.* **2017**, *73*, 745–757. [CrossRef]
31. Schillings, S.; Wanderer, T.; Cameron, L.; Tjalling van der Wal, J.; Jacquemin, J.; Veumb, K. A decision support system for assessing offshore wind energy potential in the North Sea. *Energy Policy* **2012**, *49*, 541–551. [CrossRef]
32. Kim, T.; Park, J.I.; Maeng, J. Offshore wind farm site selection study around Jeju Island, South Korea. *Renew. Energy* **2016**, *94*, 619–628. [CrossRef]
33. Murphy, J.; Lynch, K.; Serri, L.; Airdoldi, D.; Lopes, M. Site selection analysis for offshore combined resource projects in Europe. In *Report of the Off-Shore Renewable Energy Conversion Platforms—Coordination Action (ORECCA) Project*; 2011; Available online: http://orecca.rse-web.it/doc_info/Site_Selection_Analysis_Report.pdf (accessed on 13 January 2022).
34. Mekonnen, A.; Gorsevski, P. A web-based participatory GIS (PGIS) for offshore wind farm suitability within Lake Erie, Ohio. *Renew. Sustain. Energy Rev.* **2015**, *41*, 162–177. [CrossRef]
35. Jay, S. Planners to the rescue: Spatial planning facilitating the development of offshore wind energy. *Mar. Pollut. Bull.* **2010**, *10*, 493–499. [CrossRef] [PubMed]
36. Ikhennicheu, M.; Danglade, B.; Pascal, R.; Arramounet, V.; Trébaol, Q.; Gorintin, F. Analytical method for loads determination on floating solar farms in three typical environments. *Sol. Energy* **2021**, *219*, 34–41. [CrossRef]
37. Cradden, L.; Kalogeri, C.; Martinez Barrios, I.; Galanis, G.; Ingram, D.; Kallos, G. Multi-criteria site selection for offshore renewable energy platforms. *Renew. Energy* **2016**, *87*, 791–806. [CrossRef]
38. Díaz, H.; Fonseca, R.B.; Guedes Soares, C. Site selection process for floating offshore wind farms in Madeira Islands. In *Advances in Renewable Energies Offshore*; Guedes, S.C., Ed.; Taylor & Francis Group: London, UK, 2019; pp. 729–737.
39. Spyridonidou, S.; Sismani, G.; Loukogeorgaki, E.; Vagiona, D.G.; Ulanovsky, H.; Madar, D. Sustainable spatial energy planning of large-scale wind and PV farms in Israel: A collaborative and participatory planning approach. *Energies* **2021**, *14*, 551. [CrossRef]
40. Oliveira-Pinto, S.; Stokkermans, J. Marine floating solar plants: An overview of potential, challenges and feasibility. *Proc. Inst. Civ. Eng. Marit. Eng.* **2020**, *173*, 120–135. [CrossRef]
41. Doljak, D.; Stanojević, G. Evaluation of natural conditions for site selection of ground-mounted photovoltaic power plants in Serbia. *Energy* **2017**, *127*, 291–300. [CrossRef]
42. Vagiona, D.G. Comparative Multicriteria Analysis methods for ranking sites for solar farm deployment: A Case study in Greece. *Energies* **2021**, *14*, 8371. [CrossRef]
43. Government Gazette 2885/2017. Available online: https://www.greekhydrocarbons.gr/pdfs/BlockDocuments/2885B_2017_Lease_Agreement-Geographical_Delimitation.pdf (accessed on 1 October 2021).
44. Hellenic Navy Hydrographic Service. Available online: https://www.hnhs.gr/el/ (accessed on 1 October 2021).
45. Open Geospatial Data and Services for Greece. Available online: http://geodata.gov.gr/ (accessed on 1 October 2021).
46. MarineTraffic: Global Ship Tracking Intelligence. Available online: https://www.marinetraffic.com/en/ais/home/centerx:-12.0/centery:25.0/zoom:4 (accessed on 1 October 2021).
47. WWF Greece. Available online: https://www.contentarchive.wwf.gr/images/pdfs/AquacultureLow.pdf (accessed on 1 October 2021).
48. Regulatory Authority for Energy. Available online: https://geo.rae.gr/ (accessed on 1 October 2021).
49. Independent Power Transmission Generation. Available online: https://www.admie.gr/en/grid/description/lines-map (accessed on 1 October 2021).
50. Hellenic Statistical Authority. Available online: https://www.statistics.gr/el/statistics/pop (accessed on 1 October 2021).
51. Nikitidou, E.; Kazantzidis, A.; Tzoumanikas, P.; Salamalikis, V.; Bais, A.F. Retrieval of surface solar irradiance, based on satellite-derived cloud in-formation, in Greece. Renew. *Energy* **2015**, *90*, 776–783. [CrossRef]
52. Hsing-Chen, L.; Ching-Ter, C. Comparative analysis of MCDM methods for ranking renewable energy sources in Taiwan. *Renew. Sustain. Energy Rev.* **2018**, *92*, 883–896. [CrossRef]
53. Zhu, Y.; Tian, D.; Yan, F. Effectiveness of entropy weight method in decision-making. *Math. Probl. Eng.* **2020**, *2020*, 3564835. [CrossRef]
54. Dong, G.H.; Shen, J.Q.; Jia, Y.Z.; Sun, F.H. Comprehensive evaluation of water resource security: Case study from Luoyang City, China. *Water* **2018**, *10*, 1106. [CrossRef]
55. Amiri, V.; Rezaei, M.; Sohrabi, N. Groundwater quality assessment using entropy weighted water quality index (EWQI) in Lenjanat, Iran. *Environ. Earth Sci.* **2014**, *72*, 3479–3490. [CrossRef]
56. Saaty, R.W. The analytic hierarchy process—What it is and how it is used. *Math. Model.* **1987**, *9*, 161–176. [CrossRef]
57. Saaty, T.L. How to make a decision: The analytic hierarchy process. *Eur. J. Oper. Res.* **1990**, *48*, 9–26. [CrossRef]
58. Chen, S.J.; Hwang, C.L. *Fuzzy Multiple Attribute Decision Making: Methods and Applications*; Springer: Berlin/Heidelberg, Germany, 1992.

59. Hwang, C.L.; Yoon, K. *Multiple Attribute Decision Making: Methods and Applications, A State-of-the-Art Survey*; Springer: Berlin/Heidelberg, Germany, 1981.
60. Jahanshahloo, G.R.; Hosseinzadeh Lotfi, F.; Izadikhah, M. Extension of the TOPSIS method for decision-making problems with fuzzy data. *Appl. Math. Comput.* **2006**, *181*, 1544–1551. [CrossRef]
61. Opricovic, S.; Tzeng, G.H. Compromise solution by MCDM methods: A comparative analysis of VIKOR and TOPSIS. *Eur. J. Oper. Res.* **2004**, *156*, 445–455. [CrossRef]
62. Vagiona, D.G.; Karanikolas, N.M. A multicriteria approach to evaluate offshore wind farms siting in Greece. *Glob. NEST J.* **2012**, *14*, 235–243. [CrossRef]
63. Kaya, I.; Çolak, M.; Terzi, F. Use of MCDM techniques for energy policy and decision-making problems: A review. *Int. J. Energy Res.* **2018**, *42*, 2344–2372. [CrossRef]
64. Siksnelyte-Butkiene, I.; Zavadskas, E.K.; Streimikiene, D. Multi-Criteria Decision-Making (MCDM) for the assessment of renewable energy technologies in a household: A review. *Energies* **2020**, *13*, 1164. [CrossRef]
65. European Commission Factsheet, Clean Energy. Available online: https://ec.europa.eu/commission/presscorner/detail/en/fs_19_6723 (accessed on 13 January 2022).
66. López, M.; Claus, R.; Soto, F.; Cebada, A.; Hernández-Garrastacho, Z.A.; Simancas, O. Structural reliability of a novel offshore floating photo. In *Developments in the Analysis and Design of Marine Structures*, 1st ed.; CRC Press: London, UK, 2021.
67. Charles Lawrence Kamuyu, W.; Lim, J.R.; Won, C.S.; Ahn, H.K. Prediction model of photovoltaic module temperature for power performance of floating PVs. *Energies* **2018**, *11*, 447. [CrossRef]

Journal of
Marine Science and Engineering

Article

Techno-Economic Assessment of Offshore Wind Energy in the Philippines

Gerard Lorenz D. Maandal [1], Mili-Ann M. Tamayao-Kieke [2] and Louis Angelo M. Danao [3,*]

1 Energy Engineering Graduate Program, University of the Philippines, Diliman, Quezon City 1101, Philippines; gelomaandal@gmail.com
2 Department of Industrial Engineering and Operations Research, University of the Philippines, Diliman, Quezon City 1101, Philippines; mmtamayao@up.edu.ph
3 Department of Mechanical Engineering, University of the Philippines, Diliman, Quezon City 1101, Philippines
* Correspondence: louisdanao@up.edu.ph

Abstract: The technical and economic assessments for emerging renewable energy technologies, specifically offshore wind energy, is critical for their improvement and deployment. These assessments serve as one of the main bases for the construction of offshore wind farms, which would be beneficial to the countries gearing toward a sustainable future such as the Philippines. This study presents the technical and economic viability of offshore wind farms in the Philippines. The analysis was divided into four phases, namely, application of exclusion criteria, technical analysis, economic assessment, and sensitivity analysis. Arc GIS 10.5 was used to spatially visualize the results of the study. Exclusion criteria were applied to narrow down the potential siting for offshore wind farms, namely, active submerged cables, local ferry routes, marine protected areas, reefs, oil and gas extraction areas, bathymetry, distance to grid, typhoons, and earthquakes. In the technical analysis, the turbines SWT-3.6-120 and 6.2 M126 Senvion were considered. The offshore wind speed data were extrapolated from 80 m to 90 m and 95 m using power law. The wind power density, wind power, and annual energy production were calculated from the extrapolated wind speed. Areas in the Philippines with a capacity factor greater than 30% and performance greater than 10% were considered technically viable. The economic assessment considered the historical data of constructed offshore wind farms from 2008 to 2018. Multiple linear regression was done to model the cost associated with the construction of offshore wind farms, namely, turbine, foundation, electrical, and operation and maintenance costs (i.e., investment cost). Finally, the levelized cost of electricity and break-even selling price were calculated to check the economic viability of the offshore wind farms. Sensitivity analysis was done to investigate how *LCOE* and price of electricity are sensitive to the discount rate, capacity factor, investment cost, useful life, mean wind speed, and shape parameter. Upon application of exclusion criteria, several sites were determined to be viable with the North of Cagayan having the highest capacity factor. The calculated capacity factor ranges from ~42% to ~50% for SWT-3.6-120 and ~38.56% to ~48% for 6.2M126 turbines. The final regression model with investment cost as the dependent variable included the minimum sea depth and the plant capacity as the predictor variables. The regression model had an adjusted R2 of 90.43%. The regression model was validated with existing offshore wind farms with a mean absolute percentage error of 11.33%. The *LCOE* calculated for a 25.0372 km^2 offshore area ranges from USD 157.66/MWh and USD 154.1/MWh. The breakeven electricity price for an offshore wind farm in the Philippines ranges from PHP 8.028/kWh to PHP 8.306/kWh.

Keywords: offshore wind; technical analysis; economic analysis; Philippines

Citation: Maandal, G.L.D.; Tamayao-Kieke, M.-A.M.; Danao, L.A.M. Techno-Economic Assessment of Offshore Wind Energy in the Philippines. *J. Mar. Sci. Eng.* **2021**, *9*, 758. https://doi.org/10.3390/jmse9070758

Academic Editor: Liliana Rusu

Received: 15 June 2021
Accepted: 7 July 2021
Published: 9 July 2021

1. Introduction

The Philippines adopted the Republic Act 9513 or An Act Promoting the Development, Utilization, and Commercialization of Renewable Energy Resources and for Other Purposes on 16 December 2008 to pave the way for sustainable and affordable renewable energy [1]. The policies adopted in the Act are to accelerate the exploration and development of renewable energy resources, to encourage the development and utilization of such to reduce harmful emissions, and establish necessary infrastructure to carry out the mandate specified in the Act.

Under RA 9513, the National Renewable Energy Program (NREP) was formulated to provide the necessary framework for the country [2]. The main target of the NREP's energy policy is to increase the existing renewable energy capacity by 15.304 GW by 2030. The additional capacity is to come from wind, hydro, geothermal, solar, ocean, and biomass. Specific to wind energy, NREP's energy policy framework aims to achieve grid parity by 2025 through the addition of 2345 MW of capacity. However, only an additional 395 MW have been installed by 2015 and none thereafter [3]. A difference of 654 MW is required to reach the target for that same year. This backlog is indicative of the slow pace of development in the Philippines so that the target grid parity will not be realized as envisioned. Other pathways to achieve this goal are necessary and venturing into the offshore wind arena may be a viable option.

Offshore wind farms (OWF) have gained attention in the past years due to some inherent advantages present. Capacities of OWF are significantly greater than their onshore counterparts, sometimes reaching 1000 MW [4]. The wind resource in the ocean is less intermittent than on land and can be predicted with greater accuracy [5]. Furthermore, social acceptance is greater for OWF due to low visual and noise impact [4]. To assess these, the technical and economic viability of OWFs in the Philippines was studied in this work.

1.1. Statement of the Problem

As of 2015, the Philippines has not achieved its target capacity for wind energy installations by 62%. OWFs are investigated and considered in this research to possibly fill in the gap in the capacity. Specifically, the technical and economic aspects of OWFs were studied for their viability in the Philippine setting. As a caveat, there are difficulties in the assessment of OWFs in the Philippines, which are stated below:

- Methodologies on techno-economic assessment of offshore wind energy have not been applied to the Philippine setting.
- A notion exists that it is a risky investment with high cost and uncertainty in return.
- There is no readily available and reliable information for investments regarding the viability of offshore wind farms in the Philippines.
- There has been no formulation for the recommendation of the viability of offshore wind energy in the Philippines.

1.2. Objectives of the Study

The main objective of the study was to be develop a framework in assessing the technical and economic potential of offshore wind energy in the Philippines. It aims to achieve the following specific objectives:

- To develop a methodology for the techno-economic assessment of offshore wind farms in the Philippines.
- To assess the wind resource in the Philippine oceans for the potential of putting up an offshore wind farm.
- To investigate the economic viability of constructing an offshore wind farm in the Philippines through *LCOE*.
- To formulate a recommendation for the viability of OWF in the Philippines.

1.3. Research Significance

The current pace of development of wind energy in the Philippines is slow in terms of attaining grid parity in wind energy with the commissioning of 2345 MW of additional capacity by 2030. The bulk of the commissioning of wind power plants was done in 2014 under four projects with a total capacity of 303.90 MW [3]. In 2015, there were only two projects constructed with a combined installed capacity of 90 MW [6,7]. By exploring the emerging technology of offshore wind energy, higher capacity wind farms can be built with the availability of space and good wind resource. This techno-economic assessment research can help assess the viability of the construction of an offshore wind farm in the Philippines. The investors and government may benefit from this study since they will be able to see the technical and financial model of the study wherein the *LCOE* and price of electricity are included. These economic measures may give the investors an option to consider offshore wind farms as one of their investments and provide the government insights for the promotion of this emerging renewable energy technology.

The novelty of this research compared to existing technical and economic assessment of offshore wind farms is the extensive application of exclusion criteria, comparison of two offshore wind turbines, and rigorous economic analysis in the Philippines setting.

1.4. Limitations of the Study

Since offshore wind energy is an emerging technology around the world, there are various improvements that can be adopted in this research in terms of technical and economic data acquisition. The accuracy and reliability of the results in the technical potential of offshore wind is heavily dependent on the data at reference height, spatial resolution, and time interval. The available data acquired by the researcher are from the Phil-LiDAR 2 Program of the Department of Science and Technology with a spatial resolution of 4 km, and mean wind speeds for five years at 10 m, 20 m, 80 m, and 100 m elevations.

The data for the economic analysis were acquired from 4C Offshore. The offshore wind farms considered were the ones constructed from 2008 to 2018. The data acquired include the name of the power plant, country of origin, minimum and maximum sea depth, area, offshore cable length, onshore cable length, inter-array cable length, port for O&M, distance from port, turbine model, number of turbines, turbine capacity, plant capacity, and investment cost. Offshore wind farms with incomplete data were not considered in the study.

Since the different parts of the Philippines have varying wind speeds, only a selected site with the highest capacity factor will be considered for the economic analysis. In this study, the offshore wind energy will be compared to the price of onshore wind farms in terms of investment cost and price of electricity since onshore wind is a renewable energy technology with a share of 3.57% of the total installed capacity in the Philippines, respectively [3].

The stationary foundation technologies, namely, monopile, jacket, and tripod, will be the interest of this research due to its high capacity production at stationary structures. Thus, the limitation of the study was within 50 m bathymetry.

2. Review of Related Literature

2.1. Offshore Wind Farms

There are numerous advantages of offshore wind farms that address the disadvantages of putting up an onshore wind farm. Wind farms, in general, emit noise mostly from the rotation of blades [8]. In some cases, the annoyance from the noise coming from onshore wind farms can lead to psychological distress and sleep disturbance to the people living within the vicinity of wind turbines [9]. With the offshore wind farm situated in the ocean, low noise is heard [4].

Offshore wind farms have a more stable wind production compared to onshore wind farms. The power duration curve of offshore wind power plants is greater than onshore wind farms in Germany [5]. Higher and steadier offshore wind farms are more productive in the UK with a 36% capacity factor compared to onshore wind farms (27% capacity factor), which implies a higher capacity credit and smaller back-up cost [10].

Forecasting of wind energy is another concern in wind energy technology due to the intermittency in wind resources. In the case of offshore wind farms, wind energy can be predicted with a relatively smaller error compared to onshore wind farms. Offshore wind farms need fewer reserves in the balancing of forecast error [5]. The results of the study of Richts et al. [5] illustrate that offshore wind farms can have more than 50% of their power output predicted with 99.994% reliability in one-hour advance compared to onshore wind farms with only about 30% power output predicted with the same level of reliability.

A study showed that the proximity of wind turbines near residence areas negatively influences attitudes toward installations. However, with proper consultations and proper participatory siting, wind farms can receive a higher level of acceptance [11]. Through proper consultations with the affected stakeholders and participatory siting, the possibility of the construction of an offshore wind farm can also be accepted.

2.2. Offshore Wind: Current Status

2.2.1. Installed Capacity

Rapid increase in offshore wind farms has been evidently seen for the past two decades in Europe. The total installed capacity for offshore wind farms was only 50 MW in 2000 and increased up to 1471 MW by the year 2008, which means an annual growth of approximately 50% [10]. In 2016, the total installed capacity in Europe reached 12,631 MW [12].

In 2006, the average wind farm size in Europe was only 46.3 MW. After 10 years, the average size has increased up to 379.5 MW based on offshore wind farms under construction in 2016 [12]. Offshore wind farms have high potential to have larger size capacity as planned in Europe.

2.2.2. Number of Turbines and Project Area

The project average area and number of turbines of commissioned and under construction offshore wind farms in Europe was studied extensively [13]. It was seen that the project area in 2015 was lower compared to 2012, but the average size was higher than in 2012. The developments in the turbine industry can explain this event as the turbines have increased in capacity and the number of turbines has also increased.

2.2.3. Distance to Shore and Water Depths

In European installations, the distance to shore vs. water depth was studied [12]. As the distance of the offshore wind farm to the shore increases, it also translates to a higher installed capacity. There is a rapid increase in the distance to the shore as time goes by [11]. In 2013, the average distance was just 25 km, but it has rapidly increased to an average distance of 42 km to the shore.

2.2.4. Cost

There has been an increase in the CAPEX despite the technological advances in offshore wind energy [13]. The annual construction of deeper and farther offshore may contribute to the rise of CAPEX since these factors increase the grid connection, foundation, and installation costs.

2.2.5. Wind Turbines

The wind turbine is an important parameter in the construction of offshore wind farms. In Europe, the average capacity rating was 4.8 MW based on 361 offshore wind turbines that are in the progress of being constructed in 2016 [12]. It was seen that the progression of rated capacity gradually increases from 0 MW to 4.9 MW in 20 years. Among the different offshore wind turbine manufacturers in Europe, Siemens Wind Power was the biggest supplier with 67.8% installed capacity [12].

2.3. Foundation Technologies

There are proven and tested stationary foundations within 50 m of depth, namely, monopile, jacket, and tripod [14]. The floating structures are considered optimal for bathymetry greater than 50 m, namely, tension leg platform, semi-submersible, and spar buoy [15]. The stationary foundation technologies will be the interest of this research due to its high capacity production at stationary structures.

2.4. Renewable Energy Law in the Philippines

The Republic Act No. 9513, also known as the "Renewable Energy Act of 2008", is an act promoting the development, utilization, and commercialization of renewable energy resources and for other purposes. In this law, policies and regulations are enacted to favor the utilization of renewable energy technologies in the Philippines such as wind energy. A feed-in tariff system is applied to renewable energy technologies to leverage them from the wholesale electricity spot market by prioritizing their connection to the grid, prioritizing for their purchase and transmission or "priority dispatch", fixed tariff for up to 20 years, and has a mandated minimum percentage of generation.

Aside from the feed-in tariff system that promotes renewable energy technologies, tax incentives are included in this law. Renewable energy power plants can have income tax holiday for the first seven years of the power plant, duty-free importation of machinery, equipment, and materials for the first ten years, and a 10% corporate tax rate instead of 30% [1]. With these policies enacted, the chances of constructing an offshore wind farm can increase, but the technical and economic aspects should still be investigated.

2.5. Related Techno-Economic Studies

The technical and economic feasibility of OWF in Chile were studied by Mattar and Guzmán-Ibarra [16]. The Vestas V164 8.0 MW wind turbine was studied using wind data from the ERA-interim reanalysis spanning the years 1979 to 2014. Performance and capacity factor were calculated for wind speeds extrapolated from 10 m above sea level. Information on costing was obtained from previous studies involving the full life cycle of wind farms and used in the economic analysis of the study. Estimates for a useful life of 25 years were computed for the levelized cost of electricity (LCOE), payback period, internal rate of return, and net present value. Three capacities were considered to estimate the *LCOE*. These were 80 MW, 160 MW, and 240 MW.

Power density maxed at 3190 W/m^2 with the capacity factor of 70% and *LCOE* between 72 and 100 USD/MWh for locations between 45° and 56° S. Rated at 13 m/s, the best location for the selected wind turbine was shown to be between 30° and 32° S. Power density at that location was between 700 W/m^2 and 900 W/m^2 with capacity factor ranging from 40 to 60%. The *LCOE* values were seen to range from 100 to 114 USD/MWh. It was determined that an increase in discount rate by only 2% can increase the *LCOE* threefold.

The techno-economic and environmental considerations of OWF along India's coastline were studied by Nagababu et al. [17] using a geospatial information system (GIS). Cost supply curve and spatial distribution of levelized production cost (LPC) were created and sensitivity analyses were carried out to examine the effect of project lifetime, annual energy production, availability, and discount rate on cost.

Results revealed a wide range of wind power densities for both east and west coasts. The western coastline sees power densities from 13 to 294 W/m^2 while the eastern coastline gets power densities of 63 to 393 W/m^2. Theoretically, these translate to annual energy potential ranges of 54 to 823 MWh/km^2 and 107 to 1117 MWh/km^2 for the western and eastern coast, respectively. Compared to existing renewable energy resources in India within the 200 Euro/MWh FIT, a good 40% of the potential is available that is economically competitive. Bottom fixed wind farms in the western coastline will produce energy at costs of 231 and 262 Euro/MWh for 0 to 30 m and 30 to 50 m depths, respectively. The eastern coastline will see higher costs for similar installations at 308 and 334 Euro/MWh for 0 to 30 m and 30 to 50 m depths, respectively. Interest rate was seen as a very influential parameter for the calculation of the LPC.

Khraiwish Dalabeeh [18] studied the cost of energy and capacity factor of OWF at five prospective locations in Jordan by developing a simulation model using Weibull parameters and computing for annual generation of different wind turbine models. The most suitable wind turbine was matched to the locations studied and the cost of electricity and capacity factors were determined subsequently. Nine years' worth of wind data were utilized for the analysis of the prospective sites.

The analysis showed that there were five suitable sites that have good wind resources for electricity generation. The Xemc-Darwind was observed to produce the lowest cost of electricity at 25.9 USD/MWh. Aaer and Acciona produced the greatest cost at 100 to 222 USD/MWh for sites Der Alla, Amman, and Irbid due to low wind resource and high property prices. R. Monif came out to be the most suitable site for the deployment using the Xemc-Darwind rotor with the least cost of electricity at 25.9 USD/MWh and highest capacity factor of 42%. Fossil fuel power generation costs in 2013 were significantly higher than the computed wind energy generation cost with an average of 260 USD/MWh.

Schweizer et al. [19] conducted the first feasibility study of the offshore wind farm in the coast of Rimini at the Northern Adriatic Sea. The wind speed data were acquired in 2008 to 2013 through the anemometers mounted on a gas platform, and the characteristics of the wave were analyzed through the data using a wave buoy installed in 2007 (located few kilometers at the study area). The study also described the environmental external conditions that could potentially act on the offshore wind farm and are crucial for the engineering design. A commercially available wind turbine was selected in the study with a rated power of 3.6 MW, and the capacity factor was calculated to be 25% using the experimental wind probability distribution at 80 m above sea level. It was found that for layout A (15 wind turbines, 54 MW installed capacity), the best case had 97% WT availability, 9.75% internal losses while the worst case had 90% WT availability, 11% internal losses, and 0.205 €/kWh feed-in tariff. For layout B (60 wind turbines, 216 MW installed capacity, 0.160 €/kWh feed-in tariff), the best case had 97% WT availability, 9.75% internal losses, and 0.205 €/kWh feed-in tariff, while the worst case had 90% WT availability, 11% internal losses, and 0.160 €/kWh feed-in tariff. It was seen that the assumed feasibility study and commissioning will last up to four years with the offshore wind farm to be operational in 25 years, while the last year is for decommissioning. In all of the cases, positive net cash flow can be seen. The lowest net income was found in layout A in the worst case scenario, and the most adequate result in terms of finance could be found in layout B with the best case scenario.

Cavazzi and Dutton [20] studied the potential of offshore wind energy in the UK with the aid of GIS. A suitability map was produced taking into account technical and economic factors that influence the OWF potential of the study area. Certain sites were excluded based on specific constraints applied such as existing offshore installations, submarine cables, oil and gas extraction areas, among others. The *LCOE* was calculated and a sensitivity analysis was conducted to ascertain which factors greatly influence the *LCOE*. It was observed that the interest rate contributed the most in influencing the *LCOE*. Of the available offshore wind resources in the UK, 10% is potentially accessible for a 150 GW wind farm that will produce energy at a cost of less than 140 Euro/MWh.

2.6. Exclusion Criteria

The Philippines is composed of 7641 islands, and there are three main island groups, namely, Luzon, Visayas, and Mindanao. Since the Philippines has more than 7000 islands, there is a need to narrow down the sites that could have the potential for the construction of an offshore wind farm. With this premise, the study excludes sites if they do not meet the certain constraint criterion. In the study of the multi criteria decision analysis of offshore wind energy potential in Egypt, the constraints considered for site exclusion are distance from shore, depth, submerged cable paths, oil and gas extraction areas, shipping routes, nature reserves, and military areas [21]. The constraints considered in the study on the assessment of the UK's offshore wind energy potential by using geographic information systems were anchorage areas, existing wind farms, active submarine cables, protected wrecks, sand mining, IMO traffic separation schemes, and oil and gas extraction areas [20].

For this study, the constraints considered for excluding sites are active submerged cable paths, local ferry routes, marine conservation areas, reefs, oil and gas extraction areas, depth, and distance from grid. Since the Philippines is located in the Pacific Ring of Fire, it is susceptible to calamities. The other constraints that will be considered are typhoons and earthquakes.

Active submerged cables are used for transmitting data through the use of cables laid on the seabed. These are essential for the telecommunications industry in transmitting data to connect with other countries and also to connect to the Internet. These cables can reach up to 9000 km in length with a price of $300,000 USD [22]. In this study, the path of active submerged cables with a buffer of 5 km were excluded in the siting process of the offshore wind farm. The active submerged cable map was acquired from Submarine Cable Map [23].

Local ferry routes in the Philippines were generated by Thierry Crevoisier for the World Food Program on 25 February 2014 [24]. In this study, the local ferry routes in the Philippines were excluded in the siting process since these offshore wind farm may pose safety concerns along the path. A buffer of 3 km was applied when using this criterion.

According to the Republic Act 7586, also known as the "National Integrated Protected Areas System Act of 1992", protected areas are identified as areas of water and land that are set aside for preservation due to its unique biological and physical significance, managed for the improvement in biodiversity, and protected from destructive human exploitation [25]. The National Integrated Protected Areas System (NIPAS) is tasked to manage the preservation of the protected areas. The following are categorized as protected areas, namely, strict nature reserve, natural park, natural monument, wildlife sanctuary, protected landscapes and seascapes, resource reserve, and natural biotic areas. In this study, all protected areas were excluded [26]. The marine protected areas are the areas with a dark shade of blue in the map. A buffer zone of 3 km was applied based on a study that 90% of marine protected areas were less than 1 km^2 [27].

The Philippines is part of the Coral Triangle with Malaysia, Timor-Leste, Indonesia, Papua New Guinea, and the Solomon Islands. This group of countries contain one-third of the world's coral reefs [28]. Coral reefs act as the main buffer against erosion due to typhoons and waves, and it also serves as houses for the fish in the ocean [29]. The data on the location of coral reefs in the Philippines is based on ReefBase, which serves as the official database of the International Coral Reef Action Network and Global Coral Reef Monitoring Network [30]. Locations with coral reefs are excluded in the analysis. The coral reef areas are indicated by the dark brown shade in the map. Similar to marine protected areas, a buffer area of 3 km was applied in the exclusion of coral reefs.

The Department of Energy contracts private companies to extract oil and gas in the vicinity of the country. The map for the oil and gas extraction areas was acquired from the Department of Energy [31]. These oil and gas extraction areas reach up to 1,476,000 hectares such as in the case of BHP Billiton in Southwest Palawan. The areas that have already contracted and the areas up for contracting for around 5 km will be excluded from the study.

The depth of the ocean or the bathymetry affects the type of foundation to be applied and the economic cost of the offshore wind farm. The type of foundation technology to be used in this study is a stationary foundation since it can be applied within a depth of 50 m [14]. There is an increase in the CAPEX as the depth also increases [13]. The map for the depth of the ocean in the Philippines was acquired from the General Bathymetric Chart of the Ocean [32]. The bathymetry excluded in this study were depths greater than 50 m.

As the distance from the grid to the offshore wind farm increases, the transmission cost also increases related to grid connection [20]. The distance greater than 120 km from the grid will be excluded from the analysis since longer distance may yield high uncertainty with the cost. The data for the transmission lines was acquired from National Renewable Energy Laboratory [33].

In the study by Worsnop et al. [34], they predicted that offshore wind turbines cannot withstand hurricane gusts of up to 90 m/s (324 km/h)—modern wind turbines can only withstand gusts of up to 70 m/s (252 km/h) based on engineering standards. In the Philippine context, tropical cyclones reach up to a maximum of 250 km/h according to the United Nations Office for the Coordination of Humanitarian Affairs [35]. However, there are only limited studies regarding the impacts of tropical cyclones to offshore wind farms, and this could be a potential area of research for the improvement of these offshore wind turbines. Due to the limitation on the studies in tropical cyclone impacts, wind speed of gusts greater than 250 km/hour will be excluded in the study. The buffer area considered in the application of typhoon path was 50 km since the maximum wind speeds were within the eye wall with a maximum distance of 50 km [36,37].

In 2011, the Kamisu offshore wind farm in Japan was struck by an earthquake with an epicenter distance of 300 km and a magnitude of 6 [38]. The 2 MW offshore wind farm survived the disaster, and was even operational after three days. The Kamisu offshore wind farm is located 40 m from the coast with a monopile foundation technology. One of the solutions of the Niras international engineering consultancy regarding earthquake and typhoon problems is the application of a jacket foundation, which is built on a grid construction of steel placed on four piles [39]. A possible option for offshore wind farms is jacket foundations, which are more robust with regard to the size of the foundation at a greater ocean depth. In this study, a buffer of 15 km from the high historical earthquakes will be applied based on the 10 km critical damage to a dyke of a 7.8 magnitude earthquake in Japan [40].

2.7. Wind Curtailment

It is a challenge if there is a high level of penetration from wind energy to the grid due to its limits in predictability and variability. Wind energy may be curtailed when there is a lack of available transmission, system balancing issues, or back-feeding. Wind curtailment is referred to as the utilization of less wind at a certain time even though there is still more wind power potential available [41].

In 2013, China's installed wind capacity was 77.16 GW with 142 TWh generated electricity or 2.6% of the total electricity generated in the country [42]. In that same year, it curtailed 16.23 TWh of wind energy or 10.74% of the total wind energy generated. Some parts of the country had higher wind curtailment such as in Jilin and Gansu with 21.79% and 20.69%, respectively. High concentration of wind farms in northern and northwestern parts of China, but low electricity demands have contributed to the curtailment. The transmission capacity to deliver electricity to different parts of China with high demand is also insufficient. Some other factors for wind curtailment are also guaranteed minimum dispatch for coal plants and the need for combined heat and power plants during winter for heating. China's solution to wind curtailment is an improvement in forecasting and scheduling, construction of dispatch system for wind power, utilization of automatic generation control systems (AGC), and the application of electric boilers powered by wind energy.

In Denmark, there was an additional installed capacity of 4893 MW in 2014 that produced electricity of 13.1 TWh (1272 MW and 5.2 TWh are from offshore wind farms) [42]. The wind integration in this country is reinforced by the interconnection of systems to neighboring countries with an international electricity market. For instance, over-generation of wind energy is avoided through negative price signals in the day-ahead and intra-day power market. Large capacity offshore wind farms negotiate through the tendering process for compensation.

The United States of America has numerous balancing areas in different parts of the country. The wind curtailment is generally less than 4% [42]. The factors affecting wind curtailment are excessive wind energy during low demand periods and during requirements of minimum generation such in the case of Bonneville Power Administration (BPA) and the Electric Reliability Council of Texas (ERCOT).

The wind curtailment in different countries was also reported by Bird et al. [42]. The data for Germany dated to 2012. The electricity generation from wind ranges from 0.4% to 31.9%. Globally, wind curtailment ranges from 1% to 3% with the exception of 11% in China.

2.8. Data Sources

The data acquired for the wind speed are from Phil-LiDAR 2 Program of the Department of Science and Technology with a 4 km spatial resolution and distance of 10 km from the shore [43]. The offshore wind data are an annual wind speed at elevations of 10 m, 20 m, 80 m, and 100 m in 2008, 2010, 2014, 2015, and 2016, respectively.

2.9. Technical Analysis

2.9.1. Power Law

An important parameter for the assessment of wind power potential in a specific site is the reference height of the wind speed data. Nowadays, wind turbine heights range from 50–120 m [44]. More reliable results can be acquired if the reference height of the wind speed data is close enough to the hub height.

The mathematical model for power law can be seen in Equation (1) where v_1 (m/s) is the wind speed at reference height z_1 (m); v_2 (m/s) is the wind speed at height z_2; and z_o is the roughness length factor. An established value of 0.0002 m is used, which corresponds to sea surface [16,17].

$$v_2 = v_1 \frac{\ln\left[\frac{z_2}{z_0}\right]}{\ln\left[\frac{z_1}{z_o}\right]}. \quad (1)$$

2.9.2. Weibull Model

There are two known statistical approaches for calculating wind power potential, namely, the Weibull and Rayleigh Distribution models. Based on the study of Celik [45], the Weibull model gives better performance compared to the Rayleigh model based on monthly fitness coefficients in calculating the wind power potential in Iskenderun, Turkey. However, due to limitations in the type of data acquired, the Weibull model was used with a shape parameter value of 2, effectively reducing it to the Rayleigh model.

The mathematical model for the estimation of the wind power potential can be calculated from the Weibull distribution model in Equation (2) with $p(V_2)$ as the Weibull probability density; k as the shape parameter (dimensionless); c as the scale parameter (m/s); and v_2 as the wind speed (m/s) [44]. The assumed value of the shape parameter based on another study is 2 [20].

$$p(V_2) = \left(\frac{k}{c}\right)\left(\frac{v_2}{c}\right)^{k-1} exp\left[-\left(\frac{v_2}{c}\right)^k\right]. \quad (2)$$

2.9.3. Wind Turbine Power Curve

The SWT-3.6-120 manufactured by Siemens and the 6.2M126 manufactured by Senvion were used in the study. The SWT-3.6-120 with a 3.6 MW optimal output was utilized in the analysis because of its high tolerance to wind gusts of up to 70 m/s. It has a 120 m rotor diameter with a cut-in wind speed of 3–5 m/s, nominal power at 12 m/s to 13 m/s, and cut-out wind speed of 25 m/s based on the manufacturer's datasheet [46]. The data were digitized from the published power curve for the technical analysis of the annual wind energy production from 0 to 25 m/s.

The Senvion 6.2M126 has a rated output of 6.15 MW, and it is used to check the effects of having a high capacity with respect to net present value, payback period, internal rate of return, and levelized cost of electricity. It has a 126 m rotor diameter with a cut-in wind speed of 3.5 m/s, nominal wind speed at 13.5 m/s, and cut-out wind speed of 30 m/s based on the manufacturer's datasheet [47]. The data were digitized from the published power curve for the technical analysis of the annual wind energy production from 0 to 30 m/s.

2.9.4. Wind Power and Wind Power Density

The wind power density is the power per area that can be harnessed from the specified site, which can then be computed from the wind speed data and probability density curve. The formula for wind power density can be seen in Equation (3) with *P/A* as the wind power density (W/m^2); *P* as the wind power (W); *A* as the swept area (m^2); ρ as the air density (1.225 kg/m^3); $p(V_2)$ as the wind probability density; and V_z as the magnitude of the wind speed (m/s) [16].

$$\frac{P}{A} = 0.5\rho \int_0^{\infty} {V_2}^3 p(V_2) dV. \tag{3}$$

2.9.5. Annual Wind Energy Production

The annual wind energy production is the energy that is harnessed from the site. The formula for annual wind production *E* can be seen in Equation (4) with *T* as the time period of the study (8760 h); $p(V_2)$ as the Weibull probability density; and $p(V_2)$ as the wind turbine power output corresponding from the wind turbine curve [16].

$$E = T \int_0^{\infty} p(V_2) P(V_2) dV. \tag{4}$$

2.9.6. Capacity Factor

The capacity factor is the ratio between the wind power generation from the turbine and the total power generation it produces at full capacity in a period of time. The equation for capacity factor *CF* can be seen in Equation (5) with *PE* as the estimated output; *PN* as the rated output; and *T* as the time period [13]. The generally accepted values for offshore wind capacity factor are in the range between 30% and 55% [48].

$$CF = \frac{PE}{PN(T)}. \tag{5}$$

2.9.7. Performance

The annual seasonal performance is the ratio between the wind energy produced by the turbine and the energy that is possessed by the wind annually. It shows the relationship between the wind turbine and the wind energy in the location. The equation for the performance of a turbine is shown in Equation (6) with ηEST as the wind turbine's seasonal performance; *E* as the wind energy produced; and E_d as the energy possessed by the wind. The considered technically viable performance for the wind turbines in this study is greater than 10% [16].

$$\eta EST = \frac{E}{E_d}. \tag{6}$$

2.9.8. Array Spacing and Number of Turbines

The number of turbines for the areas that are viable for the technical aspect and exclusion criteria are dependent on the total available area and array spacing, as seen in Equation (7) [49,50]. The array spacing is estimated using the rotor diameter, downwind spacing factor, and crosswind spacing factor as seen in Equation (8) [21,49,50]. Values of 10 and 5 were applied to the downwind and crosswind spacing factors, respectively, to reduce the inter-turbine wake losses.

$$Number\ of\ Turbines = \frac{Total\ Available\ Area}{Array\ Spacing}. \tag{7}$$

$$Array\ Spacing = (Rotor\ diamater)^2 \times Downwind\ Spacing\ Factor \times Cross\ Wind\ Spacing\ Factor. \tag{8}$$

2.10. Economic Analysis

2.10.1. Investment Cost

The investment cost of an offshore wind farm has commercial sensitivities, which makes the estimation difficult. The components of total capital cost such as wind turbine and electrical costs are acquired from different sources. The summation of all the total capital costs C_{Total} can be seen in Equation (9) with C_{WT} as the wind turbine cost; C_F as the foundation cost; C_E as the electrical cost; and other costs such as operation and maintenance [17].

$$C_{Total} = C_{WT} + C_F + C_E + Other\ cost. \tag{9}$$

The data acquired for the cost components of offshore wind farms were from "4COffshore", which is a market research and consultancy organization that focuses on offshore energy markets, and it has been cited in numerous studies [51–53]. The offshore wind farms that are constructed from 2008 to 2018 only were considered in the study. The data acquired from "4COffshore" were the name of the power plant, country of origin, minimum and maximum sea depth, area, offshore cable length, onshore cable length, inter-array cable length, port for O&M, distance from port, turbine model, number of turbines, turbine capacity, plant capacity, and investment cost.

2.10.2. Multiple Linear Regression

In this study, different cost parameters are considered to quantify each component of the investment cost using multiple linear regression as seen in Equation (10), where Y is the investment cost component; $\beta_{j,j} = 0, 1, 2, \ldots, k$, are the regression coefficients; $x_{j,j} = 0, 1, 2, \ldots, k$, are the predictor variables; and ε is a random error term [54]. The predictor variables considered were date commissioned, type of foundation, hub height, sea depth, minimum distance from shore, useful life, offshore cable length, onshore cable length, inter-array cable length, distance from port, distance from grid, turbine size, number of turbines, and capacity. The estimated investment cost of existing offshore wind farms around the world for the last ten years were considered.

$$Y = \beta_O + \beta_1 x_1 + \beta_2 x_2 + \ldots + \beta_k x_k + \varepsilon. \tag{10}$$

The coefficient of determination (R^2) was investigated using the statistical software to acquire the best multiple linear regression model. The coefficient of determination or the R^2 is a measure of the goodness-of-fit of the regression model. The highest and greater than 80% value will be applied to the economic analysis of the study.

The mean absolute percentage error ($MAPE$) of the model was analyzed, which aims to validate results. The $MAPE$ is a measure of the quality of the regression model [55] and measures how close the predicted values of the regression model are to the target variables.

2.10.3. Multiple Regression Assumptions

There are four assumptions in multiple regression analysis for the regression model to be valid. These four assumptions are normality, linearity, homoscedasticity, and reliability of measurement. If these assumptions are not satisfied, the results may be deemed inaccurate and may also result to type I or type II errors [56].

The assumption of normality implies that the distribution of the dependent variables and independent variables are normal. This can be checked through visual inspection of the residuals. A q–q plot can be done to visually check the normality of the variables. A model is considered a good model for normality if the residuals do not deviate largely from a straight line, while a bad model would show a large deviation from a straight line [57].

The assumption of linearity indicates that there is a linear relationship between the independent and dependent variables. A further visual inspection of the scatterplot of the residuals can be done to determine whether there is a linear pattern or not. A model is considered a good model for a linearity if there is equal distribution of the residual across a horizontal line through 0, while a bad model will show a curvilinear pattern [57].

Homoscedasticity implies that the variance of errors is the same across all independent variables. In the case of heteroscedasticity, there are different variance of errors from different values of the independent variables. The assumption for homoscedasticity can be satisfied by checking if the residuals are equally distributed along the predicted values. A model is considered homoscedastic if it appears to be randomly spread out a horizontal line while a model is heteroscedastic if there is a wider spreading of residuals along the x-axis, leading to a steep line fitted through [57].

A good reliability of measurement for multiple regression indicates that the effect sizes of other variables should be minimized. It also implies that there should be no multicollinearity, or the independent variables should not be highly correlated with each other because this may lead to over estimation. A residual vs. leverage plot may help in searching for influential observations that do not follow the trend of most other observations, and the variance inflation factor (*VIF*) is a method for checking the correlation of independent variables. A reference line called the Cook's distance is added and observations outside this line are considered influential observations outside the trend [57]. Observations outside should be omitted since they influence the resulting slope coefficient and R^2 significantly. In the *VIF* method, independent variables with greater than 5 values indicate that there is multicollinearity with each other and should be omitted carefully.

2.10.4. Net Present Value

Net present value (*NPV*) is an economic metric to indicate the total benefit from an investment. It is a measure of the incoming and outgoing cash flow in the investment as seen in Equation (11), whereas the N is the number of years from zero to the last year of the investment; FC is the cash flow for an annual period reaching the last year of the investment; and r is the discount rate. A NPV greater than zero means that the investment is profitable, while a NPV less than zero is not profitable, and the investment should not be pursued. If the NPV is exactly zero, nothing has been gained or lost, which means the project should also not be pursued [16].

$$NPV = \sum_{t=0}^{N} \frac{FC_t}{(1+r)^t} = v_1 \frac{\ln\left[\frac{z_2}{z_0}\right]}{\ln\left[\frac{z_1}{z_o}\right]}. \quad (11)$$

2.10.5. Levelized Cost of Electricity

Levelized cost of electricity (*LCOE*) is a general measure for the evaluation of the life cycle cost of a power plant. It can also be used for the comparison of different power plant technologies in terms of its life cycle cost and energy production. It accounts for the total installation cost, annual operation and maintenance, and energy production [58].

The formula for *LCOE* can be found in Equation (12) with I_T is the investment at time t; M_T is the operating and maintenance cost at time t; E_T is the energy produced (MWh) at time t; r is the discount rate (%); and t is the time from base year up to the total number of years of operation [16].

$$LCOE = \frac{\sum_{t=0}^{n} \frac{I_T+M_T}{(1+r)^t}}{\sum_{t=0}^{n} \frac{E_T}{(1+r)^t}}. \quad (12)$$

3. Methodology

The methodology described in this study focused on two aspects of the potential siting of an offshore wind farm, namely, the technical and economic aspects. In the succeeding chapters, the framework of the methodology of the study will be discussed together with its detailed procedures. The methodologies were aligned with the objectives of the study found in Section 1.2.

The novelty of this research compared to existing technical and economic assessment of offshore wind farms are the application of numerous exclusion criteria, comparison of two offshore wind turbines, and rigorous economic analysis in the Philippines setting. Compared to other techno-economic assessment studies, nine exclusion criteria were applied in this study to exclude areas that are not suitable as a potential site for offshore wind farms. Calamities such as historical earthquakes and typhoons have been considered as exclusion criteria. In the technical analysis, two turbines have been considered, namely, the SWT-3.6-120 and 6.2M126 turbines. In other techno-economic studies, they have considered only one turbine. The economic analysis done in this study covers the actual costs of offshore wind farms around the world for the period 2008 to 2018. These data were then used to generate a multiple regression model to capture the investment costs of offshore wind farms. The levelized cost of electricity and the price of electricity of offshore wind farms in the Philippines were taken into account. In other techno-economic studies, they only considered the levelized cost of electricity. Finally, the novelty of this study is that analysis was done with relevant assumptions in the Philippines. It is also the first available and published technical and economic assessment of offshore wind farms in the Philippines that follows engineering methods.

3.1. Framework of Methodology

The flow diagram of the methodology is shown in Figure 1. The framework has three stages: data gathering, exclusion analysis, and technical and economic analysis. The acquisition of data included methods of the recent research work on techno-economic analysis of OWF, collection of wind speed data, and developing the simulation model using GIS. The application of the exclusion criteria not only narrowed down the locations of potential sites, but more importantly, masked out all of the unsuitable sites in the Philippines due to restrictions by local legislation and current global offshore industry practice. Subsequently, the suitable sites were then subjected to technical and economic analysis.

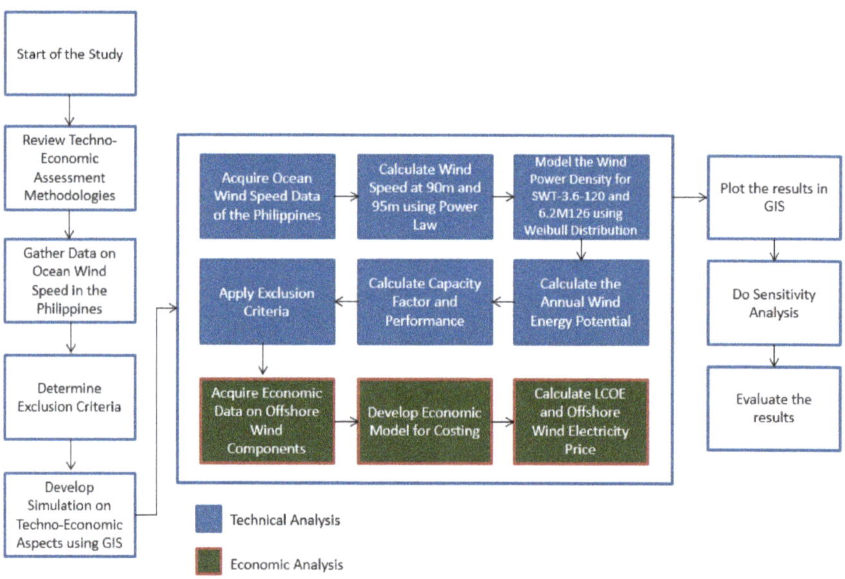

Figure 1. Framework of methodology.

3.1.1. Exclusion Criteria

The constraints considered in this study are presented in Figure 2. Using the raster method in GIS, all constraints were overlaid in the wind resource map to exclude these areas as unsuitable locations in the analysis. It should be noted that the suitability of the locations was determined purely by technical and economic considerations in the study. The exclusion step was applied to reduce the potential sites with no consideration to linearized scoring of factors that affect suitability.

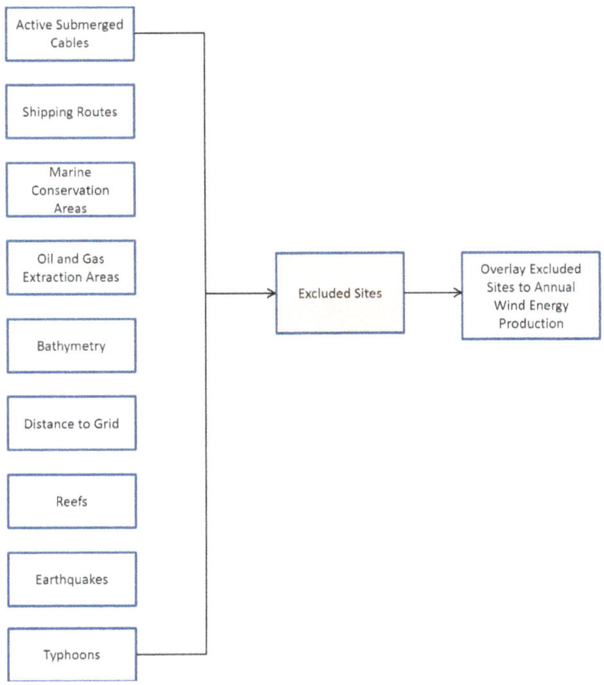

Figure 2. Detailed exclusion analysis.

Active submerged cable data were extracted from TeleGeography, and the map was projected to WGS 1984 UTM Zone 51 N, which is the projection for the Philippines. As seen in Figure 3, most of the active submerged cable is connected to the west part of Luzon near the capital of the Philippines—Manila City. This city is one of the most heavily populated areas and the center for business in the country, which explains the concentration of active submerged cables in the area. Luzon has the most active submerged cable connections that are connected from other countries, while Visayas and Mindanao are connected to Luzon for the interconnection of subsea cables. A buffer spacing of 5 km from the cables was applied for a conservative estimate that the offshore wind farm will not affect these cables.

Local ferry routes were extracted from the World Food Program. The map was projected to WGS 1984 UTM Zone 51 N, as shown in Figure 4. The path of the ferry routes in the Philippines all pass through the Visayas. Since the Philippines is composed of many islands, ferry routes are sometimes the only way to get to the other islands. The buffer spacing for the shipping route was 3 km, which is a conservative estimate considering an array spacing of 0.79 km^2 for a 126 m 6.2M126 turbine.

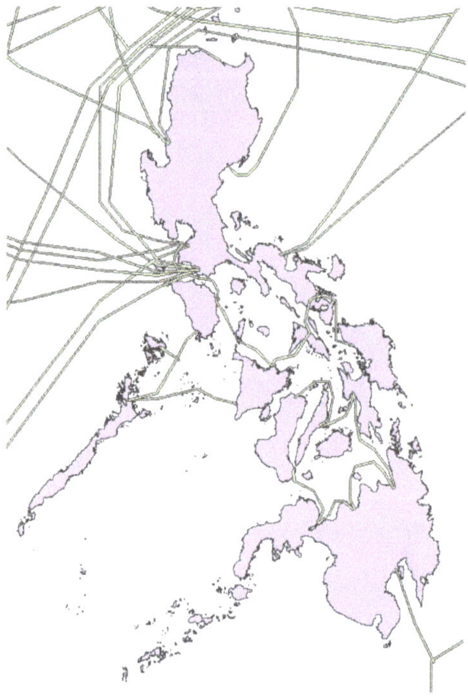

Figure 3. Active submerged cables.

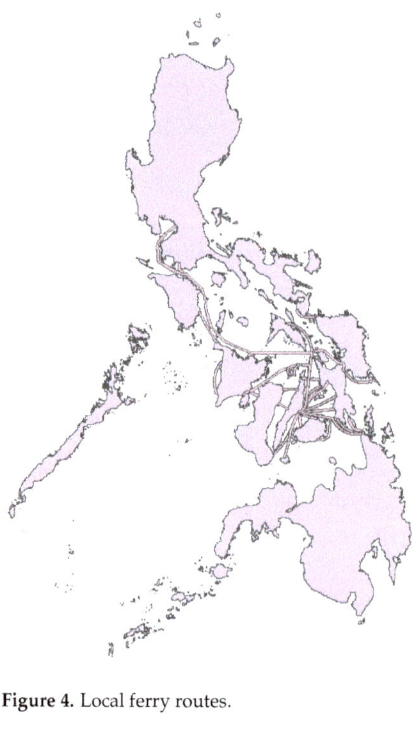

Figure 4. Local ferry routes.

The data for marine conservation areas were acquired from Protected Planet, which has an online world database of marine and terrestrial protected areas around the world. The data were used to generate a map using Arc GIS 10 with a projection of WGS 1984 UTM Zone 51 N, as shown in Figure 5. The map shows the nature reserve, natural park, natural monument, wildlife sanctuary, protected landscapes and seascapes, resource reserve, and natural biotic areas of the Philippines. Some of the evident marine protected areas in the map are the Tubbataha Reef, Turtle Islands Wildlife Sanctuary, Sarangani Bay, Siargao Protected Seascape, and Tanon Protected Seascape, Negros Occidental Coastal Wetlands Conservation Area, and Malampaya Sound. These marine protected areas were consolidated with other exclusion criteria to narrow down the potential location of the viability of offshore wind farms. The buffer spacing applied to the marine protected areas was 3 km, as already represented in the map.

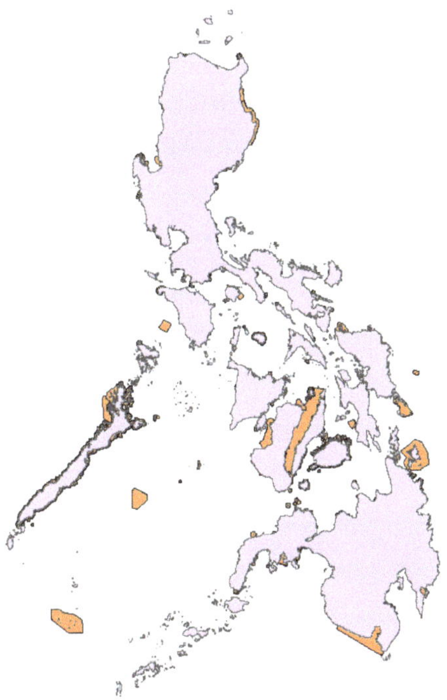

Figure 5. Marine conservation areas.

Coral reefs data in the Philippines were acquired from PhilGIS, which is a free GIS data source for educational and nonprofit use, and the reef data were from ReefBase, which is a global database for coral reefs. The data for reef were generated in a map using Arc GIS 10 with a projection of WGS 1984 UTM Zone 51 N, as shown in Figure 6. The types of reefs shown in the map are barrier, fringing, patch, and shelf reefs. The concentration of reefs is dominant in Visayas compared in Luzon and Mindanao. Similar to the marine protected areas, a buffer spacing of 3 km was applied in the exclusion of coral reefs.

Oil and gas extraction areas were acquired from the Department of Energy of the Philippines, and the map generated from Arc GIS was projected to WGS 1984 UTM Zone 51 N, as seen in Figure 7. The largest oil and gas extraction exploration in the Philippines is found at the east side of the Palawan Islands, which measures up to 1,476,000 km^2, and is headed by the Philippine National Oil Company (PNOC). Some of the other companies exploring oil and gas in the country are China International Mining Petroleum Corporation (197,000 km^2), Mindoro-Palawan Oil and Gas Inc. (724,000 km^2), Nido Petroleum

Philipines Pty. Ltd. (1,344,000 km^2), PNOC (36,000 km^2), Gas2Grid Ltd. (75,000 km^2), Polyard Petroleum International Company (684,000 km^2), Nido Petroleum Philippines Pty. Ltd. (314,000 km^2), and Otto Energy Investments Ltd. (988,000 km^2). These oil and gas exploration areas were consolidated with other exclusion criteria to narrow down the potential location of the viability of offshore wind farms. The buffer spacing considered for this criterion was 5 km, as shown in the map, so that the offshore wind farm will not affect the oil and gas extraction areas.

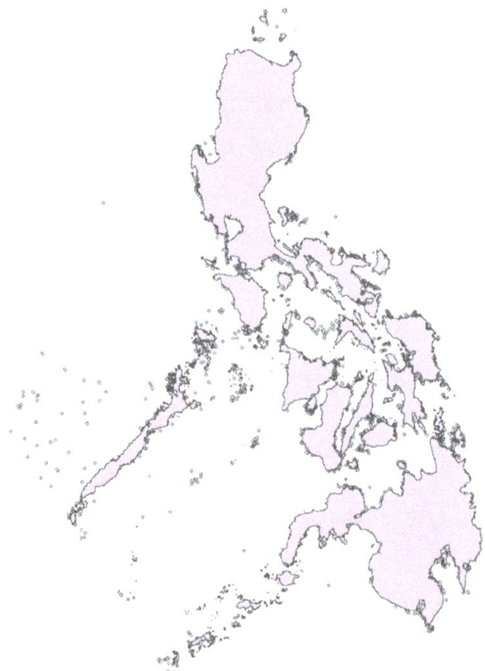

Figure 6. Coral reefs.

Bathymetry of the Philippines was acquired from the General Bathymetric Chart of the Oceans (GEBCO), and the map generated from Arc GIS was projected to WGS 1984 UTM Zone 51 N, as seen in Figure 8. The Reclassify tool in Arc GIS was used to remove the bathymetry greater than 50 m. The yellow areas represent the areas that were within the 50 m criteria. Areas greater than 50 m bathymetry were excluded in the analysis together with other exclusion criteria to narrow down the potential location of the viability of offshore wind farms.

The data from the National Renewable Energy data explorer on the grid of the Philippines were used in this study, and the map generated using Arc GIS was projected to WGS 1984 UTM Zone 51 N, as seen in Figure 9. The Euclidean Distance tool was used to generate a map with the corresponding distance from the grid. The Reclassify tool was then used to remove the values greater than 120 km distance from the grid. Areas beyond 120 km or outside of the green shade in the map were excluded in the analysis. The distance greater than 120 km and other exclusion criteria were consolidated to narrow down the potential sites for the analysis.

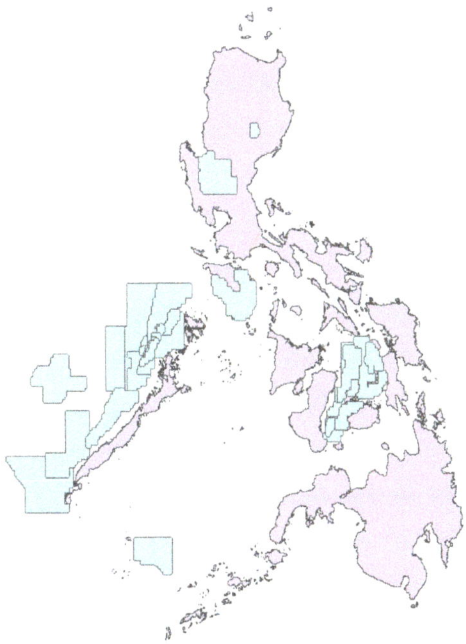

Figure 7. Oil and gas extraction areas.

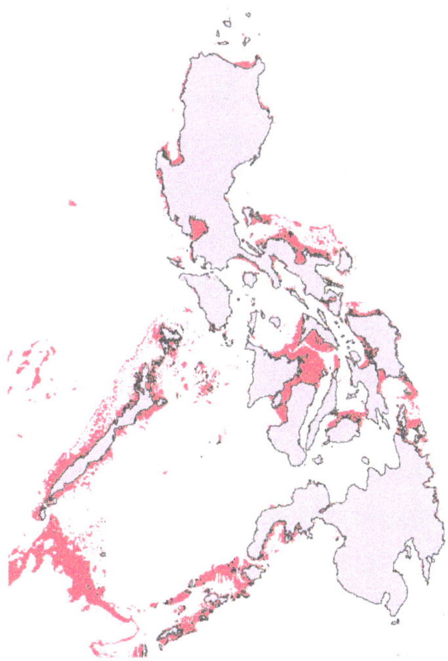

Figure 8. Bathymetry within 50 m.

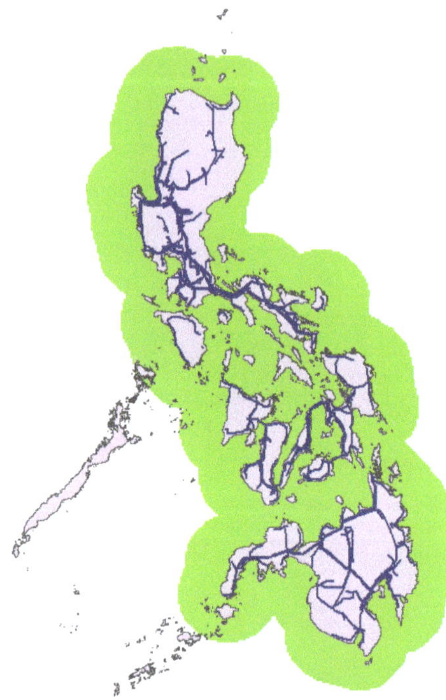

Figure 9. Distance to grid within 120 km.

Historical earthquakes were acquired from the Philippine Institute of Volcanology and Seismology (PHIVOLCS). The acquired data were composed of the recorded magnitude and location of earthquakes in the Philippines from 1908 to 2018. The earthquakes with 6.5 magnitude and above were plotted in Arc GIS with a projection of WGS 1984 UTM Zone 51 N, as seen in Figure 10. A buffer area of 15 km was applied to the study area based on the critical damage extent to a dyke of a 7.8 magnitude earthquake in Japan [40].

The typhoon path and wind speed data of the Philippines were acquired from the Philippine Atmospheric Geophysical and Astronomical Services Administration (PAGASA). The gathered data spans from 1998 to 2018. The map was generated using Arc GIS with a projection of WGS 1984 UTM Zone 51 N, as illustrated in Figure 11. The highest historical typhoon that landed in the Philippines is Typhoon Yolanda (International Name: Haiyan), with a wind speed reaching 315 km/h (indicated by the red path). The path considered were those typhoons that reached greater than 250 km/h from 1998 to 2018. A buffer area of 50 km was applied in the exclusion of historical typhoon paths.

A graph digitizer was used to produce the power curve of SWT-3.6-120 and 6.2M126, as seen in Figures 12 and 13, respectively. A total of 562 and 270 data points were plotted based on the power curve of the SWT-3.6-120 brochure and power curve of 6.2M126, respectively. The x-axis represents the wind speed in m/s, and the y-axis represents the power output in kW of the wind turbine.

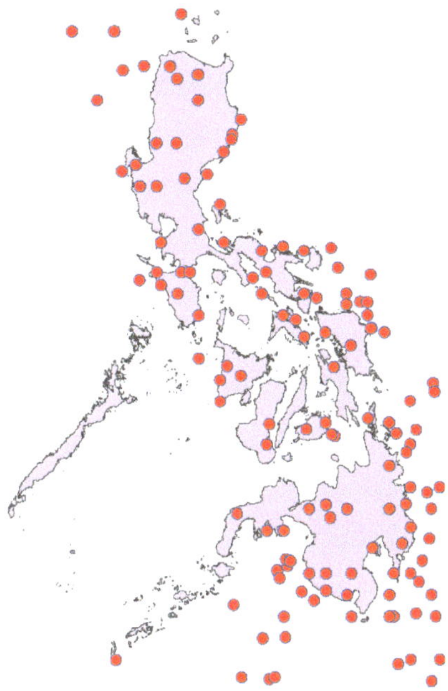

Figure 10. Earthquakes with at least a 6.5 magnitude.

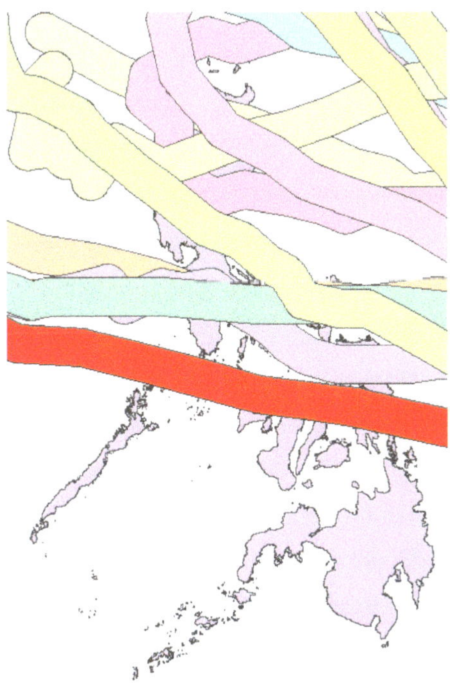

Figure 11. Typhoons with greater than 250 km/h.

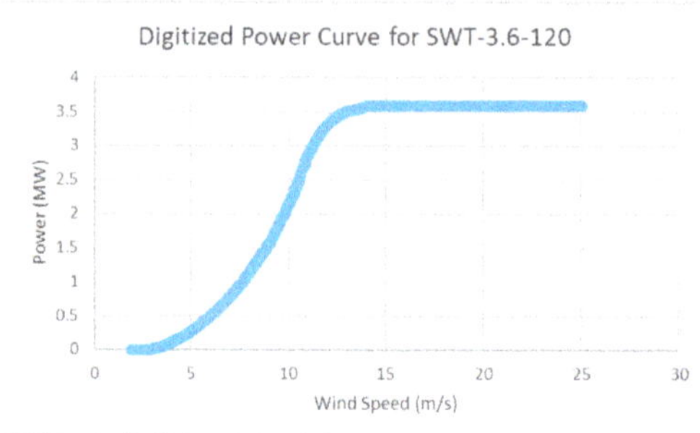

Figure 12. Power curve of Siemens SWT-3.6-120.

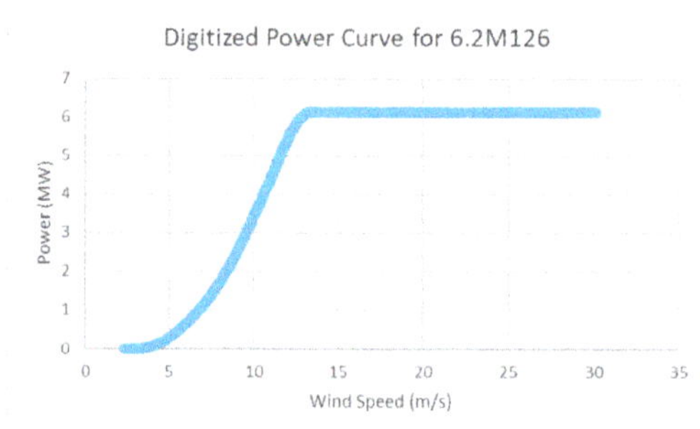

Figure 13. Power curve of Senvion 6.2M126.

3.1.2. Technical Analysis

The detailed technical analysis can be seen in Figure 14. The acquired wind speed data were at 80 m and necessitated being extrapolated to the hub height of SWT-3.6-120 at 90 m and hub height of 6.2M126 at 95 m using the Power Law in Equation (1). Equation (3) allows for the computation of the wind power density and wind power utilizing the Weibull function model of SWT-3.6-120 and 6.2M126. To compute the annual wind energy production, the power curve of the selected turbines and their corresponding Weibull functions were used as inputs to Equation (4). Capacity factor and performance were computed using Equations (5) and (6), respectively. After generating the annual capacity factor and performance maps, the exclusion criteria layer was masked over to remove the unsuitable locations from the potential sites for OWF and the investment cost from the regression model was applied to calculate the levelized cost of electricity and offshore wind electricity price in the Philippines. Finally, the two turbines were then compared with the *LCOE* at different areas and plant capacities.

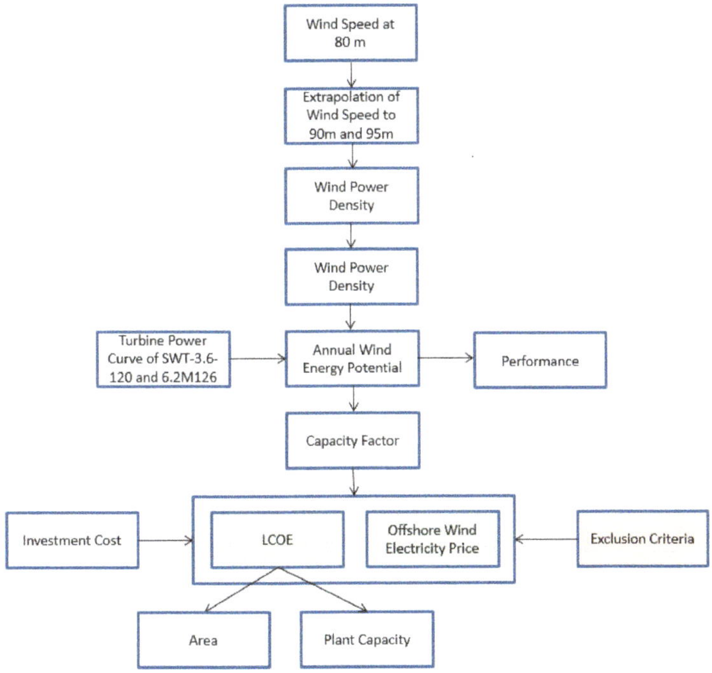

Figure 14. Detailed technical analysis.

3.1.3. Economic Analysis

The detailed economic analysis of the study can be seen in Figure 15. The independent variables considered for the electrical cost of offshore wind farms were onshore cable length, offshore cable length, inter-array cable length, and the area. For the turbine cost, the plant capacity was the considered independent variable in the analysis. The selection of these variables were based on the premise that the cables are the major components of the electrical cost and that the turbine cost is dependent on the plant capacity [17,20,51]. The independent variables for the foundation cost depend on the bathymetry of the ocean and the size of the turbine [59]. The nearest port distance was considered for the maintenance and operation [20,59].

The data for independent variables were acquired from "4COffshore", which is a market research and consultancy organization that focuses on offshore energy markets, and it has been cited in numerous studies [51–53]. The data acquired were name of power plant, country of origin, minimum and maximum sea depth, area, offshore cable length, onshore cable length, inter-array cable length, port for O&M, distance from port, turbine model, number of turbines, turbine capacity, plant capacity, and investment cost. All the values for investment cost in different years were converted to 2017 USD using the historical values of each country and the historical USD exchange rate. The nearest port distances were acquired based on the visual inspection in the 4COffshore wind map, and the distance between the nearest port and offshore wind farm was measured through the difference of their coordinates. The missing onshore cable length was estimated based on the location of the stated grid connection point and the cable landing point through the difference of their coordinates. The four regression assumptions were checked using the R statistical software. The "lm" function was used to generate the multiple linear regression model. The "plot" function was used to test the normality, linearity, homoscedasticity, and reliability of the model with the visual inspection of the normal Q–Q plot, residual vs. fitted plot, scale-location plot, and residual vs. leverage plot, respectively. The reliability of the model was further examined using the "vif" function, which determines the independent variables with multicollinearity. The independent variables with vif greater than 10 were carefully omitted. The selected regression model was validated using actual data from offshore wind farms through the measure of mean absolute percentage error. The investment cost calculated using the selected regression model was applied together with the capacity factor and exclusion criteria to acquire the *LCOE* and offshore wind electricity price in the Philippines.

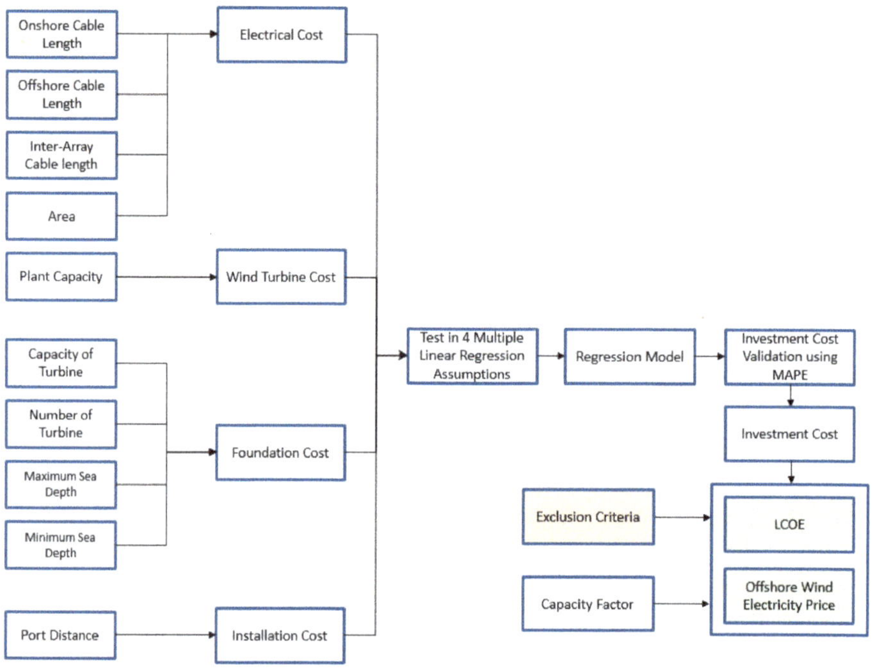

Figure 15. Detailed economic analysis.

3.1.4. Sensitivity Analysis

Sensitivity analysis was carried out in the last part of the study to determine the effect of certain parameters to the output values under a given set of assumptions. The parameters that were analyzed to check the sensitivity to the resulting *LCOE* and price of electricity were investment cost ($\pm$ mean absolute percentage error), capacity factor ($\pm$2%), weighted average cost of capital ($\pm$2%), cost of debt ($\pm$1%), cost of equity ($\pm$2%), and plant capacity ($\pm$5%). The basis in the adjustment of the parameters were small reasonable changes based on the nature of the parameters.

3.2. *R Statistical Software*

The R statistical software was used for handling the 128,000 by 26 data points for the technical analysis. A program was also implemented to calculate the wind speed at 90 m, Weibull probability density function, wind power density, wind power, annual energy production, capacity factor, and performance. The statistical tools of R were used to acquire the equation for the multiple linear regression in the economic analysis.

3.3. *GIS Software*

Geospatial information system (GIS) methodology has been widely used by different researchers in describing the technical and economic potential of offshore wind farms. It is a modeling tool that can relate the costs, location, and power plant capacity through displayed resources, spatial cost, and restrictions in areas. The main product of this is the relation of the levelized cost of electricity (LCOE) to the spatial area of study [17].

The Arc GIS 10 software requires the wind speed data resource in the location of study, bathymetry, cable distance as well as the economic factors involved in the construction of an offshore wind farm. This modeling tool can compute for various cost algorithms to get the *LCOE* [20].

4. Results and Discussions

4.1. Technical Analysis

The recommended hub heights for the Siemens SWT-3.6-120 and Senvion 6.2M126 were 90 m and 95 m, respectively. As such, the wind speed data at 80 m were extrapolated to these two elevations. The succeeding analysis will show the technical parameters corresponding to both turbines plotted in their ideal elevations.

4.1.1. Wind Speed

The data on wind speed at 80 m were extrapolated to a hub height of 90 m for the SWT-3.6-120 and to 95 m for the Senvion 6.2M126 using the power law in Equation (1). Figure 16a,b shows the average wind speeds of the Philippines for the SWT-3.6-120 and 6.2M126, respectively, from the years 2008, 2010, 2014, 2015, and 2016.

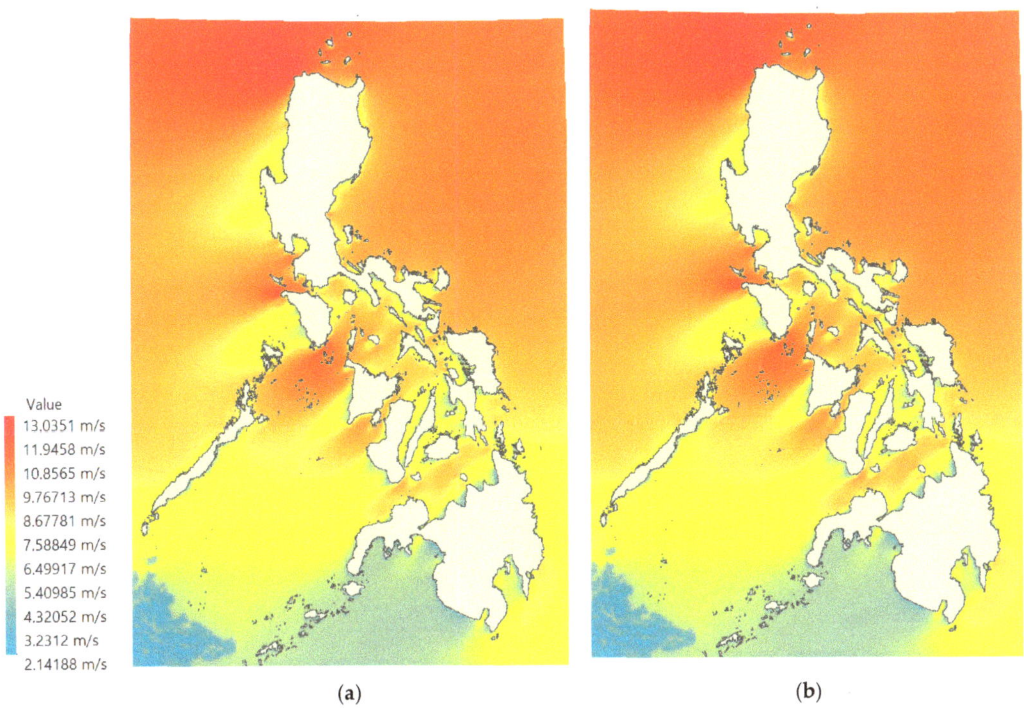

Figure 16. Wind speed at a hub height of (**a**) SWT-3.6-120 and (**b**) 6.2M126.

Shown in Figure 16a is the contour plot of wind speed for the Siemens SWT-3.6-120 turbine hub height. The values range from 2.133 m/s to 12.981 m/s with an average of 8.041 m/s. The northern parts of Ilocos Norte sees the greatest wind speeds ranging from 10.611 m/s to 12.945 m/s. Coming in next are the northern parts of Occidental Mindoro and the southeastern parts of Oriental Mindoro, which see wind speed values from 9.175 m/s to 12.091 m/s and from 9.091 m/s to 11.63 m/s, respectively. Luzon primarily sees greater wind speeds than Visayas and most especially Mindanao, as evidenced in the blue contours indicating lower values shown in the regions of the latter island.

For the Senvion 6.2M126 hub height, wind speeds vary from 2.142 m/s to 13.035 m/s with a mean of 8.096 m/s. The contours are shown in Figure 16b. As expected, the wind speed values seen by the Senvion turbine were higher than those of the Siemens case due to a slightly higher elevation. Additionally expected were the locations of greatest wind speeds since the extrapolation will result in the same trends due to a common roughness factor for all offshore areas studied. The northern region of Ilocos Norte experiences wind speeds of 10.655 m/s up to 12.991 m/s while the northern region of Occidental Mindoro sees 9.213 m/s to 12.325 m/s speeds, and the southeastern region of Oriental Mindoro experience speeds from 9.279 m/s to 11.678 m/s.

4.1.2. Wind Power Density

The average wind power density for SWT-3.6-120 and 6.2M126 in the years 2008, 2010, 2014, 2015, and 2016 can be seen in Figure 17a,b. It is worth noting that the mean wind power density is independent of the turbine, and is the power density of the wind for the area. For the wind power density of SWT-3.6-120, the lowest was at 11.539 W/m^2 and the highest was at 1777.830 W/m^2, as seen in Figure 17a. The mean of the wind power density was at 700.918 W/m^2. Corresponding to the areas of greatest wind speeds were those of the greatest wind power densities with the northern region of Ilocos Norte ranging from 1238.358 W/m^2 to 1771.519 W/m^2, the northern region of Occidental Mindoro seeing 882.755 W/m^2 to 1604.634 W/m^2, and finally, the southeastern region of Oriental Mindoro ranging from 873.324 W/m^2 to 1499.410 W/m^2. For the wind power density of 6.2M126, the lowest was at 11.683 W/m^2 and the highest was at 2257.367 W/m^2, as seen in Figure 17b. The mean of the wind power density was at 753.835 W/m^2. Compared with SWT-3.6-120, there were increases of 1.25%, 26.97%, and 7.55% in the lowest, highest, and mean wind power densities, respectively. There was a variation in the increase of these measures due to the nature of the equation, which is dependent on the cube of the wind speed and the different Weibull probability distribution model for each turbine. The wind Weibull probability distribution model for SWT-3.6-120 takes into account 26 integers from 0 to 25, which represents the wind speed it operates in, while the 6.2M126 model takes into account 31 integers from 0 to 30. Similar to the results of SWT-3.6-120, the locations with the greatest wind power densities (as shown in the red contours) were found in the northern parts of Ilocos Norte, which had values ranging from 1405.655 W/m^2 to 2241.579 W/m^2 while the northern parts of Occidental Mindoro and the southeastern parts of Oriental Mindoro experience densities ranging from 937.934 W/m^2 to 1999.059 W/m^2 and 955.461 W/m^2 to 1757.770 W/m^2, respectively.

4.1.3. Wind Power

The average annual wind power for the Siemens SWT-3.6-120 and Senvion 6.2M126 are presented in Figure 18a,b. Wind power is a function of the diameter of the rotor (see Equation (3)). The Siemens SWT-3.6-120 has a diameter of 120 m while the Senvion 6.2M126 has a diameter of 126. One can surmise that the Senvion should produce greater wind power due to greater wind speeds as well as a greater swept area.

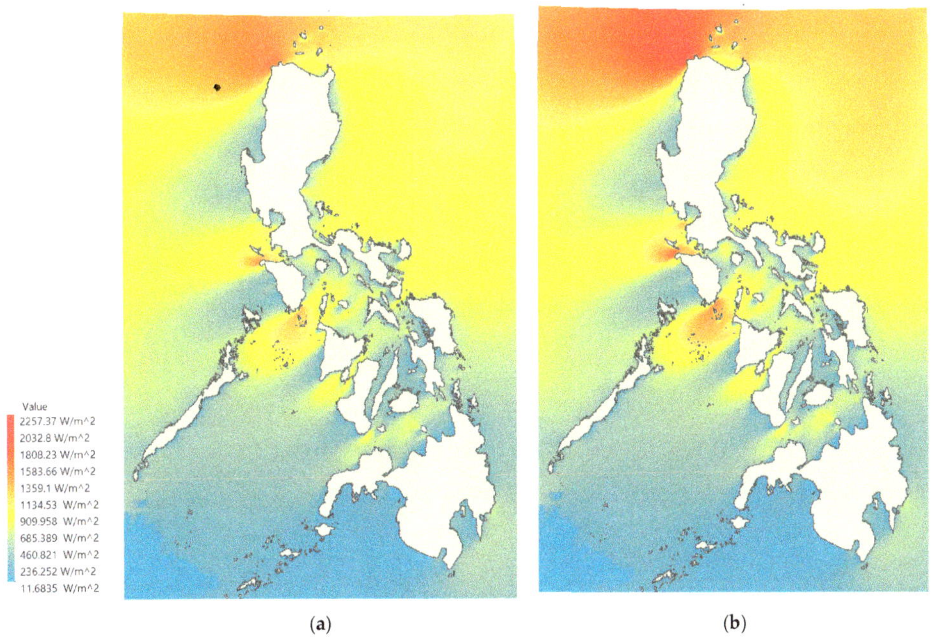

Figure 17. Wind power density for (**a**) SWT-3.6-120 and (**b**) 6.2M126.

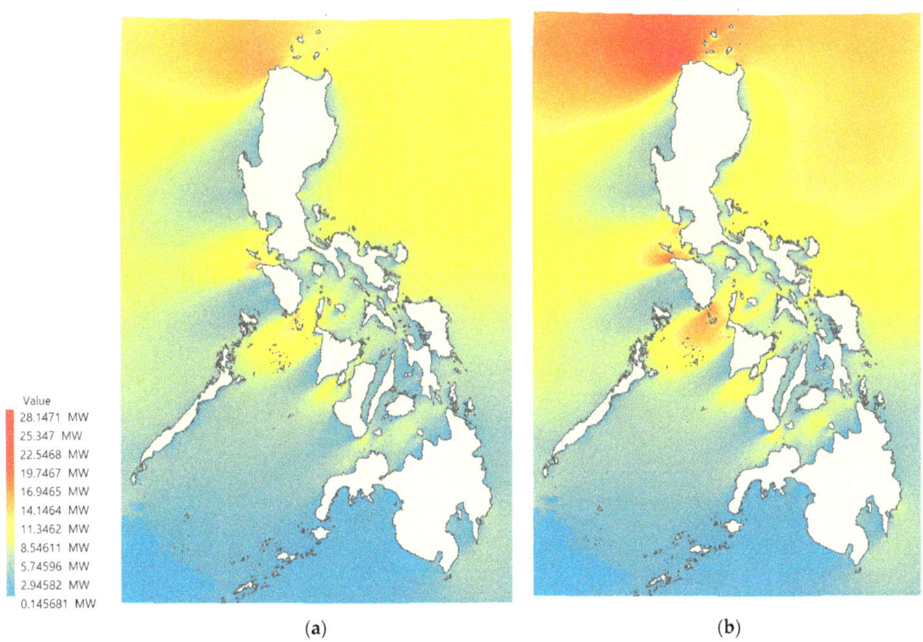

Figure 18. Wind power for (**a**) SWT-3.6-120 and (**b**) 6.2M126.

In studying SWT-3.6-120, the least wind power observed was 0.131 MW, while the greatest was 20.107 MW. The mean wind power observed was 7.927 MW. As with the wind power density, the locations that registered the greatest wind power were the ones in the northern region of Ilocos Norte, reaching low values of 14.006 MW and high values of 20.035 MW. Furthermore, the northern region of Occidental Mindoro and the southeastern region of Oriental Mindoro register wind power ranges from 9.984 MW to 18.148 MW and from 9.877 MW to 16.958 MW, respectively, and seen as orange contours in the map. Using the same analysis for Senvion 6.2M126, the least value of wind power observed was 0.146 MW while the greatest value was 28.147 MW, with the latter clearly seen in Figure 18b. The average annual wind power was 9.4 MW. Compared to the Siemens rotor, one can observe an increase in the power metrics for the Senvion case with a computed 11.45%, 39.99%, and 18.58% increase in the least, greatest, and average wind power, respectively. The wind Weibull probability distribution model for SWT-3.6-120 takes into account 26 integers from 0 to 25, which represents the operational wind speed, while the 6.2M126 model takes into account 31 integers from 0 to 30. Consistent with the previous results, the greatest wind power densities were located in the northern regions of Ilocos Norte with power values ranging from 17.527 MW to 27.950 MW. Following suit were the areas in the northern regions of Occidental Mindoro and the southeastern regions of Oriental Mindoro, registering values of 11.695 MW to 24.926 MW and 11.914 MW to 21.918 MW, respectively.

4.1.4. Annual Energy Production

Shown in Figure 19a,b are the mean annual energy production for both turbines studied. The Siemens SWT-3.6-120 produces at least 179.519 MWh and at most 19,759.275 MWh. The average annual energy production is 11,905.933 MWh. As expected, one can find the highest potential in the same locations as those of the greatest wind speed and wind power recorded. These are the northern parts of Ilocos Norte, producing 17,181.576 MWh to 19,742.881 MWh of energy per year. Likewise, the northern parts of Occidental Mindoro produced high yields at 14,763.450 MWh to 19,212.203 MWh whereas the southeastern parts of Oriental Mindoro produced around 14,379.013 MWh to 18,769.434 MWh of energy. Areas with low energy production were observed close to the shorelines of Visayas and Mindanao.

The analysis further showed that the Senvion 6.2M126 produced the least energy at a rate of 155.305 MWh and the greatest at 34,974.594 MWh. The contours of annual energy production are shown in Figure 19b. The average energy produced was 19,550.665 MWh. While there were increases in the mean and greatest energy production versus the Siemens turbine at 77% and 64.21%, respectively, a drop of 13.49% was observed in the lowest energy production. This drop was likely due to a higher cut-in wind speed of the Senvion 6.2M126 which is at 3.5 m/s. Despite the higher cut-in wind speed, the Senvion makes up in energy production as the greatest production rate and mean production rate were both significantly higher than that of the Siemens. At a higher rated capacity of 6.15 MW and higher wind speed seen, the Senvion naturally outperformed the Siemens in energy production. The sites with the greatest production of energy were in the same locations as those of wind speed and wind power. The northern areas of Ilocos Norte can potentially produce energy from 29,111.229 MWh to 34,903.898 MWh with the northern areas of Occidental Mindoro and southeastern areas of Oriental Mindoro producing 24,404.434 MWh to 33,724.617 MWh and 24,664.807 MWh to 32,284.715 MWh, respectively.

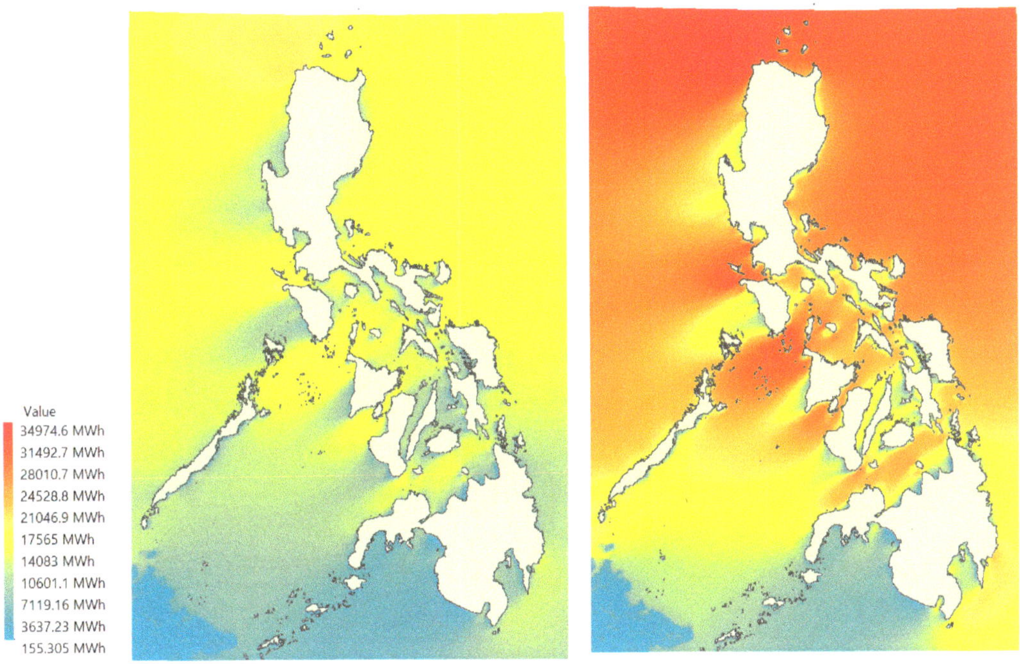

(**a**) (**b**)

Figure 19. Annual energy production of (**a**) SWT-3.6-120 and (**b**) 6.2M126.

4.1.5. Capacity Factor

Shown in Figure 20a,b are the computed capacity factors of the Siemens SWT-3.6-120 and the Senvion 6.2M126, respectively. The red contours indicate high capacity factor, while the blue contours indicate low capacity factor. The lowest, mean, and highest capacity factors computed for the Siemens SWT-3.6-120 were 0.57%, 37.75%, and 62.66%, respectively. Very high values of 54.48% to 62.60% were observed in the northern parts of Ilocos Norte. Similarly, high values were also seen around the northern parts of Occidental Mindoro ranging from 46.81% to 60.92% as well as the southeastern parts of Oriental Mindoro ranging from 45.60% to 59.52%.

The lowest, mean, and highest capacity factors computed for the Senvion 6.2M126 were 0.29%, 36.29%, and 64.92%, respectively. In the same manner that the lowest value of annual energy production was significantly lower than that of the Siemens turbine, the least capacity factor was recorded to have dropped by 49.12%. There was also a decrease in the mean capacity factor by 3.87%, while an increase in the greatest capacity factor value by 3.6% was observed. The top three areas with the greatest capacity factors were the northern region of Ilocos Norte registering values from 54.04% to 64.79%, the northern region of Occidental Mindoro with recorded values of 45.30% to 62.60%, and the southeastern region of Oriental Mindoro reaching a range of 45.78% to 59.93%. Figure 21a,b show the capacity factor with less than 30% for SWT-3.6-120 and 6.2M126, respectively, indicated by the black contours. In this study, areas that had a capacity factor less than 30% were excluded in the potential siting of offshore wind farms.

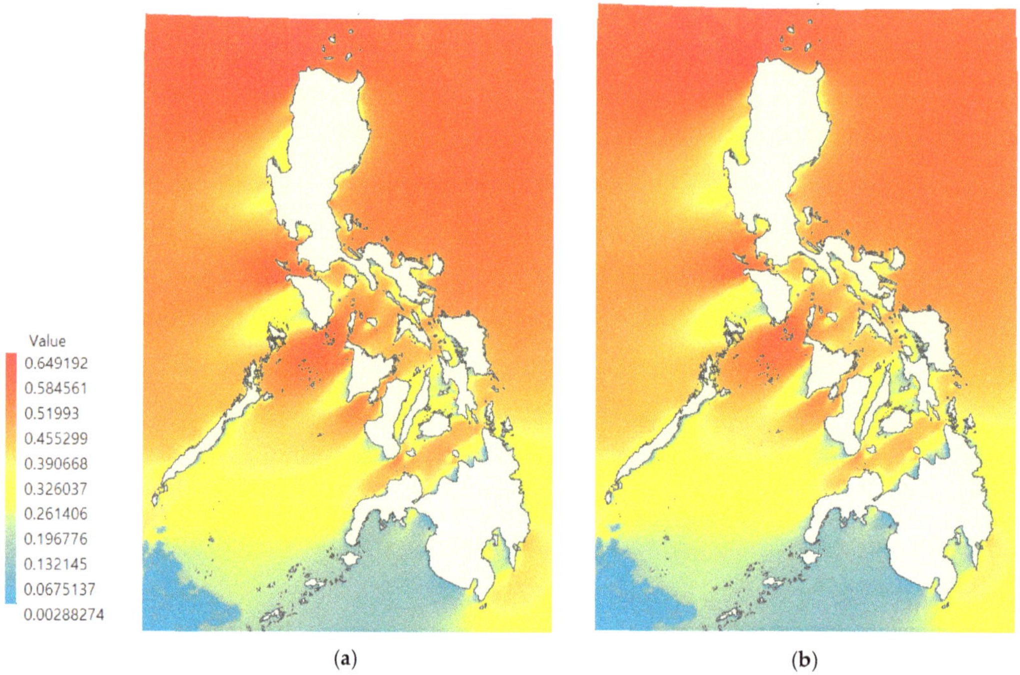

Figure 20. Capacity factor of (**a**) SWT-3.6-120 and (**b**) 6.2M126.

4.1.6. Performance

The performance and the ratio between the energy produced by the wind turbine and the wind energy available of the Siemens SWT-3.6-120 and the Senvion 6.2M126 are shown in Figure 22a,b. The worst performance computed for the Siemens SWT-3.6-120 was 11.27%, while the recorded best performance was 29.89%. On average, the performance of the Siemens turbine was 20.60%. The poorest performing areas were the northern region of Ilocos Norte, the northern region of Occidental Mindoro, and the southeastern region of Oriental Mindoro, with performance ranging from 11.30% to 14.62%, 12.13% to 17.62%, and 12.70% to 18.17%, respectively. The contrasting results between performance and the previous parameters are not surprising since performance-wise, the two turbines are able to maximize the energy conversion available when the wind speeds are lower. At higher wind speeds that translate to higher wind power and capacity factor, the wind turbine conversion is limited to the capacity of the system and is fixed at its declared rating. This means that there is more energy in the wind that is not converted due to the limitation of the energy converter (flat line power output beyond the rated wind speed of about 13 m/s).

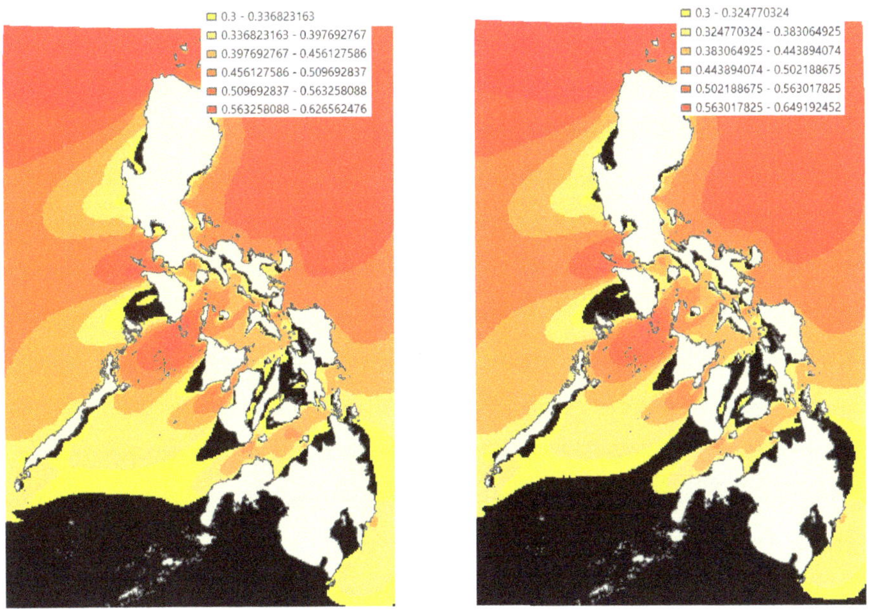

(**a**) (**b**)

Figure 21. Black contours with less than 30% capacity factor for (**a**) SWT-3.6-120, (**b**) 6.2M126.

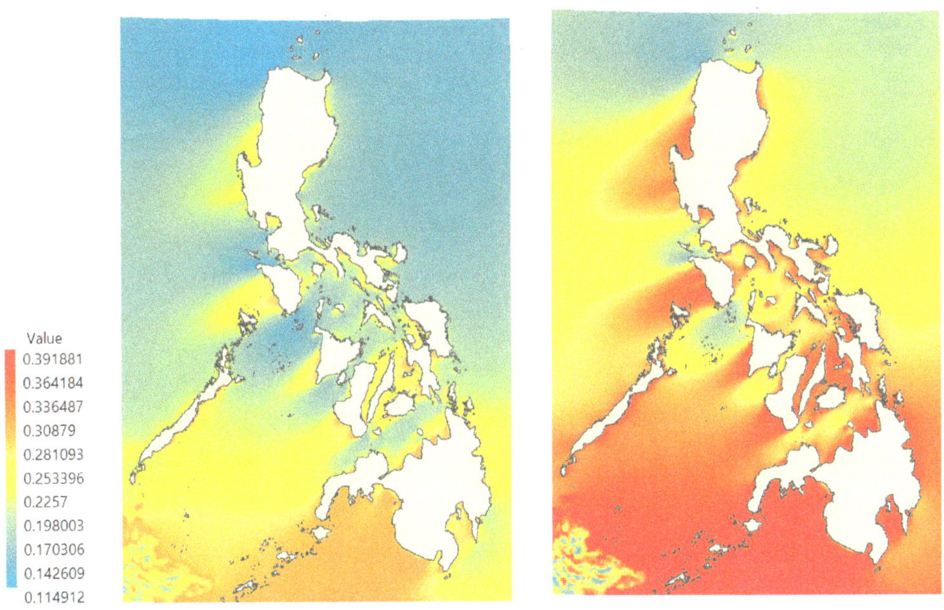

(**a**) (**b**)

Figure 22. Performance of (**a**) SWT-3.6-120 and (**b**) 6.2M126.

The worst performance of the Senvion 6.2M126 was comparable to the Siemens turbine at 11.49%, while the best performance was recorded higher at 39.19%. The average performance was also better than the Siemens at 28.26%. The improvement in performance relative to the Siemens turbine can be attributed to higher annual energy production. Probing the performance map, one can observe that the low performing regions correspond, unsurprisingly, to the areas of best wind power. These poorest performing areas were the northern parts of Ilocos Norte with rates ranging from 14.41% to 20.23%, the northern parts of Occidental Mindoro with figures of 15.89% to 25.14%, and the southeastern parts of Oriental Mindoro performed slightly better but were still very low at 16.97 to 24.85%. Again, the results revealed the inverse effect of high capacity factor to low performance since high wind power outbalances the influence of the annual energy production. Conversely, high performing areas usually have low values of capacity factor due to lower wind power values.

4.2. Application of Exclusion Criteria

Applying all exclusion criteria on the capacity factor layer, the sites that are potentially viable for deployment of wind farms for both wind turbines studied are shown in Figures 23 and 24. These locations are namely the (i) north of Cagayan, (ii) west of Rizal, (iii) north of Camarines Sur, (iv) north of Samar, (v) southwest of Masbate, (vi) Dinagat Island, (vii) Guimaras, and (viii) northeast of Palawan. These areas identified are shown in patches of green and dot the map all over the country from the north or Luzon to the northernmost tip of Mindanao. From the previous sections, one can make the conclusion that the top three areas would still be the same ones at the end of the analysis. Unfortunately, the three best areas were excluded after the exclusion mask was applied. The resulting map with the highest capacity factor shows the areas north of Luzon in the Cagayan region as the best for wind farm development. The Siemens SWT-3.6-120 capacity factor range was from 50.32% to 41.71% while the Senvion 6.2M126 capacity factor range was from 48.42% to 38.56%. This location was the same site used for the economic analysis.

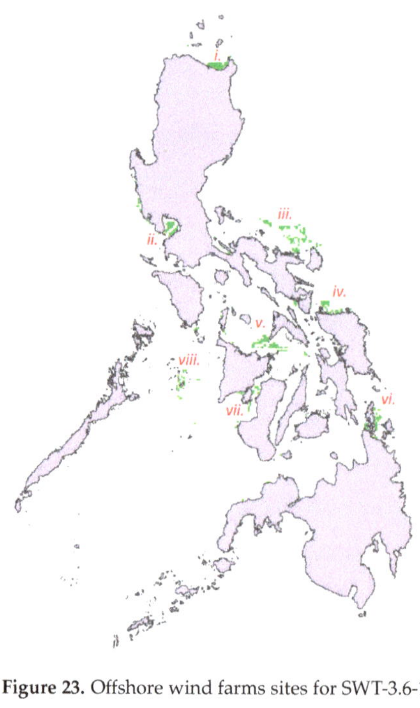

Figure 23. Offshore wind farms sites for SWT-3.6-120.

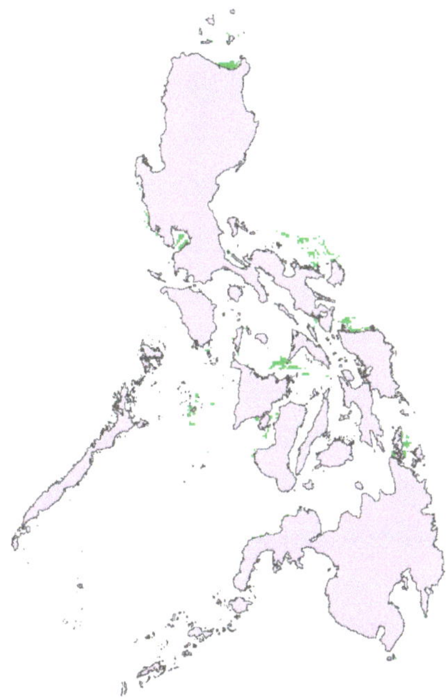

Figure 24. Offshore wind farms sites for 6.2M126.

4.3. Economic Analysis

4.3.1. Multiple Linear Regression

The data for independent variables were acquired from "4COffshore". There were 63 observations acquired for the offshore wind farms with the corresponding name of power plant, country of origin, minimum and maximum sea depth, area, offshore cable length, onshore cable length, inter-array cable length, port for O&M, distance from port, turbine model, number of turbines, turbine capacity, plant capacity, and investment cost. The first 39 observations acquired were separated in a different spreadsheet that used the multiple linear regression model, while the other 24 observations were used for validation of the model generated.

4.3.2. Regression Model Diagnostics

Figure 25a–d illustrates the four-assumption test with the exclusion of three observations. The observations followed a normal distribution, as shown in Figure 25a. It also displayed linearity and homoscedasticity, as seen in Figure 25b,c, respectively. Finally, the reliability test through the residuals vs. leverage plot showed that there was no influential observation that was outside the trend. The next step in checking the assumptions of multiple linear regression is the variance inflation factor test, which indicates if there is multicollinearity between the independent variables. The three observations were omitted due to failure in satisfying the reliability test.

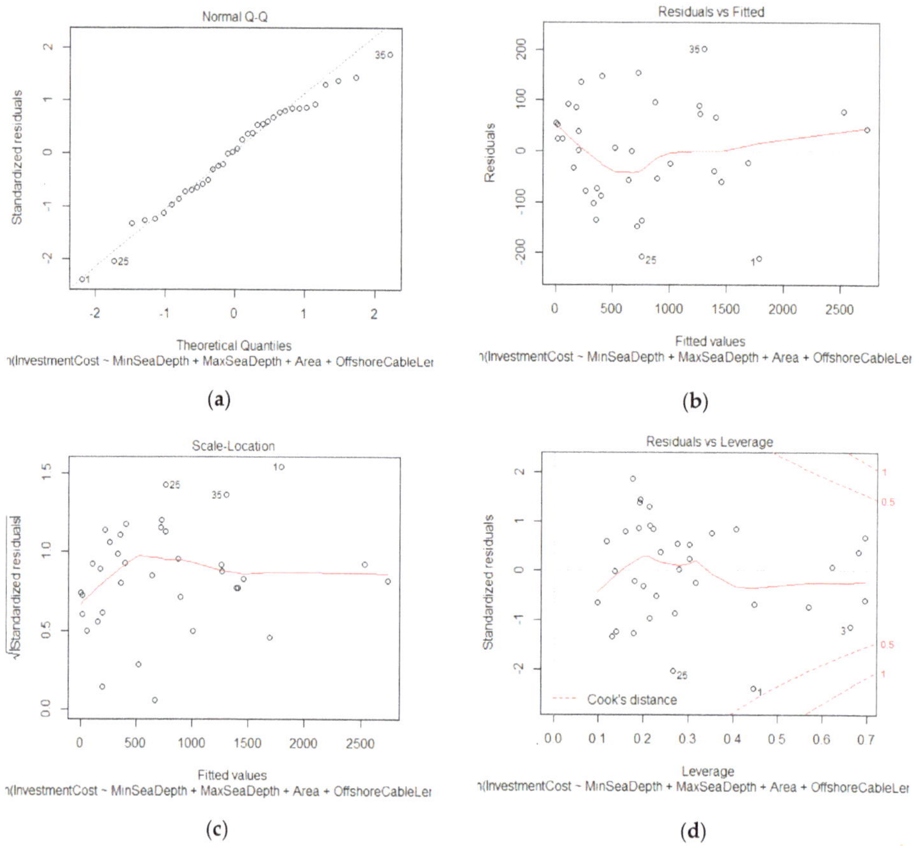

Figure 25. Multiple linear regression test for (**a**) normality, (**b**) linearity, (**c**) homoscedasticity, and (**d**) reliability of the 36 observations.

4.3.3. Model Selection

The multiple linear regression models with independent variables less than 5 *VIF* are shown in Table 1. There were 26 models generated and the other models with greater than 5 *VIF* were excluded since they indicate multicollinearity. Models 8, 9, 16, and 17 represent the plant capacity with the independent variable plant capacity itself. However, models 21, 22, 25, and 26 express the plant capacity in terms of number of turbines and capacity of the turbines. Since the models satisfied the four linear regression assumptions, the best model was chosen based on the criteria of R^2 and adjusted R^2.

4.3.4. Adjusted R^2

The coefficient of determination of each multiple linear regression models is shown in Table 2. Among the models that satisfied the multiple linear regression assumptions, model 8 had the highest R^2 and adjusted R^2 with 97.40% and 96.75%, respectively. The model includes the independent variables minimum sea depth, area, offshore cable length, onshore cable length, port distance, capacity of the turbine, and plant capacity. This result implies that the model is a good fit for the observations considered.

Table 1. Variance inflation factor of the different models of independent variables.

Name	Individual Variables	Variable Removed	VIF
Model 8	MinSeaDepth + Area + OffshoreCableLength + OnshoreCableLength + PortDistance + CapTurbine + PlantCap	MaxSeaDepth, NumTurbine, InterArrayCableLength	MinSeaDepth = 1.966903, Area = 3.014808, OffshoreCableLength = 2.154003, OnshoreCablelength = 1.248556, PortDistance = 1.377031, CapTurbine = 1.402520, PlantCap = 3.646495
Model 9	MaxSeaDepth + Area + OffshoreCablelength + OnshoreCablelength + PortDistance + CapTurbine + PlantCap	MinSeaDepth, NumTurbine, InterArrayCablelength	MaxSeaDepth = 2.474813, Area = 3.207744, OffshoreCablelength = 2.009005, OnshoreCableLength = 1.229875, PortDistance = 1.493217, CapTurbine = 1.647750, PlantCap = 3.860488
Model 16	MinSeaDepth + OffshoreCablelength + OnshoreCablelength + PortDistance + CapTurbine + PlantCap	MaxSeaDepth, Area, InterArrayCablelength, NumTurbine	MinSeaDepth = 1.966185, OffshoreCablelength = 2.113326, OnshoreCablelength = 1.229009, PortDistance = 1.363429, CapTurbine = 1.402447, PlantCap = 1.558588
Model 17	MaxSeaDepth + OffshoreCablelength + OnshoreCablelength + PortDistance + Cap Turbine + PlantCap	MinSeaDepth, Area, InterArrayCablelength, NumTurbine	MaxSea Depth = 2.325112, OffshoreCableLength = 2.008671, OnshoreCablelength = 1.210969, PortDistance = 1.440038, CapTurbine = 1.622965, PlantCap = 1.486171
Model 21	MinSeaDepth + Area + OffshoreCableLength + OnshoreCablelength + PortDistance + NumTurbine + CapTurbine	PlantCap, MaxSeaDepth, InterArrayCableLength	MinSeaDepth = 1.948631, Area = 2.866016, OffshoreCableLength = 1.920793, OnshoreCablelength = 1.247374, PortDistance = 1.370884, NumTurbine = 3.566499, CapTurbine = 1.615025
Model 22	MaxSeaDepth +Area+ OffshoreCablelength + OnshoreCablelength + PortDistance + NumTurbine + CapTurbine	PlantCap, MinSeaDepth, InterArrayCablelength	MaxSeaDepth = 2.422104, Area = 2.994082, OffshoreCablelength = 1.945860, OnshoreCablelength = 1.229641, PortDistance = 1.498961, NumTurbine = 3.730028, CapTurbine = 2.030133
Model 25	MinSeaDepth + OffshoreCablelength + OnshoreCablelength + PortDistance + NumTurbine + CapTurbine	PlantCap, MaxSeaDepth, Area, InterArrayCablelength	MinSeaDepth = 1.943687, OffshoreCablelength = 1.920234, OnshoreCableLength = 1.224317, PortDistance = 1.344159, NumTurbine = 1.603537, Cap Turbine = 1.528544
Model 26	MaxSeaDepth + OffshoreCablelength + OnshoreCablelength + PortDistance + NumTurbine + CapTurbine	PlantCap, MinSeaDepth, Area, InterArrayCablelength	MaxSeaDepth = 2.312621, OffshoreCablelength = 1.932499, OnshoreCablelength = 1.208195, PortDistance = 1.436073, NumTurbine = 1.538419, Cap Turbine = 1.841882

4.3.5. Investment Cost Regression Model 8

The results of the multiple linear regression coefficients of model 8 show the coefficients for the intercept, minimum sea depth, area, offshore cable length, onshore cable length, port distance, capacity of turbine, plant capacity at −163.0030, 4.2942, −3.9149, 0.6885, −2.0687, 0.4292, 13.3512, and 5.0368, respectively. These coefficients should be multiplied to the predictor variables to calculate the investment cost. The coefficient of the independent variable area and onshore cable length were negative, which implies that as these parameters increase, there is a decrease in investment cost. Further analysis was done to check the negative coefficients.

4.3.6. Checking of Investment Cost Regression Model 8

In the initial result of generating model 8, the normality was checked with all the independent variables combined. The normality of each independent variable was not taken into account in the analysis, and should be investigated. The Pearson correlation was added in the analysis, which checks the linear relationship between two independent variables [60].

Table 2. Coefficient of determination of the multiple linear regression models.

Name	Individual Variables	Variable Removed	Multiple R^2	Adj. R^2
Model 8	MinSeaDepth + Area + OffshoreCablelength + OnshoreCablelength + PortDistance + CapTurbine + PlantCap	MaxSeaDepth, NumTurbine, InterArrayCablelength	97.40%	96.75%
Model 9	MaxSeaDepth + Area + OffshoreCablelength + OnshoreCablelength + PortDistance + CapTurbine + PlantCap	MinSeaDepth, NumTurbine, InterArrayCablelength	97.24%	96.55%
Model 16	MinSeaDepth + OffshoreCablelength + OnshoreCablelength + PortDistance + CapTurbine+ Plantcap	MaxSeaDepth, Area, InterArrayCablelength, NumTurbine	96.42%	95.68%
Model 17	MaxSeaDepth + OffshoreCablelength + OnshoreCablelength + PortDistance + CapTurbine + PlantCap	MinSeaDepth, Area, InterArrayCablelength, NumTurbine	96.34%	95.59%
Model 21	MinSeaDepth + Area + OffshoreCablelength + OnshoreCablelength + PortDistance + NumTurbine + CapTurbine	PlantCap, MaxSeaDepth, InterArrayCablelength	91.80%	89.75%
Model 22	MaxSeaDepth +Area+ OffshoreCablelength + OnshoreCablelength + PortDistance + NumTurbine+ CapTurbine	Plant Cap, MinSeaDepth, InterArrayCablelength	91.97%	89.96%
Model 25	MinSeaDepth + OffshoreCablelength + OnshoreCablelength + PortDistance + NumTurbine + CapTurbine	PlantCap, MaxSeaDepth, Area, InterArrayCablelength	91.61%	89.87%
Model 26	MaxSeaDepth + OffshoreCablelength + OnshoreCablelength + PortDistance + NumTurbine + CapTurbine	PlantCap, MinSeaDepth, Area, InterArrayCablelength	91.85%	90.16%

Upon checking the normality of each of the independent variables and dependent variable, only the maximum sea depth had a normal distribution. The independent variables minimum sea depth, area, offshore cable length, onshore cable length, inter-array cable length, port distance, number of turbine, capacity of turbine, and plant capacity were transformed to log normal to normalize the distribution. Observations 10 and 34 were omitted due to "–infinity" result in the transformation to log normal, and observation 23 was omitted due to multicollinearity. The dependent variable investment cost was also transformed to a log normal to normalize the distribution.

The Pearson correlation of all the independent variables with a value between -0.5 to 0.5 indicate that there was a medium correlation between each other, and this was the limit considered in the study. The independent variables that satisfied this criterion were minimum sea depth, port distance, capacity of turbine, and plant capacity and was labelled as model 27.

The normality, linearity, homoscedasticity and reliability were checked for the log normal of the independent variables minimum sea depth, port distance, capacity of turbine, and plant capacity, as shown in Figure 26a–d. As shown in Figure 26a, the majority of the observations followed a normal distribution, but points 4, 26, and 29 should be taken into account if it is within the Cook's distance. The test for linearity showed that the observations resembled a linear pattern, as shown in Figure 26b. The observations followed homoscedasticity since the variance of errors were somehow similar across all independent variables, as seen in Figure 26c. For the reliability test in Figure 26d, all of the observations were within the Cook's distance and implies that there was no multicollinearity. The four linear regression assumptions were all satisfied.

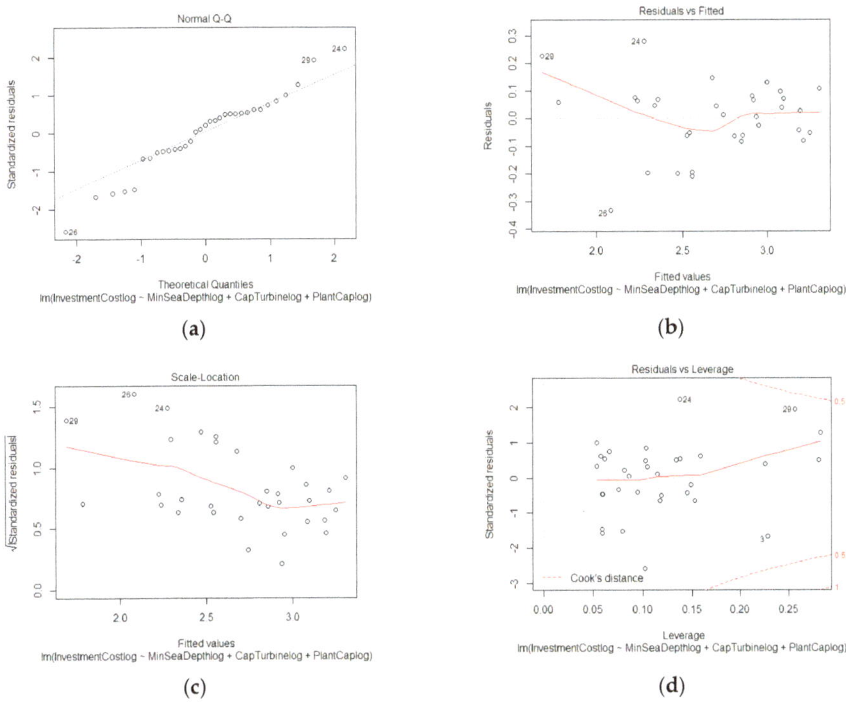

Figure 26. Multiple linear regression test for normality (**a**), linearity (**b**), homoscedasticity (**c**), and reliability (**d**) of model 27.

4.3.7. Selection of Investment Cost Regression Model

The multiple linear regression coefficients and R^2 of model 27 showed that the intercept value was 0.50223 with coefficients of 0.21401, 0.21699, and 0.88934 for the minimum sea depth, capacity of turbine, and plant capacity, respectively. Among the independent variables, the plant capacity had the highest influence since it had the highest coefficient. The coefficient of determination of the adjusted R^2 was 90.5%, which indicates that the model was still a good fit for the observations considered. On the other hand, the P-value for the turbine capacity was 0.27987, which was not less than 0.05 and means that it was not statistically significant. Therefore, the independent variable capacity of the turbine was omitted.

Model 28 is the regression model in which the capacity of the turbine was omitted from model 27. The four linear regression assumptions were satisfied by model 28. The multiple linear regression coefficients and R^2 of model 28 show that the intercept value was 0.61472 with coefficients of 0.25773 for the minimum sea depth and 0.87386 for the plant capacity. The plant capacity had the highest influence since it had the highest coefficient compared to the minimum sea depth. The coefficient of determination of the adjusted R^2 was 90.43%, which indicates that the model was still a good fit for the observations considered. This will be the regression model considered for acquiring the investment cost.

4.3.8. Multiple Linear Regression Model Validation

The multiple linear regression model 28 was validated using the other 24 observations that were not included in the model generation, as shown in Table 3. Model 28 had the least mean absolute percentage error ($MAPE$), which amounted to 11.33%. This means that there was an uncertainty of about 11.33% when predicting the investment cost using the multiple linear regression model 8. In the sensitivity analysis, ±11.33% was analyzed for the investment cost.

Table 3. Validation of multiple linear regression model 28 using mean absolute percentage error.

	Actual Investment Cost (Million USD)	Predicted Investment Cost (Million USD)	Residuals	MAPE
1	1155.0303	1272.5352	−117.504902	0.10173318
2	1610.5574	1636.7401	−26.182712	0.01625693
3	1155.0303	1272.5352	−117.504902	0.10173318
4	1443.7878	1302.7241	141.063756	0.09770394
5	2079.0545	1988.9850	90.069423	0.04332230
6	2485.6456	2534.3208	−48.675219	0.01958253
7	1389.2856	1287.8844	101.401244	0.07298805
8	463.2340	334.1906	129.043308	0.27857048
9	1501.5393	1315.9762	185.563161	0.12358195
10	1355.8067	1331.0228	24.783887	0.01827981
11	1039.5272	996.5151	43.012121	0.04137662
12	2146.6939	1997.6728	149.021163	0.06941891
13	784.7990	717.1558	67.643246	0.08619180
14	740.0445	826.3828	−86.338284	0.11666634
15	1006.9377	922.6834	84.254318	0.08367381
16	578.7908	477.7106	101.080216	0.17464032
17	415.9957	424.3203	−8.324543	0.02001113
18	297.5603	401.2918	−103.731481	0.34860655
19	1030.7567	753.9011	276.855596	0.26859451
20	517.0838	606.1924	−89.108637	0.17232921
21	3163.5490	2602.6829	560.866112	0.17729016
22	492.4285	577.0530	−84.624516	0.17185138
23	458.1789	479.7112	−21.532298	0.04699540
24	898.0267	836.7780	61.248712	0.06820367
25	NA	NA	NA	0.11331676

4.4. Levelized Cost of Electricity

The levelized cost of electricity was computed by generating a polygon area in Arc GIS. The shape of the polygon was a rectangle, which is only a representative of the area of the offshore wind farm. This step was done to calculate the number of turbines based on the downwind and crosswind spacing factors of 10 and 5, respectively. The area generated in the map was 25.0372 km^2. There were two cases considered for the said area, namely, construction of 34 SWT-3.6-120 turbines with a plant capacity of 122.4 MW, as shown in Figure 27, and the construction of 31 6.2M126 turbines with a plant capacity of 192.2 MW, as shown in Figure 28. The levelized cost of electricity for the 34 SWT-3.6-120 turbines range from USD 156.829/MWh to USD 158.498/MWh. For the case with 31 6.2M126 turbines, the resulting *LCOE* ranged from USD 153.207/MWh to USD 154.995/MWh. The results were validated against the study done by Mattar [16] wherein the computed *LCOE* in Chile was within 72 USD/MWh to 322 USD/MWh using a V164 8.0 MW turbine. Another validation was done by applying the 41.2% capacity factor and 4.14 MW turbine with *LCOE* of USD 181/MWh in the model done by [61]. The resulting *LCOE* after applying the 41.2% capacity factor and the 4.14 MW in this study was between USD 179.19/MWh and USD 183.63/MWh. The 6.2M126 turbine had lower *LCOE* because even though the investment cost was higher for this turbine, the plant capacity was greater in 6.2M126 than with SWT-3.6-120 given the same area. The plant capacity for 6.2M126 was 192.2 MW while SWT-3.6-120 had 122.4 MW.

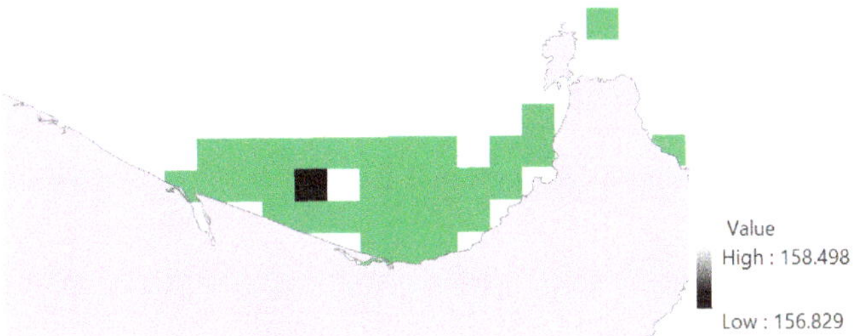

Figure 27. *LCOE* of SWT-3.6-120.

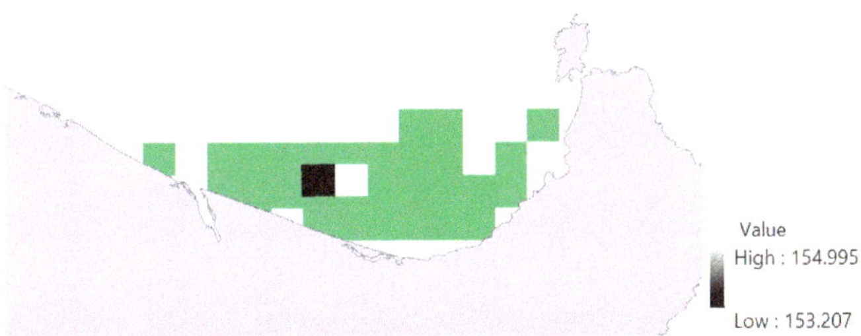

Figure 28. *LCOE* of 6.2M126.

The comparison of the *LCOE* of the SWT-3.6-120 and 6.2M126 in terms of plant capacity is shown in Figure 29. The parameters assumed for the basis of the *LCOE* were similar to the 25.0372 km^2 offshore wind farm in the north of Cagayan. The SWT-3.6-120 turbine had a lower *LCOE* compared to the 6.2M126 in all plant capacities from 0 MW to 1000 MW. The *LCOE* was lower in the case of SWT-3.6-120 since the 6.2M126 turbines have high investment cost and lower capacity factor. The *LCOE* had a similar behavior with exponential curves because even though the total life cycle cost increased with the increase in the plant capacity, the magnitude of the plant capacity as a divisor greatly affects the *LCOE*.

The comparison of the *LCOE* of the SWT-3.6-120 and 6.2M126 in terms of area can be seen in Figure 30. The parameters assumed for the basis of the *LCOE* were similar to the 25.0372 km^2 offshore wind farm in the north of Cagayan. The 6.2M126 turbine had lower *LCOE* compared to the SWT-3.120 in all areas from 0 km^2 to 150 km^2. Contrary to the results in the *LCOE* vs. plant capacity, the 6.2M126 turbine had less *LCOE* because even though its initial investment cost was higher compared with the SWT-3.6-120, the plant capacity of 6.2M126 was greater given the same area for both turbines.

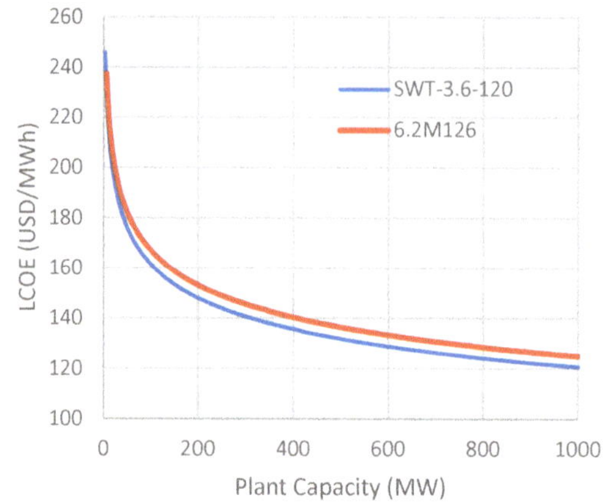

Figure 29. *LCOE* comparison of SWT-3.6-120 and 6.2M126 vs. plant capacity.

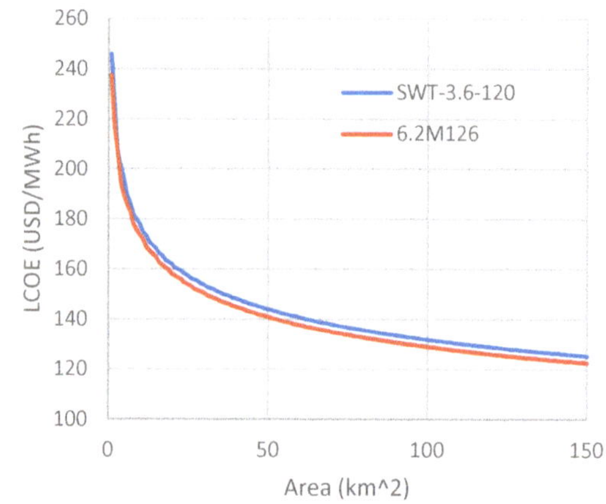

Figure 30. *LCOE* comparison of SWT-3.6-120 AND 6.2M126 vs. area.

4.5. Price of Electricity

The calculation for the price of electricity was based on the computation of the feed-in tariff for wind in the Philippines. It assumed an actual 200 MW onshore wind farm with its associated costs such as operation and maintenance cost and transmission interconnection cost. The feed-in-tariff for onshore wind farms was calculated to be PhP 7.40/kWH. The calculated offshore wind electricity price was PHP 8.028/kWh when the SWT-3.6-120 W was utilized, and a price of PHP 8.306/kWh when the 6.2M126 was used. The price of electricity from offshore wind farms is competitive against current feed-in-tariff because even though there is a higher investment cost for offshore wind farms, it was compensated for by its high capacity factor. The capacity factor assumed for the onshore wind farm was just 27.5%, while the calculated capacity factor for offshore wind farm in the study ranged from 45.80% to 47.91%.

4.6. Sensitivity Analysis

Figures 31 and 32 show the sensitivity analysis results for the *LCOE* when the SWT-3.6-120 and 6.2M126 were utilized, respectively. The base value for the *LCOE* for the SWT-3.6-120 and 6.2M126 turbines in the sensitivity analysis were USD 157.66/MWh and USD 154.1/MWh, respectively. The generated regression model had a mean absolute percentage error of ±11.33%, which was reflected in the investment cost of the sensitivity analysis. In the sensitivity analysis of the investment cost, there was a deviation of ±USD16.25/MWh to the *LCOE* for the SWT-3.6-120 turbine, and a deviation of ±USD17.91/MWh for the 6.2M126. The other parameters, namely, the capacity factor (±2%), weighted average cost of capital (±2%), cost of debt (±1%), cost of equity (±2%), and plant capacity (±5%) were also investigated to check how these parameters influence the *LCOE*. The basis in the adjustment of the parameters were small changes based on the nature of the parameters. As shown in Figures 31 and 32, the capacity factor had the highest influence among the other parameters, which had a deviation of ±USD 6.91/MWh for the SWT-3.6-120 and ±USD 6.54/MWh for 6.2M126. The weighted average cost of capital had a moderate influence on the *LCOE* with a ±USD 4.30/MWh for the SWT-3.6-120 and ±USD 4.19/MWh for the 6.2M126. The cost of debt, cost of equity, and plant capacity had less influence on the *LCOE*.

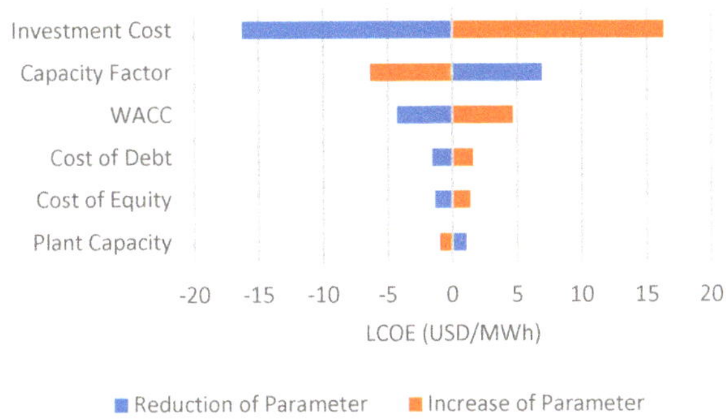

Figure 31. Sensitivity analysis of *LCOE* to different parameters (SWT-3.6-120).

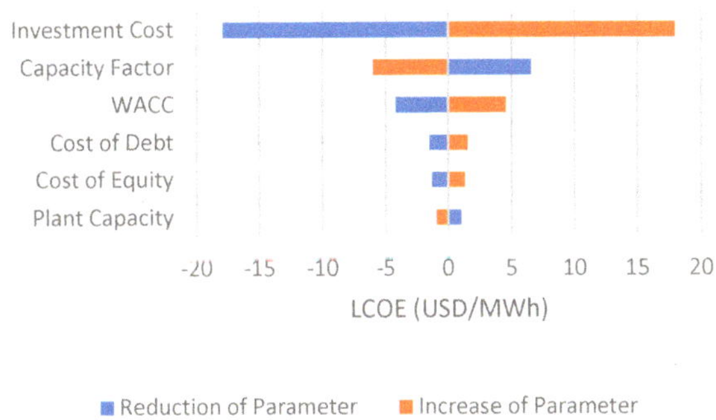

Figure 32. Sensitivity analysis of *LCOE* to different parameters (6.2M126).

The sensitivity analysis for the price of electricity from offshore wind in the Philippines using the SWT-3.6-120 and 6.2M126 turbines can be seen in Figures 33 and 34, respectively. The base offshore wind electricity prices for the SWT-3.6-120 and 6.2M126 in the analysis were PHP 8.028/kWh and PHP 8.306/kWh, respectively. The parameters considered were investment cost (±11.33%), capacity factor (±2%), plant capacity (±5%), weighted average cost of capital (±2%), cost of debt (±1%), and cost of equity (±2%). The 11.33% mean absolute percentage error of the regression model was checked in the sensitivity analysis while the other parameters were small changes based on the nature of the parameters. The ±11.33% adjustment in the investment cost resulted in a difference of ±PHP 0.91/kWh and PHP 0.94/kWh to the offshore wind electricity price when using SWT-3.6-120 and 6.2M126, respectively. The plant capacity factor, capacity factor, and weighted average cost of capital had moderate influence on the offshore wind electricity price. When adjusted to ±5% in the plant capacity, there was a difference of ±PHP 0.35/kWh and ±PHP 0.36/kWh to the electricity price when the SWT-3.6-120 and 6.2M126 were utilized, respectively. The adjustment of ±2% to the capacity factor resulted in a difference of ±PHP 0.35/kWh and ±PHP 0.38/kWh to the electricity price when utilizing the SWT-3.6-120 and 6.2M126, respectively. The adjustment of ±2% to the weighted average cost of capital resulted in a difference of ±PHP 0.18/kWh and ±PHP 0.19/kWh to the electricity price when the SWT-3.6-120 and 6.2M126 were utilized, respectively. The other parameters, namely, cost of debt and cost of equity had low influence on electricity price.

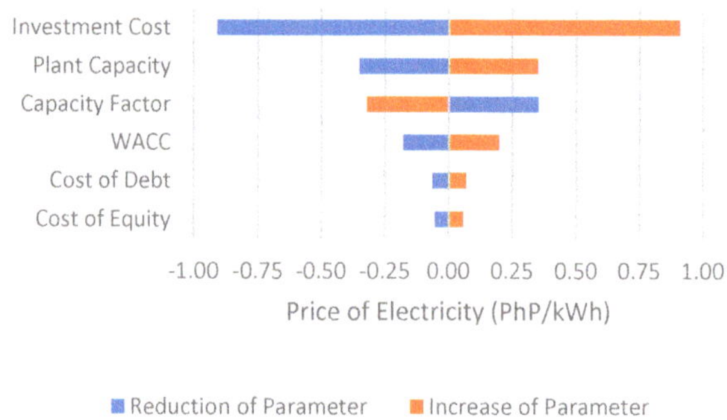

Figure 33. Sensitivity analysis of electricity price to different parameters (SWT-3.6-120).

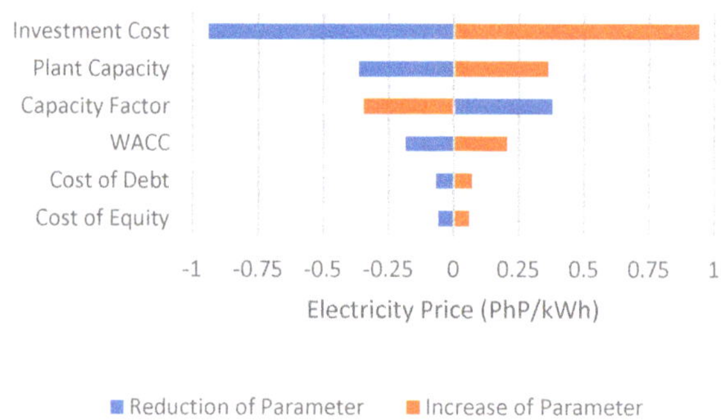

Figure 34. Sensitivity analysis of electricity price to different parameters (6.2M126).

5. Conclusions and Recommendations

The technical and economic assessments for emerging renewable energy technologies, specifically offshore wind energy, are critical for their improvement and deployment. These assessments serve as one of the main bases for the construction of offshore wind farms, which would be beneficial to the countries gearing toward a sustainable future such as the Philippines.

Two popular wind turbines were studied in the technical analysis. The capacity factor was calculated for both the Siemens SWT-3.6-120 and the Senvion 6.2M126 turbines. Results showed that the greatest values were in the northern parts of Ilocos Norte with a maximum of 62.60% for the Siemens turbine and 64.79% for the Senvion turbine. Two other locations also presented favorable values of capacity factors with the northern parts of Occidental Mindoro registering maximum values of 60.92% for the Siemens turbine and 62.60% for the Senvion turbine, while the southeastern parts of Oriental Mindoro recorded highest values of 59.52% for the Siemens turbine and 62.60% for the Senvion turbine. These locations were eventually excluded due to technical, political, and social restrictions. After masking out all exclusion criteria, the areas most suitable for development of offshore wind farms were north of Cagayan, west of Rizal, north of Camarines Sur, north of Samar, southwest of Masbate, Dinagat Island, Guimaras, and northeast of Palawan. To carry out the economic analysis, the north of Cagayan was chosen since this area recorded the greatest capacity factor from the list of potential sites.

In the economic assessment of the study, a multiple linear regression of the parameters of 33 actual offshore wind farms from 2008 to 2018 were considered. The parameters considered for the regression model were minimum and maximum sea depth, area, offshore cable length, onshore cable length, inter-array cable length, port for O&M, distance from port, turbine model, number of turbines, turbine capacity, plant capacity, and investment cost to model the investment cost of offshore wind farms. When the four linear regression assumptions were investigated, namely, normality, linearity, homoscedasticity, and reliability of measurement, the model generated consisted of minimum sea depth, turbine capacity, and the plant capacity. The regression model generated had an adjusted R^2 equal to 90.43%, which implies that it fits the model of the actual investment costs at the 95% confidence level. The regression model was also validated by applying it to 24 actual offshore wind farms, and the mean absolute percentage error of the regression model was calculated to be 11.33%, which means that there was 11.33% uncertainty when predicting the investment cost.

The results of the combined technical and economic assessment were reflected in the calculated *LCOE* and price of electricity of offshore wind farms in the Philippines. The *LCOE* calculated were USD 157.66/MWh and USD 154.1/MWh when using the SWT-3.6-120 and 6.2M126 turbines, respectively. This *LCOE* was comparable in the techno-economic study done for Chile with a range of *LCOE* from USD 72/MWh to USD 322/MWh. Another validation was done by applying the 41.2% capacity factor and 4.14 MW turbine in the model of the NREL which resulted in USD 181/MWh, and the calculated *LCOE* of the economic model resulted in USD 179.19/MWh and USD 183.63/MWh. The price of electricity was compared with the calculation for the feed in tariff of 200 MW onshore wind farm in the Philippines. The calculated price of electricity from offshore wind were PHP 7.64/kWh and PHP 8.028/kWh when the SWT-3.6-120 W was utilized, and a price of PHP 8.306/kWh when the 6.2M126 was utilized. The price calculated was competitive against the onshore wind farm feed-in tariff of PHP 7.40/kWh.

The *LCOE* of the SWT-3.6-120 and 6.2M126 were compared with each other through a scatter plot of plant capacity and area. The *LCOE* was lower for the SWT-3.6-120 given the same power plant capacity because the investment cost was greater with 6.2M126. However, the 6.2M126 had lower *LCOE* when the area for both turbines were the same due to higher plant capacity.

Based on the sensitivity analysis, the parameters that may greatly influence the *LCOE* are investment cost, capacity factor, and weighted average cost of capital for each instance when the SWT-3.6-120 and 6.2M126 were used. In the electricity price from offshore wind, the results showed that the investment cost, plant capacity, weighted average cost of capital, and the capacity factors were the most influential parameters.

However, it should be taken into account that the estimated investment cost for a 200 MW offshore wind farm is approximately USD 862.653 million while the investment cost for a 200 MW onshore wind farm amounts to USD 466 million. The offshore wind farm investment is approximately twice that of an onshore wind farm, and a possible option for pursuing offshore wind farms is through financing.

The future recommendation for this study is the consideration of the general soil characteristics of the potential sites for the construction of offshore wind farms that were not included in the study due to a lack of available data. Investigation of floating offshore wind farms in the Philippines can also be done with the methodology discussed in the study.

Author Contributions: Conceptualization, G.L.D.M. and M.-A.M.T.-K.; Methodology, G.L.D.M., M.-A.M.T.-K. and L.A.M.D.; Software, G.L.D.M.; Validation, G.L.D.M. and M.-A.M.T.-K.; Formal analysis, G.L.D.M.; Writing—original draft preparation, G.L.D.M., M.-A.M.T.-K. and L.A.M.D.; Writing—review and editing, L.A.M.D.; Visualization, G.L.D.M.; Supervision, M.-A.M.T.-K. and L.A.M.D.; Funding acquisition, L.A.M.D. All authors have read and agreed to the published version of the manuscript.

Funding: The APC was funded by the Department of Science and Technology (DOST) Engineering Research and Development for Technology (ERDT) Program.

Acknowledgments: The authors thank the Department of Science and Technology (DOST) Engineering Research and Development for Technology (ERDT) Program for the scholarship provided to the student to conduct this study.

Conflicts of Interest: The authors declare no conflict of interest.

References

1. GOV.PH. *An Act Promoting the Development, Utilization and Commercialization of Renewable Energy Resources and for Other Purposes*; Republic Act No. 9513; GOV.PH: Metro Manila, Philippines, 2008.
2. GOV.PH. *Department of Energy, Department Circular DC2009-05-0008*; Rules and Regulations Implementing Republic Act 9513; GOV.PH: Metro Manila, Philippines, 2009.
3. Department of Energy—Electric Power Industry Management Bureau. *2015 Philippine Power Situation Report*; Department of Energy: Taguig, Philippines, 2016.
4. Perveen, R.; Kishor, N.; Mohanty, S. Off-shore wind farm development: Present status and challenges. *Renew. Sustain. Energy Rev.* **2014**, *29*, 780–792. [CrossRef]
5. Richts, C.; Jansen, M.; Siefert, M. Determining the Economic Value of Offshore Wind Power Plants in the Changing Energy System. *Energy Procedia* **2015**, *80*, 422–432. [CrossRef]
6. Department of Energy. *List of Existing Power Plants in Luzon as of June 2016*; Department of Energy: Taguig, Philippines, 2016.
7. Department of Energy. *List of Existing Power Plants in Visayas as of June 2016*; Department of Energy: Taguig, Philippines, 2016.
8. Wang, S.; Wang, S. Impacts of wind energy on environment: A review. *Renew. Sustain. Energy Rev.* **2015**, *49*, 437–443. [CrossRef]
9. Bakker, R.; Pedersen, E.; Van Den Berg, G.; Stewart, R.; Lok, W.; Bouma, J. Impact of wind turbine sound on annoyance, self-reported sleep disturbance and psychological distress. *Sci. Total Environ.* **2012**, *425*, 42–51. [CrossRef]
10. Green, R.; Vasilakos, N. The economics of offshore wind. *Energy Policy* **2011**, *39*, 496–502. [CrossRef]
11. Zerrahn, A. Wind Power and Externalities. *Ecol. Econ.* **2017**, *141*, 245–260. [CrossRef]
12. Pineda, I.; Tardieu, P. *Wind in Power 2017 Annual Combined Onshore and Offshore Wind Energy Statistics*; WindEurope: Brussels, Belgium, 2018.
13. Rodrigues, S.; Restrepo, C.; Kontos, E.; Teixeira Pinto, R.; Bauer, P. Trends of offshore wind projects. *Renew. Sustain. Energy Rev.* **2015**, *49*, 1114–1135. [CrossRef]
14. Shi, W.; Park, H.; Chung, C.; Kim, Y. Comparison of Dynamic Response of Monopile, Tripod and Jacket Foundation System for a 5-MW Wind Turbine. In *Twenty-First (2011) International Offshore and Polar Engineering Conference*; International Society of Offshore and Polar Engineers: Maui, Hawaii, 2011.
15. Arapogianni, A.; Genachte, A. *Deep Water—The Next Step for Offshore Wind Energy*; European Wind Energy Association: Brussels, Belgium, 2013.
16. Mattar, C.; Guzmán-Ibarra, M. A techno-economic assessment of offshore wind energy in Chile. *Energy* **2017**, *133*, 191–205. [CrossRef]
17. Nagababu, G.; Kachhwaha, S.; Naidu, N.; Savsani, V. Application of reanalysis data to estimate offshore wind potential in EEZ of India based on marine ecosystem considerations. *Energy* **2017**, *118*, 622–631. [CrossRef]
18. Khraiwish Dalabeeh, A. Techno-economic analysis of wind power generation for selected locations in Jordan. *Renew. Energy* **2017**, *101*, 1369–1378. [CrossRef]
19. Schweizer, J.; Antonini, A.; Govoni, L.; Gottardi, G.; Archetti, R.; Supino, E.; Berretta, C.; Casadei, C.; Ozzi, C. Investigating the potential and feasibility of an offshore wind farm in the Northern Adriatic Sea. *Appl. Energy* **2016**, *177*, 449–463. [CrossRef]
20. Cavazzi, S.; Dutton, A. An Offshore Wind Energy Geographic Information System (OWE-GIS) for assessment of the UK's offshore wind energy potential. *Renew. Energy* **2016**, *87*, 212–228. [CrossRef]
21. Mahdy, M.; Bahaj, A. Multi criteria decision analysis for offshore wind energy potential in Egypt. *Renew. Energy* **2018**, *118*, 278–289. [CrossRef]
22. Hornyak, T. Here's What To Takes To Lay Google's 9000 km Undersea Cable. 2015. Available online: https://www.infoworld.com/article/2947900/networking/heres-what-to-takes-to-lay-googles-9000km-undersea-cable.html (accessed on 11 April 2018).
23. Submarine Cable Map. 2017. Available online: https://www.submarinecablemap.com/ (accessed on 15 April 2018).
24. Crevoisier, T. Philippines, Operational Ferry Routes. 2014. Available online: https://geonode.wfp.org/layers/geonode:phl_trs_waterways_wfp#more (accessed on 11 May 2018).
25. GOV.PH. *An Act Providing for the Establishment and Management of National Integrated Protected Areas System, Defining Its Scope and Coverage, and for Other Purposes*; Republic Act No. 7586; GOV.PH: Metro Manila, Philippines, 1992.

26. UNEP-WCMC. Protected Area Profile for Philippines from the World Database of Protected Areas. 2018. Available online: https://www.protectedplanet.net/country/PH (accessed on 15 April 2018).
27. Weeks, R.; Russ, G.; Alcala, A.; White, A. Effectiveness of Marine Protected Areas in the Philippines for Biodiversity Conservation. *Conserv. Biol.* **2010**, *24*, 531–540. [CrossRef]
28. Cabral, R.; Geronimo, R. How important are coral reefs to food security in the Philippines? Diving deeper than national aggregates and averages. *Mar. Policy* **2018**, *91*, 136–141. [CrossRef]
29. Gomez, E.; Aliño, P.; Yap, H.; Licuanan, W. A review of the status of Philippine reefs. *Mar. Pollut. Bull.* **1994**, *29*, 62–68. [CrossRef]
30. ReefBase. A Global Information System for Coral Reefs. 2018. Available online: http://www.reefbase.org (accessed on 1 May 2018).
31. 5th Philippine Energy Contracting Round (PERC5). 5Th Philippine Energy Contracting Round Petroleum Areas for Offer Figures and Maps. 2014. Available online: https://www.doe.gov.ph/figures-and-maps-petroleum (accessed on 15 May 2018).
32. Philippine Bathymetry Grid. General Bathymetric Chart of the Oceans (GEBCO). 2018. Available online: https://www.gebco.net/ (accessed on 20 April 2018).
33. Philippine Transmission Lines. Renewable Energy Data Explorer, National Renewable Energy Laboratory. 2018. Available online: https://www.re-explorer.org/ (accessed on 27 March 2018).
34. Worsnop, R.; Lundquist, J.; Bryan, G.; Damiani, R.; Musial, W. Gusts and shear within hurricane eyewalls can exceed offshore wind turbine design standards. *Geophys. Res. Lett.* **2017**, *44*, 6413–6420. [CrossRef]
35. Philippines: Hazard Profile. UN Office for the Coordination of Humanitarian Affairs. 2017. Available online: https://reliefweb.int/map/philippines/philippines-hazard-profile-jan-2017 (accessed on 1 March 2018).
36. Black, M.; Willoughby, H. The Concentric Eyewall Cycle of Hurricane Gilbert. *Mon. Weather Rev.* **1992**, *120*, 947–957. [CrossRef]
37. Takagi, H.; Wu, W. Maximum wind radius estimated by the 50 kt radius: Improvement of storm surge forecasting over the western North Pacific. *Nat. Hazards Earth Syst. Sci.* **2016**, *16*, 705–717. [CrossRef]
38. Matsunobu, T.; Inoue, S.; Tsuji, Y.; Yoshida, K.; Komatsuzaki, M. Seismic Design of Offshore Wind Turbine Withstands Great East Japan Earthquake and Tsunami. *J. Energy Power Eng.* **2014**, *8*, 2039–2044.
39. Truelsen, C. Offshore Wind Turbines Must Withstand Typhoons and Earthquakes in Taiwan. Available online: https://www.niras.com/projects/offshore-jacket-design/ (accessed on 10 April 2018).
40. Chiaradonna, A.; Tropeano, G.; Donofrio, A.; Silvestri, F. Interpreting the deformation phenomena of a levee damaged during the 2012 Emilia earthquake. *Soil Dyn. Earthq. Eng.* **2019**, *124*, 389–398. [CrossRef]
41. Bird, L.; Cochran, J.; Wang, X. *Wind and Solar Energy Curtailment: Experience and Practices in the United States*; NREL/TP-6A20-60983; National Renewable Energy Laboratory: Golden, CO, USA, 2014.
42. Bird, L.; Lew, D.; Milligan, M.; Carlini, E.; Estanqueiro, A.; Flynn, D.; Gomez-Lazaro, E.; Holttinen, H.; Menemenlis, N.; Orths, A.; et al. Wind and solar energy curtailment: A review of international experience. *Renew. Sustain. Energy Rev.* **2016**, *65*, 577–586. [CrossRef]
43. Ang, M.R.C.O.; Blanco, A.C. *Philippine Renewable Energy Resource Mapping from LiDAR Surveys (REMap) Project Terminal Report*; UP TCAGP: Quezon City, Philippines, 2017.
44. Murthy, K.; Rahi, O. A comprehensive review of wind resource assessment. *Renew. Sustain. Energy Rev.* **2017**, *72*, 1320–1342. [CrossRef]
45. Celik, A. A statistical analysis of wind power density based on the Weibull and Rayleigh models at the southern region of Turkey. *Renew. Energy* **2004**, *29*, 593–604. [CrossRef]
46. Siemens AG. Thoroughly Tested, Utterly Reliable Siemens Wind Turbine SWT-3.6-120. 2011. Available online: https://www.siemens.com.tr/i/Assets/Enerji/yenilenebilir_enerji/E50001-W310-A169-X-4A00_WS_SWT_3-6_120_US.pdf (accessed on 16 March 2018).
47. Senvion GmbH. 3.2M126 Wind Turbine 6.2MW. 2017. Available online: https://www.senvion.com/global/en/products-services/wind-turbines/6xm/62m126/ (accessed on 17 March 2018).
48. Tegen, S.; Lantz, E.; Hand, M.; Maples, B.; Smith, A.; Schwabe, P. *2011 Cost of Wind Energy Review*; NREL/TP-5000-56266; National Renewable Energy Laboratory: Golden, CO, USA, 2013.
49. Nagababu, G.; Kachhwaha, S.; Savsani, V. Estimation of technical and economic potential of offshore wind along the coast of India. *Energy* **2017**, *138*, 79–91. [CrossRef]
50. Sheridan, B.; Baker, S.; Pearre, N.; Firestone, J.; Kempton, W. Calculating the offshore wind power resource: Robust assessment methods applied to the U.S. Atlantic Coast. *Renew. Energy* **2012**, *43*, 224–233. [CrossRef]
51. Gonzalez-Rodriguez, A. Review of offshore wind farm cost components. *Energy Sustain. Dev.* **2017**, *37*, 10–19. [CrossRef]
52. Li, D.; Geyer, B.; Bisling, P. A model-based climatology analysis of wind power resources at 100-m height over the Bohai Sea and the Yellow Sea. *Appl. Energy* **2016**, *179*, 575–589. [CrossRef]
53. Schwanitz, V.; Wierling, A. Offshore wind investments—Realism about cost developments is necessary. *Energy* **2016**, *106*, 170–181. [CrossRef]
54. Montgomery, D.; Runger, G. *Applied Statistics and Probability for Engineers*, 3rd ed.; John Wiley & Sons, Inc.: Danvers, MA, USA, 2003.
55. De Myttenaere, A.; Golden, B.; Le Grand, B.; Rossi, F. Mean Absolute Percentage Error for regression models. *Neurocomputing* **2016**, *192*, 38–48. [CrossRef]

56. Osborne, J.W.; Waters, E. Four assumptions of multiple regression that researchers should always test. *Pract. Assess. Res. Eval.* **2002**, *8*, 2.
57. Kim, B. Understanding Diagnostic Plots for Linear Regression Analysis University of Virginia Library Research Data Services + Sciences. Data.library.virginia.edu. 2015. Available online: https://data.library.virginia.edu/diagnostic-plots/ (accessed on 3 May 2018).
58. Ioannou, A.; Angus, A.; Brennan, F. Stochastic Prediction of Offshore Wind Farm LCOE through an Integrated Cost Model. *Energy Procedia* **2017**, *107*, 383–389. [CrossRef]
59. Möller, B.; Hong, L.; Lonsing, R.; Hvelplund, F. Evaluation of offshore wind resources by scale of development. *Energy* **2012**, *48*, 314–322. [CrossRef]
60. Cohen, J. *Statistical Power Analysis for the Behavioral Sciences*, 2nd ed.; Erlbaum: Hillsdale, NJ, USA, 1987.
61. Mone, C.; Hand, M.; Bolinger, M.; Rand, J.; Heimiller, D.; Ho, J. *2015 Cost of Wind Energy Review*; NREL/TP-6A20-66861; National Renewable Energy Laboratory: Golden, CO, USA, 2017.

Journal of
Marine Science and Engineering

MDPI

Article

Assessment of the Offshore Wind Energy Potential in the Romanian Exclusive Economic Zone

Florin Onea, Eugen Rusu and Liliana Rusu *

Department of Mechanical Engineering, Faculty of Engineering, "Dunarea de Jos" University of Galati, 800008 Galati, Romania; florin.onea@ugal.ro (F.O.); eugen.rusu@ugal.ro (E.R.)
* Correspondence: liliana.rusu@ugal.ro

Abstract: The European offshore wind market is continuously expanding. This means that, together with significant technological developments, new coastal environments should be considered for the implementation of the wind farms, as is the case of the Black Sea, which is targeted in the present work. From this perspective, an overview of the wind energy potential in the Romanian exclusive economic zone (EEZ) in the Black Sea is presented in this work. This is made by analyzing a total of 20 years of wind data (corresponding to the time interval 2000–2019) coming from different sources, which include ERA5 reanalysis data and satellite measurements. Furthermore, a direct comparison between these datasets was also carried out. Finally, the results of the present work indicate that the Romanian offshore areas can replicate the success reported by the onshore wind projects, of which we can mention the Fantanele-Cogealac wind farm with an operating capacity of 600 MW.

Keywords: Romanian nearshore; EEZ; ERA5; water depth; energy resources; offshore wind turbines

Citation: Onea, F.; Rusu, E.; Rusu, L. Assessment of the Offshore Wind Energy Potential in the Romanian Exclusive Economic Zone. *J. Mar. Sci. Eng.* **2021**, *9*, 531. https://doi.org/10.3390/jmse9050531

Academic Editor: Bang-Fuh Chen

Received: 29 March 2021
Accepted: 13 May 2021
Published: 15 May 2021

1. Introduction

The human evolution and the energy demand go hand in hand. It is estimated that only during the time interval 1995–2015, the worldwide use of energy has increased from 8589 million tons of oil equivalent (Mtoe) to 13,147 Mtoe (+53%). A significant part of this market is covered by fossil fuels (80%), which represent a major problem from the perspective of the Paris agreement that is aiming to keep global warming below 2.0 °C [1]. Various strategies have been proposed to reduce the carbon emissions, among them being energy efficiency, fuel switching or renewable energy sources [2–4]. Looking at the projections promoted by the European Union, we can see that the future should be green, and it is expected to gradually increase the use of renewable sources to 45% by 2030 and almost 66% by 2050 (reported to gross electricity generation), and these values exclude the biomass production. For the year 2050, it is expected to gradually attenuate the importance of the conventional resources (e.g., oil or gas) down to zero [5]. Even the Gulf Cooperation Council (GCC) countries, which have access to important fossil fuel deposits, are committed to reduce the dependency on these sources, in the conditions when only 0.6% of their electricity production comes now from renewable resources [6].

The wind energy sector can be characterized as a very successful one, expanding continuously through new technologies and environments, such as the marine areas. Compared to the land, it is estimated that over the sea surface the wind conditions are stronger due to the lower friction, but on the other hand, the CAPEX (capital expenditure) and OPEX (operational expenditure) costs are considerably higher. The current trend is to develop large wind turbines (close to 10 MW) in deep-water areas that frequently reach hub heights of 100 m, and in order to reduce the construction costs, large projects need to be developed [7–9]. Europe is a representative area for the offshore wind industry, with a total installed capacity of 25 GW, of which 2.9 GW (11.60%—365 turbines) were installed during 2020. Among some other key features, we can mention that the average turbine size is closer to 8.20 MW, two-thirds of the projects developed in 2020 involved even larger systems. The sizes of the new wind farms are estimated to be close to 788 MW, which compared to 2019 indicates an increase of 26%, while the distance to the shore and water depth are now increased to 52 km and 44 m, respectively [10].

Europe is defined also by multiple water areas, which means that there are wide opportunities to develop marine renewable projects. In fact, the world's first operational offshore wind farm (Vindeby, Denmark [11]) was installed in the European nearshore (in the Baltic Sea) in 1991. At that time, it was difficult to predict the future of the wind sector, but looking now at the current trends some ambitious targets are set by the European Union (EU). These involve the use of new technologies, such as floating platforms, that can be installed in deeper water areas or to consider some new regions that were never taken into account before, such as the Black Sea [12,13]. Furthermore, it is estimated that by the end of the year 2050, the European offshore renewable sector will be defined by an installed capacity of 300 GW, a significant share being covered by the wind market [14].

A more complete picture of the Black Sea wind energy potential is provided in Onea and Rusu [15], taking into account some representative locations like Constanta (Romania), Odessa (Ukraine) or Trabzon (Turkey). Regarding the wind conditions (U_{80}), it was found that the Romanian sites and some sites from Russia (Sevastopol and Kerch) indicate higher resources than expected during the wintertime maximum average wind speeds of 9 m/s. As for the wind turbines, the capacity factor may reach a maximum value of 35% (the highest capacity factor) for this region, which can be obtained if the turbines operate at their full capacity for at least 10% of the time. A more detailed analysis of the wind conditions (U_{80}) in the Romanian coastal areas is carried out in Onea and Rusu [16], by considering various distances from the shore (up to 80 km) and multiple sites covering the Romanian nearshore (from north to south). From the analysis of the wind speed, it was found that it increases from the shoreline to offshore, reaching an average value of about 7.20 m/s at 60–80 km from the shore, with the mention that the sites from the northern and southern extremities indicate higher values close to the shore. Furthermore, for the sites located between 0 and 20 km from the shoreline, a water depth below 50 m is characteristic, which is suitable for the implementation of fixed wind turbines [16]. Although the research topic is quite similar to the one provided in Onea and Rusu [16], the present work is defined by some significant advances in relationship with the previous one. First, the dataset and the number of values per day are different (ERA5 in this case with 24 values per day) than the ERA-Interim, which was previously considered. Furthermore, in the previous work the performance evaluation of a turbine was made by adjusting the wind data from 10 m to a height of 80 m, while in this version the wind turbine is evaluated by considering the ERA5 wind speed directly reported to a 100 m hub height. Furthermore, taking also into account that the offshore sector evolves very fast, in the present work a 9.5 MW wind turbine is evaluated, compared to the previous work where the maximum rated power was limited to 5 MW. Also, the grid points defined in this work cover the entire Romanian EEZ (0 to 220 km offshore), compared to Onea and Rusu [16] where the points cover only the 0 to 80 km area.

Another important aspect is related to the future evolution of the Black Sea wind, in the context of climate change. From this perspective, in Rusu [17] the wind conditions were evaluated for the 30-year near future period (2021–2050) considering two RCP (representative concentration pathway) scenarios, RCP4.5 and RCP8.5. The results indicate an increase in the extreme wind conditions that may lead to an enhancement of the wind power by at least 30%, being estimated that the western part of the Black Sea may become a major source of wind energy [17]. At this point, it can be also highlighted that another preliminary study related to the implementation of the offshore wind farms in the Romanian coastal environment was made in [18], and that the present study represents in fact a development of that previous work.

In this context, the aim of the present work is to assess the Romanian offshore wind resources considering a standard turbine height (100 m) within the EEZ coastal region and to identify the performance of a large capacity wind turbine. This can be considered an opportune study since there is currently significant interest both from the Romanian authorities and from the investors to develop offshore wind projects in this coastal environment. Therefore, an important part of this work is dedicated to the analysis of the wind conditions from a meteorological point of view by evaluating some specific parameters, such as the average conditions or the seasonal variability, and to make a direct comparison with the satellite measurements. Some specific maps have been designed to show the distribution of the wind resources within the limit of the Romanian EEZ, including also the expected performances of a wind turbine of 9.5 MW. In the final part, the results of this study are compared with similar ones, including the accuracy of the ERA5 data and specifying also some limitations of the present work.

2. Materials and Methods

2.1. Study Area

The Romanian EEZ is defined by an area of 22,486 km^2, of which only 4084 km^2 represents the territorial sea, which is divided into two parts. The northern part covers 162 kilometers of coastline (Musura Bay and Cape Midia) and includes the Danube Delta Biosphere Reserve, while the southern part has an extent of only 83 kilometers and is delimited by the Vama Veche village.

The first sector is more important from an ecological point of view, while the second is related to economical activities, such as ports, but also significant touristic activities. It is estimated that almost 4.50% of Romania's population is located here, which per total covers a share of 5.30% from the entire Black Sea coastline [19]. Figure 1 presents the target area that is delimited by the EEZ borders, including the water depth details processed from GEBCO (general bathymetric chart of the oceans) [20]. The Romanian EEZ is delimited by the latitude lines 43.4398° N and 45.2128° N, while on the longitude the interval goes from 28.5278° E to 31.4097° E [21].

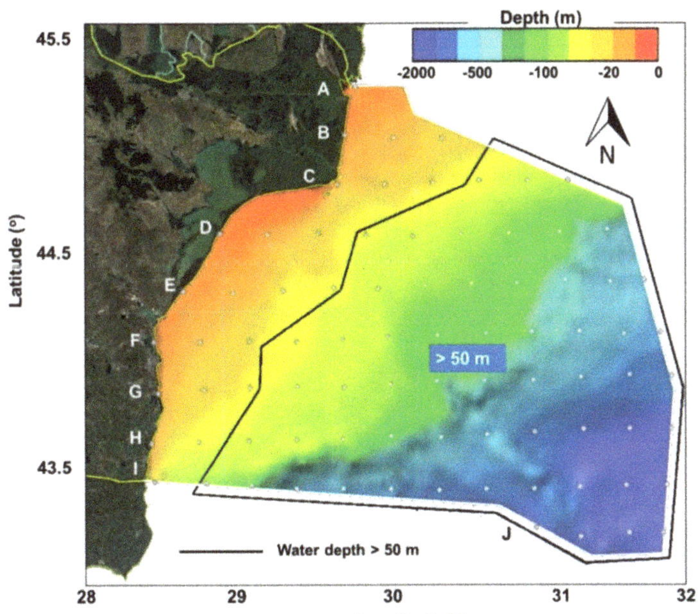

Figure 1. Representation of the Romanian exclusive economic zone (EEZ) considering the spatial map and the grid points defined. The horizontal reference lines (denoted from A to J from north to south) include points distributed along a grid defined by an approximate length of 20 km (along x and y directions).

A total of 84 points were defined in a grid (lines from A to J), the distance between each point being constant, approximately 20 km along the x and y directions. In addition, a water depth of 50 m was considered as a reference, since this represents the threshold that separates the fixed offshore wind farms from the floating ones [22,23]. By considering these points it is possible to obtain a general picture of the energy potential and the performances of a particular turbine, which can be further extended to a more detailed investigation. As a limitation of this work, the restricted areas were not taken into account, but from the information provided by Văidianu and Ristea [19], we can notice that the entire water strip located close to the shore (<50 m) is defined by shipping activities, while on the north we notice some large protected areas. Figure 1 illustrates also the water depth, and we can notice that the points located close to the shore indicate values in the range 2–24 m. The values start to increase as we go further into the deeper sea, being expected for each step of 20 km (along x axis) an increase in the water depth by 10 m for each reference line until the 100 km limit, while after that the values jump to a maximum of 1747 m (line I—Vama Veche).

2.2. Wind Data

Two different datasets are considered in the present work. The first is related to ERA5 [24]. This is a state-of-the-art global atmospheric product delivered by the European Centre for Medium-Range Weather Forecasts (ECMWF), which includes multiple climate variables such as rainfall, land surface temperature or wind data over the ocean environment [25–27]. This dataset covers the interval from 1979 to the present and incorporates some new techniques such as the ECMWF Integrated Forecast System model (IFS 41r2). This update aimed to increase both the temporal output (to 24 values per day), and the horizontal and vertical resolutions (to 0.25° and 137 vertical levels starting from the surface to 0.01 hPa). Some other improvements were introduced, this being the case of a new data assimilation scheme or the update of the parametrization schemes. More details about this project are provided in Hersbach et al. [24], while some other relevant information related to the ERA5 wind data can be found in Soares et al. [28], Ulazia et al. [29] or Carreno-Madinabeitia et al. [30]. Table 1 presents the characteristics of the ERA5 data processed in the framework of this work, where the following two types of data are considered: (a) wind speed at 10 m height above the sea level (U_{10}); and (b) wind speed at 100 m height (U_{100}). These values are directly obtained from the ECMWF database, without any further adjustment. The second wind dataset comes from AVISO (archiving, validation and interpretation of satellite oceanographic) data, which is a project that combines multiple altimeter missions. The TOPEX/POSEIDON mission was initiated in 1992, being designed to measure the wind speed. Over time, some new missions were launched (Jason-3, Sentinel-3A, Saral, Cryosat-2, Jason-1&2, T/P, Envisat, GFO, ERS-1 & 2 and Geosat), so a consistent dataset was produced [31]. Most of the satellites were placed on near-polar or sun-synchronous orbits, being defined by nadir-looking instruments having a footprint of the beam of 10 km wide. Each mission is defined by a ground track separation that has an orbit geometry that can go up to 400 km at the equator, each satellite repeating the same ground track on an interval of 3 to 10 days [32]. In Table 1, more details regarding this dataset are provided.

Table 1. Characteristics of the datasets processed for assessing the wind conditions in the Romanian coastal area.

Organization	ERA5	AVISO
	ECMWF	**Archiving, Validation and Interpretation of Oceanographic Satellite Data**
Type	Reanalysis data	Satellite measurements
Spatial coverage	Global	Global
Horizontal resolution	0.25° × 0.25°	1.00° × 1.00°
Reference height (m)	10, 100	10
Time interval	2000–2019	2013–2019
Time resolution	24 data per day	1 value per day (daily mean value)

2.3. Methods

The offshore wind industry is evolving very fast and at this moment it seems that there is interest in assessing the wind energy potential at a hub height of 100 m [28,33] and therefore the ERA5 (U_{100}) wind dataset is a suitable candidate.

Going further from this, the annual electricity production (AEP) of a wind turbine, can be estimated as in Equation (1) [34]:

$$AEP = T \cdot \int_{cut-in}^{cut-out} f(u)P(u)du \quad (1)$$

where *AEP*—in MWh; *T*—number of hours per year (8760 h/year); *f(u)*—Weibull distribution; *P(u)*—power curve of a turbine; and *Cut-in*/*Cut-out*—turbine characteristics.

The Weibull probability density function, can be defined as in Equation (2) [35]:

$$f(u) = \left(\frac{k}{c}\right)\left(\frac{u}{c}\right)^{k-1} exp\left[-\left(\left(\frac{u}{c}\right)^{k}\right)\right] \quad (2)$$

where *u*—the wind speed; *k*—the shape parameter; and *c*—the scale parameter (in m/s). Looking at some previous studies, we find that this method is not the most accurate one, since it may generate large uncertainties. For example, in the work of Elfarra and Kaya [36], two sites from the Aegean Sea and Mexico were evaluated, and it is noticed that for the sites located in the Mediterranean Sea the Weibull distribution overestimated the AEP by more than 10% while for the other site, the AEP production was underestimated by more than 6%. Also in the work of Jaramillo and Borja [37], it is mentioned that the AEP may be underestimated by at least 12%, which is caused mainly by the underestimation of the wind speeds from the intervals 12 and 20 m/s.

A single type of wind turbine will be considered for evaluation in this work. This is the Vestas V164-9.5 that is already implemented in various European offshore projects (fixed or floating). This is one of the largest wind generators available on the market (9.5 MW). Its performance will be evaluated at a hub height of 100 m. More details about this system and some of the related projects are provided in Figure 2.

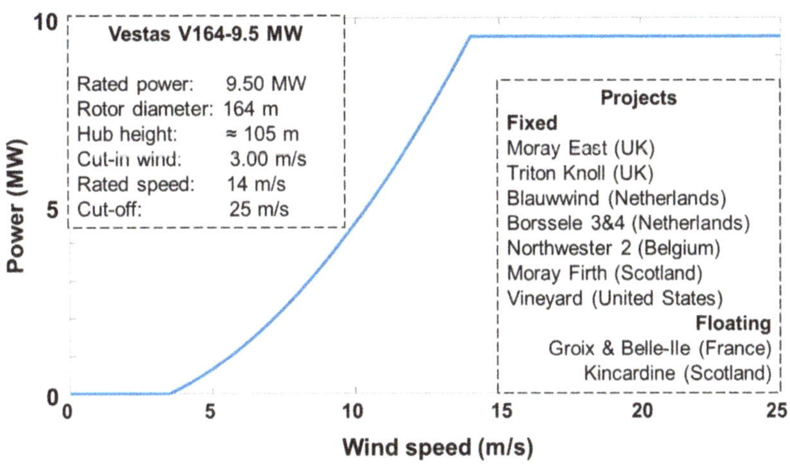

Figure 2. Power curve of the Vestas V164-9.5 turbine, including some technical data (left side) and planned offshore wind projects (right side) [38,39].

3. Results

3.1. Comparisons between the Datasets

ERA5 is used as the main source of data in this work, and therefore it is important to understand the differences between some alternative data sources, in this case, AVISO. Figure 3 presents the distribution of the U_{10} values (50th and 95th percentiles) as provided by ERA5. For the 50th percentiles (50th), the wind speed is in the range 4.11–5.97 m/s for the points located at approximately 40–60 km from the shore, and this gradually expands to 6.09 m/s close to the extremity of the EEZ zone (in the south). Per total, the points located in the southern part (from line G down) seem to be a less attractive option since they are defined by a higher water depth and lower wind resources (e.g., 5.72 m/s—line J). However, the installation of floating wind turbines can be a viable solution in areas with depths that do not allow the installation of fixed turbines. As for the 95th percentiles, the maximum values are close to 11.50 m/s, corresponding mainly to the offshore sites. In contrast, the points located close to the shore indicate a maximum value of 9.93 m/s (Sacalin Peninsula—point C1). This clearly shows that the wind energy potential in the target area increases as we go from onshore to offshore, the most promising sites for developing a wind project being located along the C-line.

Figure 4 presents a direct comparison between the AVISO and ERA5 data (U_{10}), where the results are expressed in percentages. The relative differences between ERA5 and AVISO are quantified as in Equation (3):

$$E = (X_{ERA5} - X_{AVISO})/X_{ERA5} \quad (3)$$

The daily average values of ERA5 U_{10} were computed in order to be compared with the daily mean values provided by AVISO. Excepting some points located close to the shore (lines F, G and H), all the values are positive which means that the ERA5 data indicate higher values than AVISO. A well-known issue related to the satellite measurements concerns the accuracy of the measurements on the land–sea interface, in this case being noticed as the *NaN* (not a number) values along the A-line. The *NaN* indicator is used to show that there are no valid data reported by the AVISO dataset. For the 50th percentiles (Figure 4a), the points located in the central and southern parts (close to the shore) show a better agreement, while the maximum differences of 36% occur in the center of the target area. These differences are better quantified in Table 2, where besides the 50th and 95th percentiles, the 25th and 75th percentiles are also computed, where for example D0 indicates a point located along the line D at 0.00 distance from the coast, while D80 is used to identify a similar point located at 80 km from the shore. For the 95th percentile, the best agreement corresponds to the nearshore points (1.57% and 4.26%), compared to some other points located at 80 or 120 km from the shore, where the differences reach 28.40%.

Table 2. Percentile comparisons between AVISO and ERA5 wind speeds at 10 m height above the sea level. The wind data are processed for the 7-year time interval 2013–2019, and a positive value indicates that the ERA5 wind speed is higher than that provided by AVISO.

Percentile Differences (in %)	ERA5 (U_{10})–AVISO (U_{10})									
	Top 5—Best Agreement Value (Reference Point)					Top 5—Lowest Agreement Value (Reference Point)				
25th	−0.31 (D0)	3.95 (E0)	−4.39 (I0)	−5.44 (G0)	−8.04 (F0)	38.10 (D80)	36.40 (E100)	35.90 (F100)	35.60 (G100)	35.50 (F120)
50th	−0.94 (G0)	2.07 (I0)	−3.65 (H0)	−3.90 (F0)	4.87 (D0)	37.50 (D80)	36.50 (E100)	35.80 (F100)	35.20 (G120)	35.00 (E80)
75th	−0.87 (D0)	−2.68 (H0)	3.85 (I0)	−4.17 (F0)	4.57 (D0)	33.50 (D80)	33.10 (E100)	32.90 (F120)	32.50 (E80)	32.30 (G120)
95th	−1.57 (G0)	1.77 (D0)	−2.84 (H0)	3.92 (I0)	−4.26 (F0)	28.40 (E80)	28.10 (D100)	27.90 (D80)	27.80 (D120)	27.70 (H80)

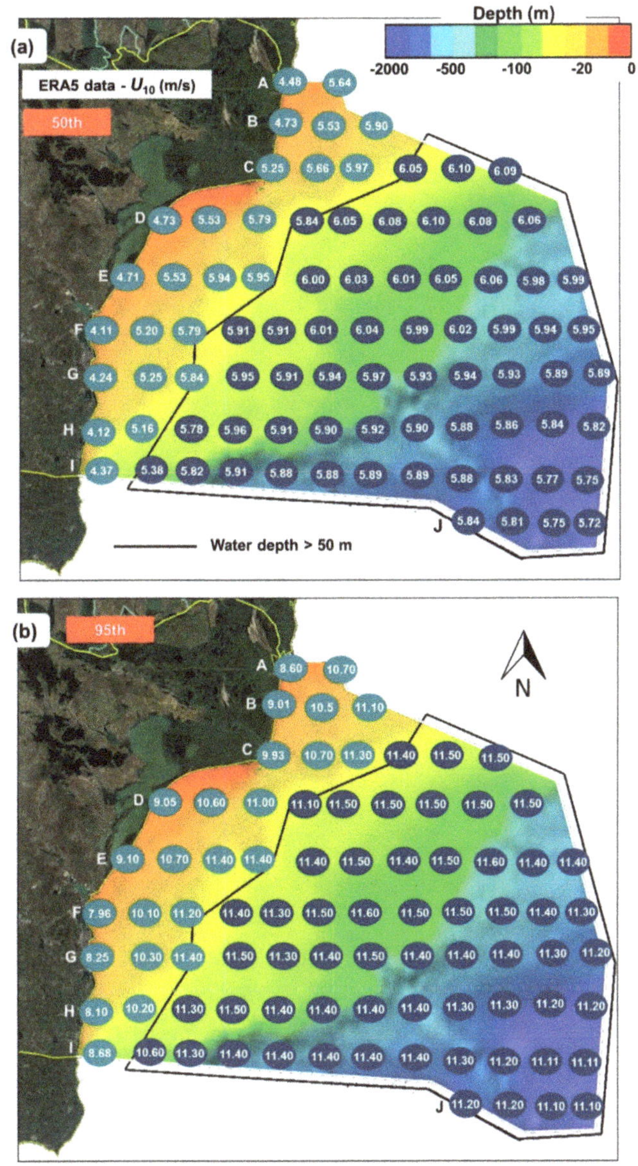

Figure 3. Percentiles distribution corresponding to the ERA5 (U_{10}) reanalysis data for the 20-year time interval 2000–2019. The results are indicated in terms of: (**a**) 50th percentile; and (**b**) 95th percentile.

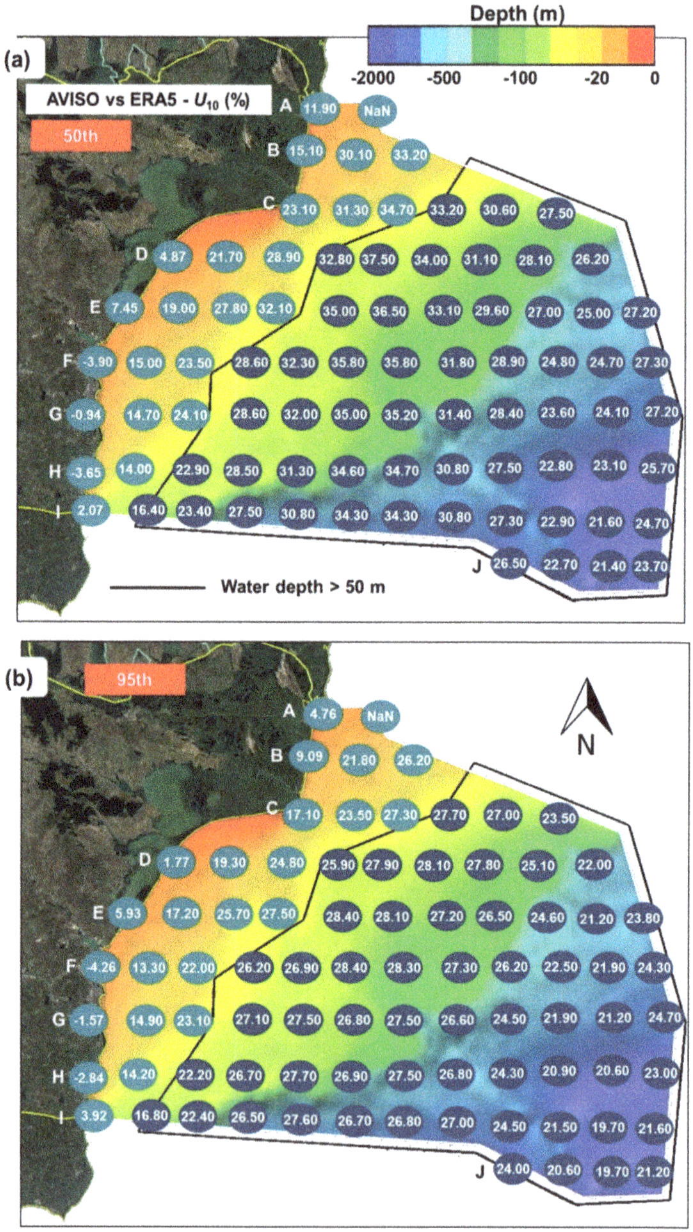

Figure 4. Comparisons between the AVISO (U_{10}) measurements and the ERA5 (U_{10}) reanalysis data for the 7-year time interval 2013–2019. The results are indicated in terms of: (**a**) 50th percentiles; (**b**) and 95th percentiles. A positive value indicates that the ERA5 wind speed is higher than that provided by the AVISO data.

3.2. Assessment of the Wind Energy Potential

Figure 5 provides the first representation of the wind energy distribution over the Romanian coastal area, including also the land. The results clearly show that the wind speed increases from the land (~5.00 m/s) to the shoreline, and further to the offshore regions. Near the shoreline the values can go up to 6.00 m/s, while from the marine areas the northern part indicates higher values, reaching a maximum of 7.48 m/s outside the Romanian national waters (in the Ukrainian area). Regarding the wind direction, several patterns are noticed. For the wind fields over the land, the dominant direction is from the land to the sea (oriented to the southeast), while near the shoreline the wind usually follows the local coastline. As we go further offshore, the wind field starts to curve, being oriented to the southwest direction.

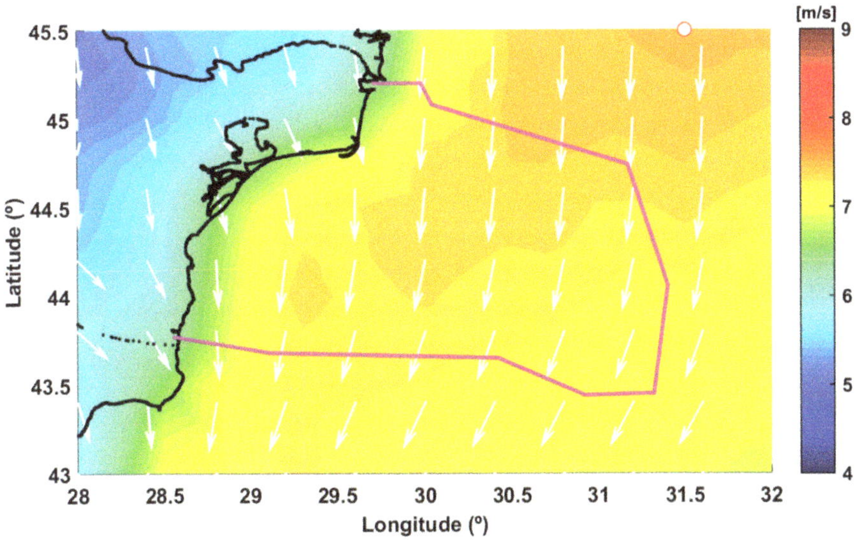

Figure 5. Average wind speed computed based on the ERA5 (U_{100}) dataset for the Romanian coastal area over the time interval 2000–2019 (full-time distribution). The wind direction is represented by arrows, while the point defined by the highest wind speed is indicated with a white circle. The Romanian EEZ is indicated by the magenta line.

A similar analysis is presented in Figure 6, considering this time the seasonal distribution (spring–summer–autumn–winter) of the parameter U_{100}. Regardless of the season, we notice that the wind conditions increase as we go from land to sea. The maximum values are located outside the Romanian coastal waters, excepting the wintertime when a maximum value of 8.76 m/s occurs in the center of the target area. This suggests that during the most energetic season (December–January–February), a wind project located in this area will obtain the best performances. More than this, the point indicated is located outside the restricted areas where the harbor activities usually take place [19], so this location may be considered a suitable candidate for a wind project. The summer season indicates lower resources, reaching a maximum value of 6.41 m/s, while the spring and autumn seasons are about the same level, reaching values of 7.44 and 7.57 m/s, respectively.

Regarding the wind directions, the patterns corresponding to the summer and autumn seasons are similar to the prevailing wind direction corresponding to the total time (Figure 5). In the case of spring, the dominant pattern is from sea to land (from east to west), being noticed close to the Danube Delta a tendency to shift in the northern sector. During the winter, the wind is moving to blow from the northwest and small changes in the orientation of the wind vectors are observed. The wind direction represents an important aspect of the design of a wind farm, considering that the energy produced is significantly affected by the wake effect. It is estimated that for the offshore projects, these losses are in the range 10.00–25.00% [40,41].

As mentioned above, wind turbines are designed to operate within a specific range of wind speeds, known as the cut-in and cut-out values. The turbine considered in this study has wind speed limits in the range of 3.00–25.00 m/s. That is why the percentages of U_{100} that were inside the functional range of the wind turbine were computed for each season, and the results are illustrated in Figure 7. It must be noticed that along the Romanian coast, percentages higher than 80% are observed in all seasons. As expected, percentages above 90% are found in the winter season.

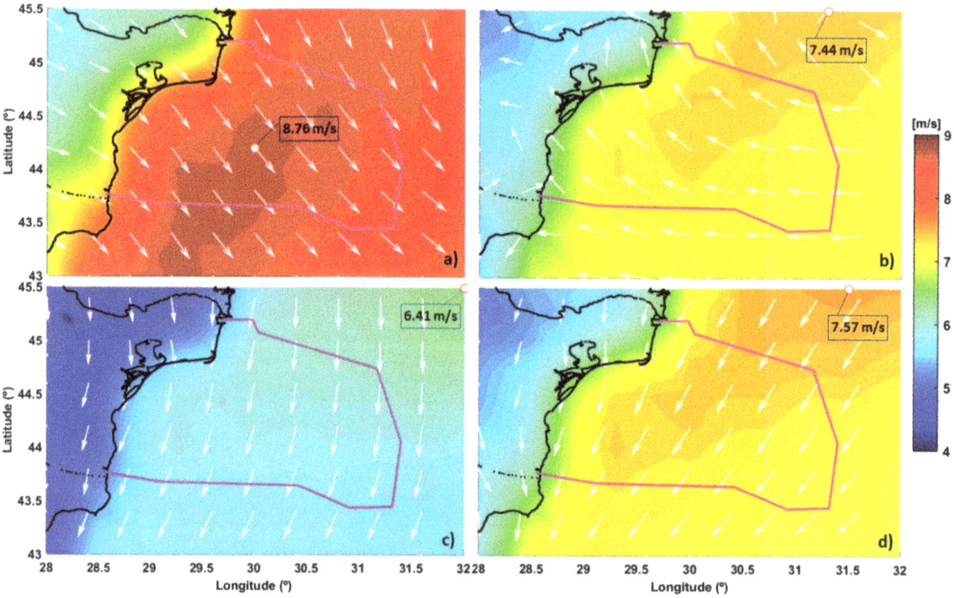

Figure 6. The seasonal averaged wind speed as indicated by the ERA5 (U_{100}) dataset for the Romanian coastal area over the time interval 2000–2019. Data corresponding to the following seasons are presented: (**a**) winter—December–January–February; (**b**) spring—March–April–May; (**c**) summer—June–July–August; and (**d**) autumn—September–October–November. The results include the wind direction vectors and the locations of the maximum wind speed, while the Romanian EEZ is indicated by the magenta line.

An important role in choosing a location for exploiting wind energy is the variability of the wind speed over the years. The *IAV* (interannual variability) index is often used in climatologic analyses to quantify the year-to-year variability of a given parameter. This can be expressed as in Equation (4) [26,42]:

$$IAV = \sqrt{\frac{1}{m}\sum_{j=1}^{m}\left[\left(\frac{1}{n}\sum_{k=1}^{n}x_{jk}\right) - \left(\frac{1}{m}\sum_{j=1}^{m}\left(\frac{1}{n}\sum_{k=1}^{n}x_{jk}\right)\right)\right]^2}\left(\frac{1}{nm}\sum_{j=1}^{m}\sum_{k=1}^{n}x_{jk}\right)^{-1} = \frac{\sigma_{\overline{x_j}}}{\overline{x}} \quad (4)$$

where *IAV*—inter-annual variability (%); x denotes the time series of the wind speed over a period of m years with n records each; $\overline{x}$—mean value of the time series of a parameter over a period of years; σ—standard deviation; the indices j and k refer to the year and record, respectively; and overbar denotes an average.

Thus, the year-to-year variability of U_{100} from the ERA5 database over the 20-year time interval 2000–2019 was quantified here and the results are illustrated in Figure 8. It can be observed that *IAV* increases as we move away from the shore. At this point, it has to be noticed that if for the coefficient of the variation in the wind speed there might be expected higher values on land than offshore, this inter-annual variability index shows very often higher values offshore, as can be noticed in the Figure 8 results which are also consistent with those presented in [42].

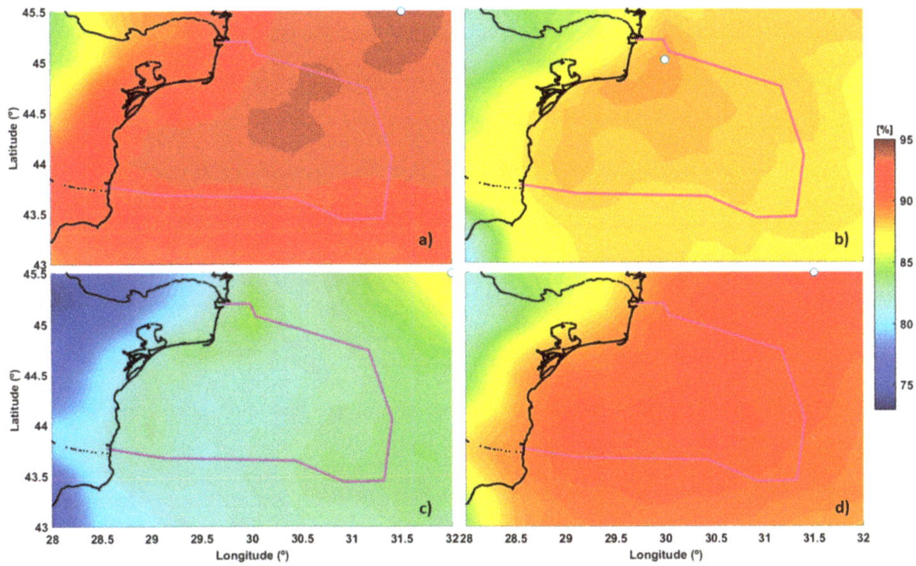

Figure 7. Seasonal percentages of the wind speed (U_{100}) in the range between the cut-in and cut-off values (standard functional range 3–25 m/s): (**a**) winter—December–January–February; (**b**) spring—March–April–May; (**c**) summer—June–July–August; and (**d**) autumn—September–October–November. The locations of the maximum values are marked by white circles, while the Romanian EEZ is indicated by the magenta line.

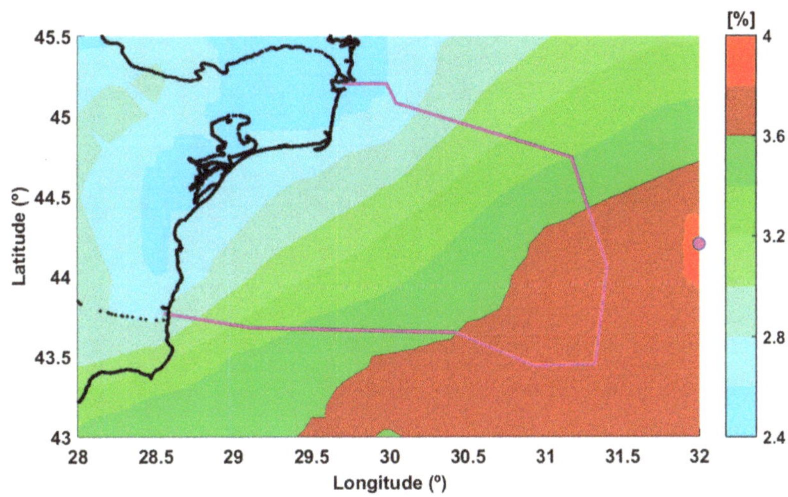

Figure 8. Interannual variability (*IAV*) of U_{100} from ERA5, where the Romanian EEZ is indicated by the magenta line.

The capacity factor is frequently used to assess the performances of a wind turbine, this is defined as in Equation (5) [43]:

$$Cf = \frac{P_e}{P_t} \tag{5}$$

where Cf—capacity factor (%); P_e—expected power to be generated by a turbine (MW); and P_t—rated power of a turbine (9.50 MW from Figure 2).

Figure 9 presents such an analysis considering the performances of the Vestas V164-9.5 wind turbine for the full-time distribution. Although the wind speed increases as we go further from the shore, this is not the case for the *Cf* indicator that for example indicates values of 28.50% for the point C40 (40.00 km from the shore), compared to values of 27.00% for the eastern boundary of EEZ. For this turbine a maximum value of 29.12% is observed near the point C80, being followed by C100 and D100/D120 with values close to 29.00%. From the points located close to the shoreline, better performances are expected for the Danube Delta (e.g., 24.62%), while from the southern part, the point I0 (Vama Veche) looks more promising with 21.22%. The points located near the interface that separates the shallow and deep waters (50 m) can be considered a promising candidate for a wind project since (a) they are defined by a higher *Cf* ~28.00%; (b) they are located close to the shore and in a lower water depth, which is in the line with the current European trends; and (c) it is possible to develop either fixed or floating projects.

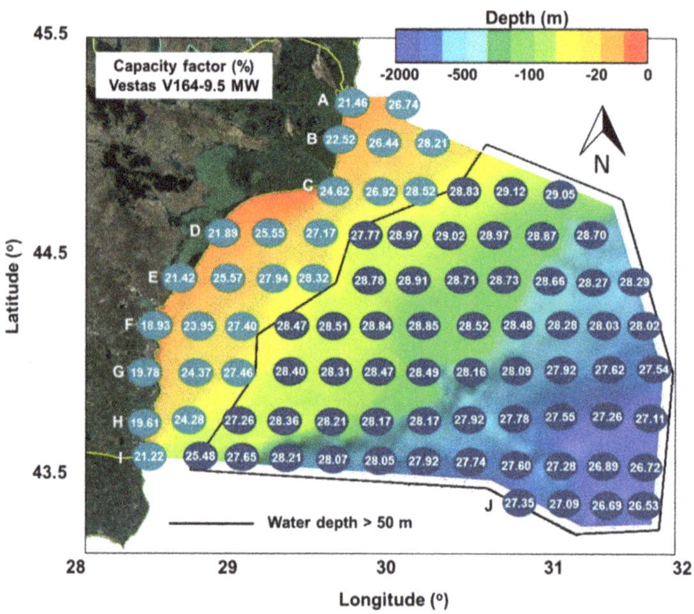

Figure 9. Capacity factor (%) of the Vestas V164-9.5 turbine for the total time, corresponding to the grid points located inside the Romanian EEZ. The results are evaluated at a hub height of 100 m, for the time interval 2000–2019, by processing the ERA5 (U_{100}) data.

The annual energy production (GWh) is illustrated in Figure 10, these values correspond to a single wind turbine, Vestas V164-9.5, which may operate in a particular point of the grid. The top values are in the range 23.00–24.00 GWh, while a significant drop occurs near the shoreline to a minimum of 16.32 GWh (approximately a 35.00% difference), these values corresponding to the point H0. From the shallow–deep water interface, we can highlight the points C40 (23.73 GWh) and E60 (23.57 GWh), while from the offshore the C points located at the extremity of the EEZ indicate better results (e.g., C80—24.23 GWh).

The *IAV* index is also used to assess the energy production of the turbine for this target area and the parameter analyzed over the 20 years is now AEP. In Figure 11 is provided such an analysis, by considering the ERA5 wind data. The *IAV* increases from the onshore to offshore areas, reaching a maximum value of 7.55% in the southeastern part of the target area. These values also increase from north to south, reaching a minimum of 5.56% close to the site A20, while near the shoreline the values go from 5.72% (A0) to 6.72% (I0). From all the sites, the J line indicates higher values (7.45 to 7.53%), a possible explanation being related to the distance from the shore (>120.00 km) or the water depth (>1200.00 m).

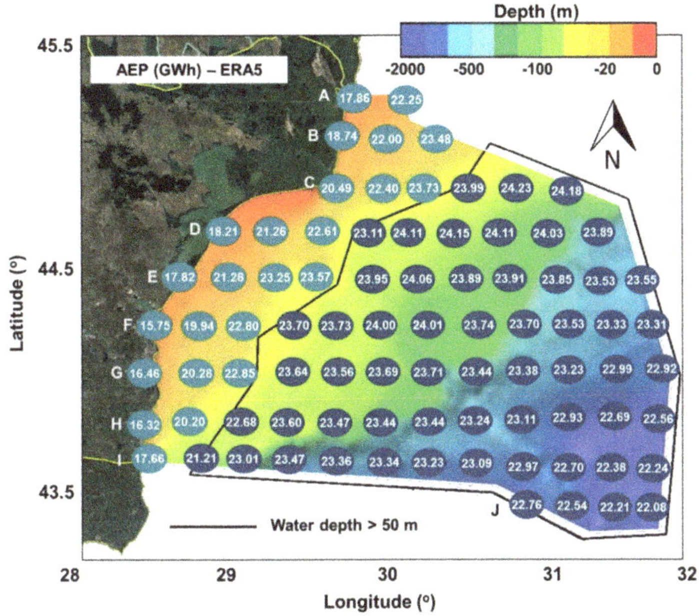

Figure 10. Vestas V164-9.5—annual energy production (in GWh) computed for the total time interval considering the ERA5 (U_{100}) reanalysis data.

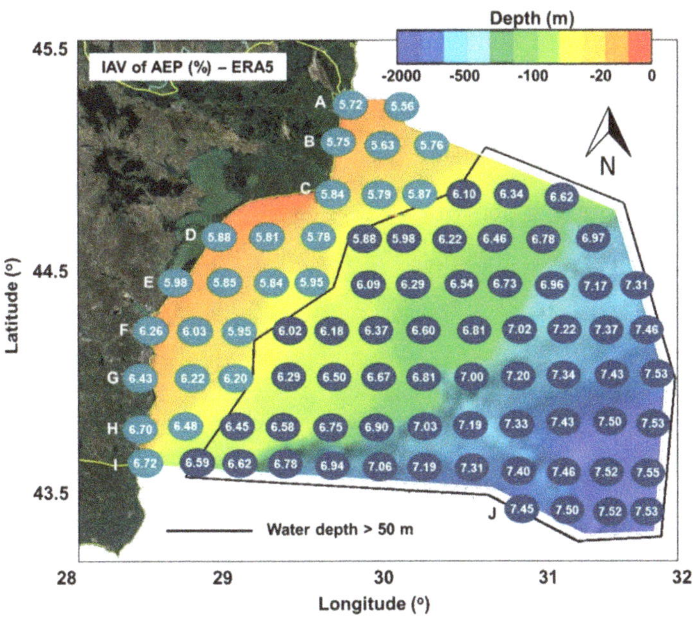

Figure 11. Inter-annual variability (*IAV*) in the annual energy production for the Vestas V164-9.5 wind turbine. The results are computed for the total time interval considering the ERA5 (U_{100}) reanalysis data.

4. Discussion

A preliminary step in the development of a renewable project consists in the analysis of the environmental conditions, the reanalysis data being capable to provide a more complete picture of the wind conditions [44–46]. Compared to the general wind patterns in the Black Sea, the Romanian coastal region is defined by more consistent wind energy resources, at which we can add a lower water depth that is related to the continental shelf [47]. Furthermore, the structure of the local seabed in the upper layers is mostly made by sand and clay deposits [48], an aspect that needs to be considered in the selection of the right type of wind turbine foundation [49].

The results presented in this work are built around the ERA5 data, and from the analysis of the wind resources (U_{100}) we notice that the wind speed increases from onshore to offshore, regardless of the time interval considered. These results are encouraging, taking also into account that near the Romanian coastline (on land) some important wind projects already operate successfully. The most representative example is the Fantanele-Cogealac wind farm (17.00 km from the Black Sea coast) that has an operational capacity of 600 MW. More than this, if we look at the wind farms of Romania, more than 78.00% of the projects are operating in the Dobrogea Plateau, which is next to the sea [50]. According to the IEC classification [51], the wind turbines from classes II and III are more suitable for this region, with better performances being expected during the wintertime when an average wind speed of 9.00 m/s is usually expected.

In relationship with some other works [15,16,52], this represents one of the first studies that evaluates the Romanian offshore wind energy potential using the new ERA5 data, which is defined by 24 values per day, and furthermore the evaluation is linked to a hub height of 100.00 m.

In the work of Cornett [53], several parameters are used to describe the temporal variability in the marine resources. One of them is the coefficient of variation (COV), which is defined as in Equation (6):

$$COV\ (U_{10}) = \frac{\sigma(U_{10}(t))}{\mu(U_{10}(t))} = \frac{\left[\overline{\left(U_{10} - \overline{U_{10}}\right)^2}\right]^{0.5}}{\overline{U_{10}}} \tag{6}$$

where σ is standard deviation; μ denotes the mean and the overbar is related to the time-averaging. A zero value of COV indicates that there is no variability, COV (U_{10}) = 1 denotes that the standard deviation is equal to the mean value, while COV (U_{10}) = 2 indicates that the standard deviation is twice the mean.

The seasonal variability (SV) is another indicator, which is defined as in Equation (7):

$$SV = \frac{U_{10_{S1}} - U_{10_{S4}}}{U_{10_{year}}} \tag{7}$$

where $U_{10_{S1}}$ and $U_{10_{S4}}$ represent the mean wind speed being related to the most and least energetic seasons. In this case, winter (December–January–February) and summer (June–July–August) represent the two extreme seasons, the first being the most energetic. In the case of $U_{10_{year}}$, this was approximated as $U_{10_{year}} \sim \frac{1}{2}(U_{10_{S1}} + U_{10_{S4}})$, similar to the one mentioned by Cornett [53].

Table 3 shows the distribution of the COV values, considering the ERA5 wind dataset. In this case, the values increase from the nearshore points (0.46) to the offshore (0.50), being observed some values of 0.51 in the southern part. The SV indicator is presented in Table 4, where we can notice smaller differences. The values go from 0.30 to 0.41, much higher ones being found in the offshore area, especially for the sites located in the southern part.

Table 3. Distribution of the COV values indicated for the ERA5 data (U_{100}), considering the time interval 2000–2019.

Line	Distance to Shore (km)											
	0	20	40	60	80	100	120	140	160	180	200	220
A	0.47											
B	0.47	0.48	0.49									
C	0.48	0.49	0.49	0.50	0.50	0.50						
D	0.47	0.49	0.49	0.50	0.50	0.50	0.50	0.50	0.50			
E	0.47	0.49	0.50	0.50	0.50	0.50	0.50	0.50	0.50	0.50	0.50	
F	0.46	0.48	0.50	0.50	0.50	0.50	0.50	0.50	0.50	0.50	0.50	0.50
G	0.46	0.48	0.50	0.51	0.50	0.50	0.50	0.50	0.50	0.50	0.50	0.50
H	0.46	0.49	0.50	0.51	0.51	0.51	0.51	0.50	0.50	0.50	0.50	0.50
I	0.47	0.50	0.51	0.51	0.51	0.51	0.51	0.51	0.50	0.50	0.50	0.50
J									0.50	0.50	0.50	0.50

Table 4. Distribution of the SV values indicated for the ERA5 data (U_{100}), considering the time interval 2000–2019.

Line	Distance to Shore (km)											
	0	20	40	60	80	100	120	140	160	180	200	220
A	0.30	0.32										
B	0.32	0.33	0.34									
C	0.34	0.35	0.35	0.36	0.36	0.35						
D	0.33	0.34	0.35	0.36	0.37	0.37	0.37	0.36	0.36			
E	0.34	0.35	0.37	0.38	0.38	0.38	0.38	0.37	0.37	0.36	0.36	
F	0.33	0.35	0.37	0.38	0.39	0.39	0.39	0.38	0.38	0.38	0.37	0.36
G	0.34	0.36	0.38	0.39	0.39	0.39	0.39	0.39	0.39	0.38	0.38	0.37
H	0.35	0.37	0.38	0.40	0.40	0.40	0.40	0.40	0.39	0.39	0.38	0.38
I	0.36	0.38	0.39	0.40	0.41	0.40	0.40	0.40	0.40	0.39	0.39	0.38
J									0.39	0.39	0.39	0.39

Although the ERA5 data are frequently used to assess the renewable resources, due to the lower resolution of the global model it is possible that for some seas surrounded by complex orography (as for example is the Black Sea) underestimation of the wind speeds to occur. Consequently, such underestimations can propagate as regards the wind energy potential. The long-term variability in the resources estimated using COV and SV shows that ERA5 data indicate more stable conditions compared to the satellite measurements.

However, it has to be highlighted at this point that ERA5 U_{10} winds were verified by Belmonte and Stoffelen [54], showing rather large model errors near the coast. The model wind data do not appear very reliable in coastal areas, particularly for enclosed seas. This is probably due to the large diurnal cycle and the occurrence of low-level jets (LLJ), which are both not well captured by ERA5 [55]. On the other hand, the AVISO dataset is not a perfect daily mean, like the average of the 24 values for ERA5. In fact, altimeter winds over a day are determined by the local solar times of the contributing altimeter sample.

As the land–sea effects near the Romanian coast are substantial, the diurnal cycle will determine the mean daily wind. Hence, this AVISO product appears not suitable to verify the simulated gridded products. An altimeter does not measure the wind at a 10 m height, but rather the ocean roughness, which may be related to the 10 m stress-equivalent wind [56]. Furthermore, the altimeter speeds are sensitive to the mean squared slope (MSS) of the ocean, and hence sensitive to the sea state which may be quite detrimental for coastal validation. Some other limitations of the ERA5 data for the energy assessments in the coastal area are further presented in Belmonte and Stoffelen [54], and Kalverla [55].

Most of the reanalysis datasets are providing data at the standard height of 10 m, which is suitable for meteorological applications, but to translate these values to a hub height, an adjustment is required (e.g., logarithmic law, power law, etc). One element of innovation that defines the ERA5 data is that it directly provides wind data at a 100 m height, considering that this is a reference height for the offshore industry [28,57].

In the work of Onea et al. [18], a similar analysis was carried out for the Romanian EEZ, considering this time for evaluation an MHI Vestas V174 9.5 MW turbine. For the present work, we considered to use the V164-9.5 MW system since all the information needed to assemble the power curve is available in the public domain (e.g., rated wind speed), which is not the case for the other system [58]. The present turbine is defined by a lower capacity factor compared to the V174 9.5, which may reach a maximum of 35.00% in the upper part of the EEZ, which is up to 6.00% compared to the one presented in this study. The main explanation for this is related to the fact that the V174 9.5 wind turbine has a lower rated speed, which makes it more competitive. These values are relatively close to the capacity factor (36%) estimated for Romania and Bulgaria, exceeding those corresponding to countries such as Italy (25–30%), Greece (29%) or Spain (31%), but far under regions like the UK (53%) or Denmark (44%), where significant offshore projects already operate [59].

5. Conclusions

In the present work, an assessment of the wind energy potential was carried out in the Romanian EEZ, considering multiple datasets that cover a total of 20 years (2000–2019). Special attention was given to the ERA5 data since these are frequently used for renewable applications. The results clearly show an increase in the wind conditions as going from the land to the sea, which means that an offshore project could become a reality in this coastal environment in the near future.

During the more energetic seasons, the average wind speed can frequently reach 8–9 m/s, values that are related to the IEC classes II and III. Taking also into account the fact that in the southern part of the Romanian nearshore some important seaports are located (for example Constanta harbor), this means that the logistic support required to develop a wind project will be more accessible. On the other hand, higher wind resources are characteristic in the northern part of the target area, where the capacity factor of the wind turbine considered for evaluation (V164-9.5 MW) frequently indicates a value of about 30%.

From this perspective, it can be concluded that the Romanian offshore wind sector represents a viable alternative to the Romanian coal power plants that during recent years gradually reduced their production capacity. The present work also opens some new research ideas that need to be tackled soon, such as the assessment of the wind energy potential by considering restricted areas, and to identify suitable wind farm layouts and consider the fluctuations in the wind direction.

Per total, the Romanian coastal area can be considered a viable candidate for the development of an offshore wind project having good wind energy potential, a large continental shelf, and also Romania is a part of the European Union that aims to significantly expand the offshore wind sector. It can be also highlighted that various projects are targeting the development of offshore wind farms very soon in this particular coastal environment, and for them the present study should be of great interest. The western part of the Black Sea has important wind resources that need to be accurately quantified to establish the expected performances of the offshore wind turbines. The first step is to understand the limitations of the global reanalysis dataset and of the satellite measurements, by making a direct comparison with in situ measurements. Regarding the Romanian EEZ, it is expected that in future works we will also include some restricted areas (e.g., shipping routes, protected areas), in order to identify the most suitable areas capable to support a marine project. Finally, it is important to mention also that at this moment the ERA5 wind data are capable to provide a general perspective of the wind conditions over a long time interval, being processed on a global scale. Nevertheless, there are now some other wind datasets available with a higher resolution, which include meso- and micro-scaling that can be used to evaluate the wind resources in the coastal environment. This is the case of the Global Wind Atlas [60] that is defined by a grid spacing of 3 km. For regional simulations, the ERA5 boundary conditions are used to force the mesoscale model, but at this moment the data available cover only the 10-year time interval 2008–2017. The present work is ongoing and this type of data represents a valuable tool for the assessment of the sites located near the coastline or onshore, where the topographic details become more important in wind modelling. At this point it has to be also highlighted that aspects of space and time sampling are relevant for observations, while they provide the only absolute source of wind information. Model winds are affected by simulation artefacts, such as varying observation input over time, diffusion of land–sea effects, diurnal cycle, convection, etc. Extensive inter-calibrated and documented satellite records of winds are publicly available in the EU Copernicus Marine Environment Monitoring Service, but not considered in the present work. From this perspective, an important objective of the future work is to complete the study related to the assessment of the offshore wind energy potential in the Romanian exclusive economic zone with an extended analysis of these scatterometer winds.

Author Contributions: F.O. performed the literature review, processed the data, and wrote the manuscript. E.R. designed and supervised the present work. L.R. contributed to the design and implementation of the research. All authors have read and agreed to the published version of the manuscript.

Funding: This work was carried out in the framework of the research project DREAM (Dynamics of the REsources and technological Advance in harvesting Marine renewable energy), supported by the Romanian Executive Agency for Higher Education, Research, Development and Innovation Funding—UEFISCDI, grant number PN-III-P4-ID-PCE-2020-0008.

Acknowledgments: The data used in this study are openly available. ERA5 data used in this study were obtained from the ECMWF data server. The altimeter products were generated and distributed by AVISO (http://www.aviso.altimetry.fr/, accessed on 29 March 2021) as part of the Ssalto ground processing segment.

Conflicts of Interest: The authors declare no conflict of interest.

References

1. Ahmad, T.; Zhang, D. A Critical Review of Comparative Global Historical Energy Consumption and Future Demand: The Story Told so Far. *Energy Rep.* **2020**, *6*, 1973–1991. [CrossRef]
2. Kesicki, F.; Anandarajah, G. The Role of Energy-Service Demand Reduction in Global Climate Change Mitigation: Combining Energy Modelling and Decomposition Analysis. *Energy Policy* **2011**, *39*, 7224–7233. [CrossRef]
3. Brockway, P.E.; Sorrell, S.; Semieniuk, G.; Heun, M.K.; Court, V. Energy Efficiency and Economy-Wide Rebound Effects: A Review of the Evidence and Its Implications. *Renew. Sustain. Energy Rev.* **2021**, *141*, 110781. [CrossRef]
4. Erdiwansyah; Mahidin; Mamat, R.; Sani, M.S.M.; Khoerunnisa, F.; Kadarohman, A. Target and Demand for Renewable Energy across 10 ASEAN Countries by 2040. *Electr. J.* **2019**, *32*, 106670. [CrossRef]
5. Brugger, H.; Eichhammer, W.; Mikova, N.; Dönitz, E. Energy Efficiency Vision 2050: How Will New Societal Trends Influence Future Energy Demand in the European Countries? *Energy Policy* **2021**, *152*, 112216. [CrossRef]
6. Elrahmani, A.; Hannun, J.; Eljack, F.; Kazi, M.-K. Status of Renewable Energy in the GCC Region and Future Opportunities. *Curr. Opin. Chem. Eng.* **2021**, *31*, 100664. [CrossRef]
7. Cheng, K.-S.; Ho, C.-Y.; Teng, J.-H. Wind Characteristics in the Taiwan Strait: A Case Study of the First Offshore Wind Farm in Taiwan. *Energies* **2020**, *13*, 6492. [CrossRef]
8. Cottura, L.; Caradonna, R.; Ghigo, A.; Novo, R.; Bracco, G.; Mattiazzo, G. Dynamic Modeling of an Offshore Floating Wind Turbine for Application in the Mediterranean Sea. *Energies* **2021**, *14*, 248. [CrossRef]
9. Costa, Á.M.; Orosa, J.A.; Vergara, D.; Fernández-Arias, P. New Tendencies in Wind Energy Operation and Maintenance. *Appl. Sci.* **2021**, *11*, 1386. [CrossRef]
10. Offshore Wind in Europe—Key Trends and Statistics 2020. Available online: https://windeurope.org/data-and-analysis/product/offshore-wind-in-europe-key-trends-and-statistics-2020 (accessed on 11 March 2021).
11. Topham, E.; McMillan, D.; Bradley, S.; Hart, E. Recycling Offshore Wind Farms at Decommissioning Stage. *Energy Policy* **2019**, *129*, 698–709. [CrossRef]
12. Onea, F.; Rusu, L. A Long-Term Assessment of the Black Sea Wave Climate. *Sustainability* **2017**, *9*, 1875. [CrossRef]
13. Onea, F.; Rusu, E. Sustainability of the Reanalysis Databases in Predicting the Wind and Wave Power along the European Coasts. *Sustainability* **2018**, *10*, 193. [CrossRef]

14. Pasimeni, F.; Fiorini, A.; Georgakaki, A. Assessing Private R&D Spending in Europe for Climate Change Mitigation Technologies via Patent Data. *World Pat. Inf.* **2019**, *59*, 101927. [CrossRef]
15. Onea, F.; Rusu, L. Evaluation of Some State-Of-The-Art Wind Technologies in the Nearshore of the Black Sea. *Energies* **2018**, *11*, 2452. [CrossRef]
16. Onea, F.; Rusu, L. A Study on the Wind Energy Potential in the Romanian Coastal Environment. *J. Mar. Sci. Eng.* **2019**, *7*, 142. [CrossRef]
17. Rusu, E. A 30-Year Projection of the Future Wind Energy Resources in the Coastal Environment of the Black Sea. *Renew. Energy* **2019**, *139*, 228–234. [CrossRef]
18. *Romania's Offshore Wind Energy Resources: Natural Potential, Regulatory Framework, and Development Prospects*; EPG: Bucharest, Romania, 2020.
19. Văidianu, N.; Ristea, M. Marine Spatial Planning in Romania: State of the Art and Evidence from Stakeholders. *Ocean Coast. Manag.* **2018**, *166*, 52–61. [CrossRef]
20. GEBCO_08 Grid—Version 20100927. Available online: https://www.gebco.net/data_and_products/gridded_bathymetry_data/version_20100927/ (accessed on 25 April 2021).
21. Flanders Marine Institute (VLIZ). *Belgium Maritime Boundaries Geodatabase: Maritime Boundaries and Exclusive Economic Zones (200NM)*; Flanders Marine Institute: Ostend, Belgium, 2019.
22. Lin, Z.; Liu, X.; Lotfian, S. Impacts of Water Depth Increase on Offshore Floating Wind Turbine Dynamics. *Ocean Eng.* **2021**, *224*, 108697. [CrossRef]
23. Díaz, H.; Guedes Soares, C. Review of the Current Status, Technology and Future Trends of Offshore Wind Farms. *Ocean Eng.* **2020**, *209*, 107381. [CrossRef]
24. Hersbach, H.; Bell, B.; Berrisford, P.; Hirahara, S.; Horányi, A.; Muñoz-Sabater, J.; Nicolas, J.; Peubey, C.; Radu, R.; Schepers, D.; et al. The ERA5 Global Reanalysis. *QJR Meteorol. Soc.* **2020**, qj.3803. [CrossRef]
25. Onea, F.; Rusu, L.; Carp, G.B.; Rusu, E. Wave Farms Impact on the Coastal Processes—A Case Study Area in the Portuguese Nearshore. *J. Mar. Sci. Eng.* **2021**, *9*, 262. [CrossRef]
26. Rusu, L.; Rusu, E. Evaluation of the Worldwide Wave Energy Distribution Based on ERA5 Data and Altimeter Measurements. *Energies* **2021**, *14*, 394. [CrossRef]
27. Olauson, J. ERA5: The New Champion of Wind Power Modelling? *Renew. Energy* **2018**, *126*, 322–331. [CrossRef]
28. Soares, P.M.M.; Lima, D.C.A.; Nogueira, M. Global Offshore Wind Energy Resources Using the New ERA-5 Reanalysis. *Environ. Res. Lett.* **2020**, *15*, 1040a2. [CrossRef]
29. Ulazia, A.; Sáenz, J.; Ibarra-Berastegi, G.; González-Rojí, S.J.; Carreno-Madinabeitia, S. Global Estimations of Wind Energy Potential Considering Seasonal Air Density Changes. *Energy* **2019**, *187*, 115938. [CrossRef]
30. Carreno-Madinabeitia, S.; Ibarra-Berastegi, G.; Sáenz, J.; Ulazia, A. Long-Term Changes in Offshore Wind Power Density and Wind Turbine Capacity Factor in the Iberian Peninsula (1900 to 2010). *Energy* **2021**, 120364. [CrossRef]
31. Home: Aviso+. Available online: https://www.aviso.altimetry.fr/en/home.html (accessed on 31 May 2019).
32. Ribal, A.; Young, I.R. 33 Years of Globally Calibrated Wave Height and Wind Speed Data Based on Altimeter Observations. *Sci Data* **2019**, *6*, 77. [CrossRef]
33. Majidi Nezhad, M.; Shaik, R.U.; Heydari, A.; Razmjoo, A.; Arslan, N.; Astiaso Garcia, D. A SWOT Analysis for Offshore Wind Energy Assessment Using Remote-Sensing Potential. *Appl. Sci.* **2020**, *10*, 6398. [CrossRef]
34. Rusu, E.; Onea, F. An Assessment of the Wind and Wave Power Potential in the Island Environment. *Energy* **2019**, *175*, 830–846. [CrossRef]
35. Salvação, N.; Guedes Soares, C. Wind Resource Assessment Offshore the Atlantic Iberian Coast with the WRF Model. *Energy* **2018**, *145*, 276–287. [CrossRef]
36. Ali Elfarra, M.; Kaya, M. Estimation of Electricity Cost of Wind Energy Using Monte Carlo Simulations Based on Nonparametric and Parametric Probability Density Functions. *Alex. Eng. J.* **2021**, *60*, 3631–3640. [CrossRef]
37. Jaramillo, O.A.; Borja, M.A. Wind Speed Analysis in La Ventosa, Mexico: A Bimodal Probability Distribution Case. *Renew. Energy* **2004**, *29*, 1613–1630. [CrossRef]
38. V-V164-9.5 MWTM. Available online: https://www.vestas.com/en/products/offshoreplatforms/v164_9_5_mw (accessed on 21 March 2021).
39. MHI Vestas Offshore V164/9500—Manufacturers and Turbines—Online Access—The Wind Power. Available online: https://www.thewindpower.net/turbine_en_1476_mhi-vestas-offshore_v164-9500.php (accessed on 19 August 2018).
40. Xie, K.; Yang, H.; Hu, B.; Li, C. Optimal Layout of a Wind Farm Considering Multiple Wind Directions. In Proceedings of the 2014 International Conference on Probabilistic Methods Applied to Power Systems (PMAPS); IEEE: Durham, UK, 2014; pp. 1–6.
41. Wu, Y.-T.; Porté-Agel, F. Modeling Turbine Wakes and Power Losses within a Wind Farm Using LES: An Application to the Horns Rev Offshore Wind Farm. *Renew. Energy* **2015**, *75*, 945–955. [CrossRef]
42. Stopa, J.E.; Cheung, K.F.; Tolman, H.L.; Chawla, A. Patterns and Cycles in the Climate Forecast System Reanalysis Wind and Wave Data. *Ocean Model.* **2013**, *70*, 207–220. [CrossRef]
43. Rusu, E.; Onea, F. A Parallel Evaluation of the Wind and Wave Energy Resources along the Latin American and European Coastal Environments. *Renew. Energy* **2019**. [CrossRef]

44. Kamranzad, B.; Etemad-Shahidi, A.; Chegini, V. Sustainability of Wave Energy Resources in Southern Caspian Sea. *Energy* **2016**, *97*, 549–559. [CrossRef]
45. Lavidas, G.; Blok, K. Shifting Wave Energy Perceptions: The Case for Wave Energy Converter (WEC) Feasibility at Milder Resources. *Renew. Energy* **2021**, *170*, 1143–1155. [CrossRef]
46. Jourdier, B. Evaluation of ERA5, MERRA-2, COSMO-REA6, NEWA and AROME to Simulate Wind Power Production over France. *Adv. Sci. Res.* **2020**, *17*, 63–77. [CrossRef]
47. Onea, F.; Rusu, E. Wind Energy Assessments along the Black Sea Basin. *Meteorol. Appl.* **2014**, *21*, 316–329. [CrossRef]
48. Dinu, C.; Wong, H.K.; Tambrea, D.; Matenco, L. Stratigraphic and Structural Characteristics of the Romanian Black Sea Shelf. *Tectonophysics* **2005**, *410*, 417–435. [CrossRef]
49. Sánchez, S.; López-Gutiérrez, J.-S.; Negro, V.; Esteban, M.D. Foundations in Offshore Wind Farms: Evolution, Characteristics and Range of Use. Analysis of Main Dimensional Parameters in Monopile Foundations. *J. Mar. Sci. Eng.* **2019**, *7*, 441. [CrossRef]
50. Dragomir, G.; Șerban, A.; Năstase, G.; Brezeanu, A.I. Wind Energy in Romania: A Review from 2009 to 2016. *Renew. Sustain. Energy Rev.* **2016**, *64*, 129–143. [CrossRef]
51. Roach, S.; Park, S.M.; Gaertner, E.; Manwell, J.; Lackner, M. Application of the New IEC International Design Standard for Offshore Wind Turbines to a Reference Site in the Massachusetts Offshore Wind Energy Area. *J. Phys. Conf. Ser.* **2020**, *1452*, 012038. [CrossRef]
52. Raileanu, A.B.; Onea, F.; Rusu, E. Implementation of Offshore Wind Turbines to Reduce Air Pollution in Coastal Areas—Case Study Constanta Harbour in the Black Sea. *J. Mar. Sci. Eng.* **2020**, *8*, 550. [CrossRef]
53. Cornett, A.M. A Global Wave Energy Resource Assessment. In *Proceedings of the Eighteenth (2008) International Offshore and Polar Engineering Conference*; Chung, J.S., Ed.; International Society Offshore & Polar Engineers: Cupertino, CA, USA, 2008; Volume 1, pp. 318–326. ISBN 978-1-880653-70-8.
54. Belmonte Rivas, M.; Stoffelen, A. Characterizing ERA-Interim and ERA5 Surface Wind Biases Using ASCAT. *Ocean Sci.* **2019**, *15*, 831–852. [CrossRef]
55. Kalverla, P.C.; Holtslag, A.A.M.; Ronda, R.J.; Steeneveld, G.-J. Quality of Wind Characteristics in Recent Wind Atlases over the North Sea. *Q. J. R. Meteorol. Soc.* **2020**, *146*, 1498–1515. [CrossRef]
56. de Kloe, J.; Stoffelen, A.; Verhoef, A. Improved Use of Scatterometer Measurements by Using Stress-Equivalent Reference Winds. *Ieee J. Sel. Top. Appl. Earth Obs. Remote Sens.* **2017**, *10*, 2340–2347. [CrossRef]
57. ERA5: How Are the 100m Winds Calculated?—Copernicus User Support Forum—ECMWF Confluence Wiki. Available online: https://confluence.ecmwf.int/pages/viewpage.action?pageId=155343870 (accessed on 30 September 2020).
58. Bauer, L. MHI Vestas Offshore V174-9.5—9,50 MW—Wind Turbine. Available online: https://en.wind-turbine-models.com/turbines/2138-mhi-vestas-offshore-v174-9.5 (accessed on 25 March 2021).
59. European Commission Joint Research Centre. *Wind Potentials for EU and Neighbouring Countries: Input Datasets for the JRC EU TIMES Model*; Publications Office: Brussels, Belgium, 2018.
60. Global Wind Atlas. Available online: https://globalwindatlas.info (accessed on 26 April 2021).

MDPI
St. Alban-Anlage 66
4052 Basel
Switzerland
www.mdpi.com

Journal of Marine Science and Engineering Editorial Office
E-mail: jmse@mdpi.com
www.mdpi.com/journal/jmse

www.ingramcontent.com/pod-product-compliance
Lightning Source LLC
Chambersburg PA
CBHW040826250926
51850CB00007BA/1274
* 9 7 8 3 0 3 6 5 9 2 3 1 2 *